Series Preface

The field of indoor air science is of growing interest and concern given that modern society spends the better part of each day indoors. Since the indoor air environment is a major, continual exposure medium for occupants, it is important to study what is present and if and how it affects the health and comfort of occupants.

Volumes in this Indoor Air Research Series are intended to provide state-of-the-art information on many areas germane to indoor air science including Chemical and Biological Sources, Exposure Assessment, Dosimetry, Engineering Controls, and Perception of Indoor Air Quality. In each volume, authors known for their expertise on the topic will present comprehensive and critical accounts of our current understanding in the area.

It is hoped that the series will advance knowledge and broaden interest among the scientific community at large in the indoor air science field.

Max Eisenberg, Ph.D.
Series Editor

Preface

Biological aerosols are, and have always been, a factor in human health. Although their existence is well recognized, interest in specific relationships between the aerosols themselves and human health, as well as synergistic effects between biological aerosols and other air pollutants, has sharply increased in recent years.

This work was commissioned by the Center for Indoor Air Research as a state-of-the-art review of bioaerosols. Bioaerosol research has, in the past several years, risen to the top of research agendas for several federal funding agencies with the result that new data appear daily. The practicalities of writing and publication schedules are an inevitable challenge to timeliness.

This book is offered as a synthesis of the information available at the time of publication. It is intended to provide background for students and practitioners of air pollution research and should be a valuable resource from which to begin new efforts in this fascinating field.

Harriet A. Burge, Ph.D.
Editor

Foreword

We share the earth's biosphere with an amazingly abundant and varied array of microorganisms that have existed since near the beginning of life on earth. They were here first, mankind has evolved in their presence, and they will be here long after the experiment of mankind has concluded. Microorganisms have developed complex and wondrous survival strategies to permit them to carpet the surface of the earth, from the tainted pink snow melting high on mountain hillsides, to the self-supporting sulfur-based ecosystems along mid-oceanic rifts.

However, natural does not mean benign. Human exposure to these organisms in their natural environment has resulted in allergic, toxic, and infectious diseases. Our nomadic forebears encountered microorganisms in their caves, on putrefying food, and in stagnant waters. Disease resulted whenever sight and smell failed to protect our prehistoric ancestors by warning them to avoid those things that experience taught were dangerous. In manipulating the natural environment, humanity has changed its relationship to these microorganisms. At least until the advent of agriculture, human interference in the natural outdoor ecosystem was minimal, and human disease related to natural exposures to microorganisms was also relatively mild or of low incidence. The advent of agriculture led to more permanent settlements and a transformation of how man perceived the environment. We now speak in terms of environmental management, modifying our landscapes, designing our foods, chemically enhancing commercial products, and constructing tight shelters with materials that have existed only for the past generation or two. In so doing, we have changed the ecology of both man and the lower organisms, creating situations where disease-causing exposures are likely, and, in some cases, inevitable.

The indoor environment has always been essentially man made, and has played a large role in human health. The indoor environment has been one where microorganisms proliferate, and where the resultant aerosols are confined to the point where dangerous disease-causing exposures can occur. Contagious diseases, Legionnaires' disease, hypersensitivity pneumonitis, and dust mite asthma, are only a few of the bioaerosol-related diseases that are primarily transmitted indoors. Advances in medical science gradually lessen the impact of the contagious diseases. However, contagious disease for which immunization is not possible, as well as environmental source infections and the hypersensitivity diseases, are increasing. This is partly due to the recognition that energy is limited (and expensive) and that sealing interior spaces limits energy usage. We forget that, in preventing access to the outdoor air, we prevent the escape of indoor-source pollutants and the entry of dilution air.

This book brings together information on some important aspects of the relationship between man and biological contaminants of indoor air. It has been edited by Dr. Burge and written by her and an illustrious group of experts for use by those of us for whom microbiology and aerobiology are foreign. If you are a facilities manager, an industrial hygienist, an occupational physician, a municipal or state health official, an architect, an environmental consultant, a teacher, or an HVAC engineer, you will find this informative and concisely written book beneficial. Beyond the pragmatic consideration of job enhancement, you will be fascinated to learn what is in your basement, in your bed, under the refrigerator, behind the walls, and in the ventilation systems of your homes and workplaces.

John D. Spengler, Ph.D.
Harvard School of Public Health

Acknowledgments

This book would not have been possible without the understanding of my family, the long-term support of Dr. William Solomon at the University of Michigan Medical School, and Dr. Max Eisenberg of the Center for Indoor Air Research, who conceived the idea in the first place, and who has been patient and encouraging throughout the process.

The Editor

Harriet A. Burge, Ph.D., is Associate Professor of Environmental Microbiology at the Harvard School of Public Health. She received her baccalaureate and master's degrees from San Francisco State University in 1960 and 1962, respectively, and a Ph.D. in Botany (mycology) from the University of Michigan in 1966. Current research projects include the study of factors affecting exposure to mycotoxins in indoor environments, the relationship between fungi and asthma, and the nature of the outdoor bioaerosol, including the development and operation of a network of outdoor air sampling stations for the collection of biological particulates that can be identified microscopically. Dr. Burge chairs the ACGIH (American Conference of Governmental Industrial Hygienists) Committee on Bioaerosols which is developing standards for the sampling and analysis of airborne biological agents. She was vice chair of the National Academy of Science Institute of Medicine Committee on Indoor Allergens, a member of the National Academy of Sciences Committee on Airliner Cabin Air Quality and Safety, and has written more than 50 publications concerning health effects, sampling and the nature of airborne fungi.

Contributors

Stuart A. Batterman, Ph.D.
Department of Environmental and Industrial Health
University of Michigan
School of Public Health
Ann Arbor, Michigan

Harriet A. Burge, Ph.D.
Department of Environmental Health
Harvard School of Public Health
Boston, Massachusetts

Martin D. Chapman, Ph.D.
Division of Allergy/Clinical Immunology
Department of Medicine
University of Virginia
Charlottesville, Virginia

Cory E. Cookingham, M.D.
Health Plus Building
Flint, Michigan

Lynn Eudey, Ph.D.
Department of Mathematics
East Carolina University
Greenville, North Carolina

Carl B. Fliermans, Ph.D.
Westinghouse Savannah River Co.
Savannah River Technical Center
Aiken, South Carolina

Estelle Levetin, Ph.D.
Faculty of Biological Science
The University of Tulsa
Tulsa, Oklahoma

Christina M. Luczyska, Ph.D.
Department of Public Health Medicine
United Medical and Dental Schools of Guy's and St. Thomas' Hospitals
University of London
London, England

Donald K. Milton, M.D., Dr.P.H.
Department of Environmental Health
Harvard School of Public Health
Boston, Massachusetts

Michael L. Muilenberg, M.S.
Department of Environmental Health
Harvard School of Public Health
Boston, Massachusetts

William R. Solomon, M.D.
Department of Internal Medicine/Allergy
The University of Michigan Medical School
Ann Arbor, Michigan

Susan Pollart Squillace, M.D.
Department of Family Medicine and Internal Medicine
University of Virginia School of Medicine
Charlottesville, Virginia

H. Jenny Su, Sc.D.
Department of Environmental and Occupational Health
National Cheng Kung University Medical College
Tainan, Taiwan

Richard L. Tyndall, Ph.D.
Health Safety Research Division
Oak Ridge National Laboratory
Oak Ridge, Tennessee

A. A. Vass, Ph.D.
Health Safety Research Division
Oak Ridge National Laboratory
Oak Ridge, Tennessee

Table of Contents

Dedication

This book is dedicated to William S. Benninghoff, Jr., 1918–1993, who was instrumental in establishing the International Aerobiology Association, developed the series of Gordon Research Conferences on Aerobiology, and was a constant source of support and encouragement for the international aerobiology community.

1 BIOAEROSOL INVESTIGATIONS

Harriet A. Burge

CONTENTS

I. INVESTIGATIVE APPROACHES

Current interest in the health effects of indoor air pollutants has placed the primary focus for bioaerosol investigations on characterizing the indoor bioaerosol with the goal of discovering a cause for ongoing disease or discomfort (e.g., Croft

0-87371-724-4/95/$0.00+$.50

et al., 1986). Bioaerosols are also studied to make connections between specific health effects (e.g., house dust mite allergen and asthma; Chapman et al., 1987, and Chapter 13, this volume) and to increase understanding of the ecology and physiology of the source organisms and the fate of their airborne effluents (Wood et al., 1993). Much of the confusion that currently exists today in bioaerosol science is in respect to these different kinds of studies, and a frequent absence of focus on those methods most suited to the problem under investigation. This chapter is an overview designed to focus attention on problem-solving approaches to bioaerosol investigations.

A. Hypothesis Development

The first step in any bioaerosol investigation is the development of a hypothesis to test. A hypothesis is no more than a carefully formulated logical answer to a question. For example, in many bioaerosol investigations the question is, "What is causing complaints in this building?" A number of different answers (or hypotheses) could be considered:

1. The building has too little outside air ventilation.
2. Inadequate ventilation is leading to the accumulation of a specific agent causing complaints.
3. Biological aerosols are present.
4. A specific bioaerosol is causing the complaints.

Each of these hypotheses would be tested in a different way. In order to provide an answer to the original question, hypothesis 1 requires that complaint rates be measured before and after increasing ventilation, and that the increase in ventilation rate be documented (measured) (e.g., Menzies et al., 1993). Hypothesis 2 includes measurement of a specific contaminant before and after ventilation changes and (sometimes) reference to a standard that indicates the likelihood of adverse health effects or complaints. Note that ventilation measurements are also required. Hypothesis 3 is not useful for answering the original question, since biological aerosols are always present in buildings. Hypothesis 4 requires measurement of the specific bioaerosol of concern, or observation or measurement of some factor that allows assumption of exposure (e.g., observation of visible growth, observation of known reservoirs, reservoir sampling, etc.) and a logical and compelling argument for the exposure/disease association (i.e., previous data, or exposure response data in the environment) (Strachan et al., 1990).

"What is causing complaints in this building?" is an example of a question that leads to the investigation of specific outbreaks of discomfort or illness. There are at least four kinds of investigations that could result from bioaerosol-related questions, each requiring different hypotheses and testing methods. These include:

1. Investigating building-related complaints.
2. Studying the significance in a population of exposure to an agent (epidemiological studies).
3. Studying the ecology, physiology, or effect of the airborne state on organisms and their bioaerosols, and studies seeking to establish cause/effect relationships between specific agents and specific diseases (experimental aerobiology).
4. Evaluating methods for studying bioaerosols.

B. Investigating Building-Related Complaints

Figure 1.1 outlines a series of steps that could be used to develop hypotheses to explain possible bioaerosol-related building complaints. A few examples will illustrate possible uses of this flow chart.

Example 1.

Nature of the complaints. Three cases of documented Legionnaires' disease occurred in one building area. The organism was *Legionella pneumophila* serotype I. The outbreak followed recommissioning of an old cooling tower near openable windows in the area of affected occupants.

Bioaerosols? Yes; Legionnaires' disease is always caused by bioaerosols.

Building related? Highly probable. Three cases of infection with the same organism in one building area make a building-related exposure likely.

Related to specific activity? Yes. Outbreak was associated with activation of old cooling tower near openable windows.

Kind of agent? Specific: *L. pneumophila* serotype I.

Hypothesis. *L. pneumophila* serotype I was amplified in the tower and released into the building air during startup, causing the disease outbreak.

Testing.

1. Verify presence of *L. pneumophila* serotype I in the tower by collecting a bulk sample to be both cultured and serotyped.
2. Document a pathway for exposure for each affected person. This step can be done by observation of ambient conditions or by using tracer aerosols.
3. Check for and test other potential reservoirs for which there is a logical transmission chain.

Conclusions. If the appropriate *Legionella* strain is recovered from the tower, and there is a logical transmission route for each patient, and no other reservoir fulfills these criteria, then the tower can be assumed to be the source.

Comments. The testing described does not provide conclusive proof that the tower was the source. However, the presence of the organism is probably

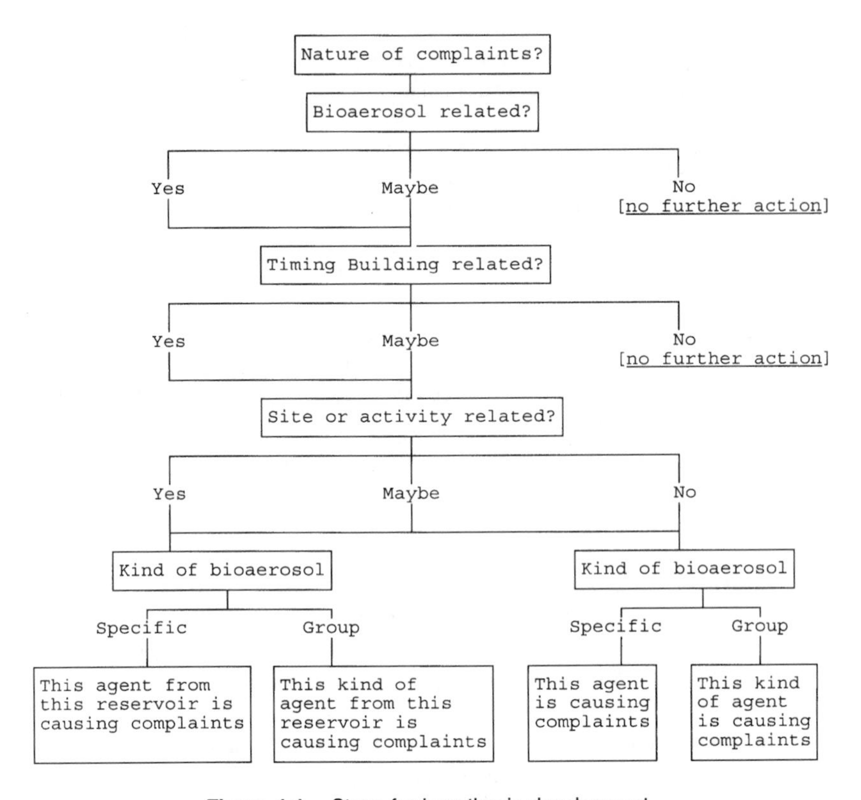

Figure 1.1. Steps for hypothesis development.

sufficient for ordering remedial actions, and absence of other potential reservoirs makes a good case for considering the environment acceptable with respect to *L. pneumophila* after remediation. Absence of *L. pneumophila* supports but does not prove the negative hypothesis. The organism could have disappeared between exposure and sampling.

Example 2.

Nature of the complaints. Hypersensitivity pneumonitis (HP) has been diagnosed in six building occupants, whose symptoms become worse when the ventilation system operates.

Bioaerosols? Possibly. HP is usually caused by bioaerosols but also can result from exposure to some highly reactive chemicals.

Building related? Yes. Six cases in one building are highly unlikely to result from chance clustering.

Related to specific reservoir or activity? Yes. Symptoms appear to be related to ventilation system operation.

Kind of biological agent? Group. A variety of small-particle antigenic aerosols can cause hypersensitivity pneumonitis.

Hypothesis. Exposure to small fungus or actinomycete spores released during operation from active or recent growth in ventilation system has caused sensitization and disease.

Testing.

1. Visual observation of ventilation system for obvious contamination.
2. Bulk sampling to isolate/identify agents including use of assay methods appropriate for identification of both fungi and actinomycetes.
3. Air sampling before and during ventilation system operation using sample collection methods appropriate for small particle aerosols, and assay methods appropriate for identification of both fungi and actinomycetes.
4. Serology to document exposure in cases and matched asymptomatic building occupants, and (possibly) matched controls from a different building.

Comments. In this case, once again, testing does not provide proof either for or against the hypothesis. Visual observation of growth proves only that some agent is present, not that it caused disease. If visible growth turns out to be of a known HP agent (e.g., *Thermoactinomyces vulgaris*) the case becomes more convincing. Positive serology also strengthens the evidence that exposure occurred, and, in some instances, serology might demonstrate differences between cases and controls. On the other hand, absence of any positive result does not mean that the hypothesis is not true. Proving the negative case is rarely possible.

Example 3.

Nature of the complaints. About 60% of building occupants are complaining of symptoms commonly associated with sick building syndrome (500 occupants in the building). Complaint rates are highest on the basement level and in offices adjacent to the elevator shafts.

Biological agent? Possibly; the cause of sick building syndrome is not clear, but microbial irritants are one hypothetical cause.

Building related? Probably, although air quality may not be involved.

Reservoir/activity related? May be related to something on the basement level, since complaints are highest there.

Kind? Group; many different kinds of biological and other agents can produce irritants that could cause sick building syndrome.

Hypothesis. Microbial growth is producing a volatile toxin that is causing complaints.

Testing. Look for visible growth, odors, and/or potential reservoirs in the basement area and the elevator shaft; visible fungus growth is likely to produce irritating volatile organic compounds. If visible growth is extensive, sample with tape or scraping to identify the specific agent. If no growth but obvious reservoirs are identified, collect bulk samples for analysis by culture.

Comments. The presence of visible extensive fungus growth or noticeable fungal odors is presumptive evidence of a contamination problem that is likely to lead to complaints, although the mechanism that relates exposure to effect is currently unknown. Once again, absence of visible contamination does not rule out bioeffluents as a cause, but reduces the chances.

C. Epidemiological Studies

The kinds of questions that one asks in an epidemiological study are somewhat different from those for a building investigation, although some of the same kinds of information are desired. Examples of bioaerosol questions that might be answered in an epidemiological investigation include:

1. Is exposure to house dust mite effluents a significant factor in the development of asthma?
2. What is the relationship between exposure to basidiospores and emergency room visits for asthma?
3. Is the presence of *Stachybotrys atra* on building surfaces associated with symptoms of nervous system dysfunction in building occupants?

These kinds of questions also require hypothesis development, but the hypothesis must be tested in large populations. The size of the populations that must be tested depends on a variety of factors, including the variability of each environmental measure to be taken, the number of cases of disease likely to appear in any selected population, etc. (see Chapter 14). Because the population size must be large (often considerably more than 100), each unit of the population cannot be studied as intensively as in the case of a single building. Also, one must decide in advance specifically what kind of sampling is to be done so that each unit can be studied in exactly the same way.

For example, question 3 might be the subject of both a building investigation and an epidemiological study. For the building investigation, we have already decided that the symptoms might be related to a specific bioaerosol. We then must discover the extent of *S. atra* on building surfaces and in the air, discover the amount of toxin present in the air (either by direct measurement or by extrapolation from the amount of toxin/spore), calculate the amount of toxin exposure each person receives, and compare that dose to studies in the literature that provide dose/response information. If any one of these steps is not possible, another possible approach is to designate a control building that matches the "sick" building in all respects but the *S. atra* contamination, and conduct a questionnaire survey of occupants of both buildings. Unfortunately, it is almost impossible to match buildings, and some doubt will always exist with respect to causality of symptoms. Since *S. atra* is a toxigenic fungus, the most likely approach in a building investigation is to recommend remediation without further study. A good example of a well-done case study is described by Croft et al. (1986).

For an epidemiologic study designed to answer the same question, one would first decide on an exposure measure for *S. atra* that is feasible for use in many buildings. The exposure measure selected would depend on the specific hypothesis. For example, if the hypothesis is that increasing levels of exposure to satratoxin produce increasing levels of complaints, then a quantitative measure of exposure to satratoxin must be chosen. Ideally, personal volumetric air sampling for the toxin itself would be used. Surrogates for such exposure include volumetric air sampling for the presence of *S. atra*, or quantitative bulk or surface sampling for the toxin or the organism. If a surrogate measure is chosen, a hypothesis must be developed to provide the link between the measure and exposure. If the hypothesis merely states that the presence of *S. atra* in the building is likely to lead to higher complaint rates, then visual observation and bulk sampling can be used to document presence of the organism.

Next, criteria for selection of buildings must be developed. Possibilities include random selection, selection of buildings with complaints, or selection of buildings with known *S. atra* contamination. If random selection or selection based on complaints is to be used, one must know the frequency of *S. atra* contamination in the building population in order to determine the number of buildings that must be sampled. In other words, some presampling will be necessary to determine either the rate of *S. atra* contamination or to allow selection of contaminated buildings. Once a set of buildings that do contain *S. atra* has been identified, one can either evaluate complaint rates related to levels of exposure, or match each contaminated building with a "clean" building and evaluate relative levels of complaints in the two building sets. These are only a few of the many kinds of epidemiological studies that could be designed. Examples of epidemiological studies relating bioaerosols to symptoms include Peat et al. (1993).

D. Experimental Aerobiology

Experimental aerobiology asks different questions. For example:

1. How does relative humidity affect the viability of specific bacteria?
2. Does a specific virus remain infective in air?
3. What is the relationship between surface growth and airborne spore levels for a specific fungus?

A number of hypotheses can be developed that relate to each of these questions. For example, one hypothesis for the third question might be:

> *Stachybotrys atra* spore levels in air are related to the number of spores available for dispersal, relative humidity at the surface, and air velocity at the surface.

Experiments to test this hypothesis would probably be done in a controlled situation where relative humidity and air velocity can be controlled. Specific humidities and air velocities would be chosen for the tests, and pilot studies would be conducted to determine the range in numbers of spores produced by *S. atra* on the given substrate under specific conditions. Sampling and analysis modes are chosen specifically to optimize collection of *S. atra* spores (probably a spore trap or filter device, since *S. atra* often does not grow well in culture). Assuming that you choose 3 levels for each parameter, then 27 different kinds of experiments would have to be conducted, each with enough replicates so that important differences between conditions can be detected. To discover the required number of replicates, pilot studies using one or more condition sets would have to be performed so that variance could be measured.

Alternatively, this hypothesis could be tested in the field. A situation where *S. atra* is growing on an exposed surface would be identified, and small samples of the growth would be collected so that the number of spores (and the variance in this number) available for dispersal could be measured. Relative humidity and air velocity could be measured close to the surface, and air samples collected. The limitations in this kind of study include the inability to control spore numbers on the substrate, relative humidity, and velocity (i.e., only those conditions present at the chosen site can be studied). Also, other variables (including temperature, light, air mixing in the space, dilution air, aerosol decay, etc.) are more difficult to control in the field.

E. Methods Studies

Methods studies include testing assay and sampling methods as well as testing protocols (i.e., sampling plans). Methods studies can be relatively simple and straightforward field studies where ambient aerosols are used to compare two or more samplers (e.g., Burge et al., 1977). Factors that need to be considered in such studies are potential interactive effects between the devices (i.e., does one disturb the collection characteristics of the other), and the differential effects of ambient conditions on the devices. Assay method comparisons can also be done, but should be conducted using identical sampling devices or, better, single samples divided for assay. Note that whenever ambient aerosols are used to test devices or analytical methods, conclusions are relevant only to the aerosols encountered and may not be relevant to other kinds of aerosols.

Laboratory-based methods studies are easily controlled, and provide more specific information than field studies. For example, a laboratory-based study might be used to test the recoverability of aerosolized endotoxin from different sample collection media (Milton et al., 1990). Use of a defined aerosol eliminates many unknowns that are always present in field-collected samples, and allows isolation of specific variables (within limits). Needless to say, any method designed for use in the field must also be tested under field conditions.

II. ANALYTICAL METHODS FOR BIOLOGICAL AGENTS

A. Overview

Analytical methods are presented before sampling methods because the sampling approach must be chosen to facilitate the analytical method. Methods of analysis in current use for bioaerosols include culture and microscopy (Muilenberg, 1989), immunoassays (see Chapter 12; Dorner et al., 1993), and the *Limulus* bioassay (see Chapter 4; Milton et al., 1992). In addition, new methods such as probes based on the polymerase chain reaction (PCR) (Palmer et al., 1993), gas chromatography/mass spectroscopy (White, 1983; Elmroth et al., 1993; Fox et al., 1990), and other chromatographic techniques show promise for specific agents (Hansen, 1993). Which method is chosen depends on the kind of bioaerosol of interest and the kind of health effect that is expected. It is always best to use the analytical method that most closely approximates the disease process. Ideal methods for selected bioaerosols are presented in Table 1.1. The *Limulus* amebocyte lysate test (LAL) and immunoassays are covered in Chapters 4 and 12, respectively.

B. Infection

Infection assays are, for the most part, impractical. However, an elegant procedure for monitoring aerosols of *Mycobacterium tuberculosis* has been reported by Riley et al. (1962), in which guinea pigs are exposed to air from rooms housing tuberculosis patients. The guinea pigs, which are very sensitive to infection (a single cell is known to result in disease), act as both the sampler and the analytic method.

C. Culture

Culture is, by far, the most commonly used analytical method for assessment of exposure to fungi and bacteria, and the most popular air samplers (the culture-plate impactors) depend on culture for analysis. At present, culture is the only means by which the common saprophytic bacteria and fungi can be accurately identified to the species level.

Table 1.1. Analytic methods for some bioaerosols related to the disease process.

Bioaerosol	Disease	Ideal analytic method	Usual analytic method
Bacterial cells	Endotoxicosis	LAL	LAL
	Tuberculosis	Infection	Culture
Fungus spores	Asthma	Immunoassay	Culture, microscopy
	HP	Immunoassay	Culture, microscopy
	Infection	Infection	Culture
	Toxicosis	Chemical assay	Culture
Animal effluents	Asthma	Immunoassay	Immunoassay

Note: LAL = *Limulus* amebocyte assay; HP = hypersensitivity pneumonitis.

Cultural analysis essentially provides information on the living organisms in a sample that are able to grow under the conditions provided. For each aerosol or bulk sample to be evaluated, a combination of conditions must be chosen that either provides the broadest coverage of the most organisms or provides optimum conditions for recovery of a single kind of particle. Note that there are *no* conditions that are optimal for both fungi and bacteria, and that any combination of conditions will select against some organisms. Conditions that can be controlled include (1) the mechanism by which the sample was collected (see below); (2) characteristics of the culture medium (pH, water activity, nutrient content, and toxin content); (3) incubation conditions, including temperature, wavelengths, intensity, and patterns of exposure to light and aeration, and (4) length of time under these conditions.

pH. In general, the fungi grow best at relatively low pH (<7) while bacteria usually require pH in the neutral range. The nutrients in most broad-spectrum media for both fungi and bacteria are readily utilized by members of either group. The reason few bacteria appear on, for example, malt extract agar and few fungi on nutrient agar is probably due to the respective pH values for these media (Table 1.2).

Water activity. Water activity (a_w) is a measure of the amount of available water in a substrate (Kendrick, 1992). It is usually measured by allowing the substrate material to come to equilibration in a small chamber and measuring relative humidity in the chamber. Most organisms (including all bacteria) require a substrate a_w in excess of 0.90. Fungi, however, have varying requirements for available water and can be loosely classed according to these requirements (Table 1.3). The xerophilic fungi require low water activity to grow and sporulate and are not recovered on standard culture media (e.g., *Aspergillus restrictus* and members of the *Aspergillus glaucus* group). On the other hand, some mesophilic fungi that are important components of indoor air (e.g., *Stachybotrys atra,* members of the genus *Fusarium*) require media of high water activity. Therefore, when surveying any reservoir for unknown fungal populations, it is advisable to use both a mesophilic medium (e.g., malt

Table 1.2. Some commonly used culture media for bacteria and fungi.

	Carbon/nitrogen sources	pH	Water activity	Toxins
Fungus media				
2% Malt	Malt extract	4.7	High	—
MEA	Malt extract Peptone Dextrose	4.7	High	—
RBS	Dextrose Soytone	7.2	High	Rose bengal Streptomycin
Saboraud's	Dextrose Peptone	5.6	High	—
Czapek's	Saccharose Sodium nitrate	7.3	High	—
DG-18	Glucose Peptone		Low	Glycerol Dichloran Streptomycin
Bacterial media				
Nutrient	Beef extract Peptone	6.8	High	—
TSA	Tryptone Soytone	7.3	High	—
R2A	Yeast extract Peptone Dextrose Starch Casamino acids Sodium pyruvate	7.2	High	—
MacConkey's	Lactose Peptone	7.1	High	Neutral red Crystal violet

Table 1.3. Some common fungal saprophytes classified by water requirement.

Hydrophilic	Mesophilic	Xerotolerant	Xerophilic
Fusarium	*Alternaria*	*A. versicolor*	*A. glaucus gr.*
Phialophora	*P. purpurogenum*	*C. sphaerospermum*	*A. restrictus*
Sporobolomyces	*C. herbarum*	*P. chrysogenum*	*Wallemia*
	S. atra		*Sterigmatomyces*

Note: P = *Penicillium*; A = *Aspergillus*; C = *Cladosporium*.

extract agar) and a low water activity medium (e.g., DG-18) (Hocking and Pitt, 1980; Verhoeff et al., 1990).

Nutrient content. Although most fungi and bacteria normally encountered in indoor air have broad nutrient requirements, some can utilize specialized substrates that can be used for selective isolation. For example, the provision of cellulose as the sole carbon source will select for those organisms that produce cellulase (e.g., *S. atra, Chaetomium globosum,* etc.) and will prevent or drastically slow the growth of many common environmental isolates, including common species of *Cladosporium* and *Alternaria.*

Table 1.4. Optimum temperature ranges for fungi and bacteria.

	Psychrophiles	Mesophiles	Thermotolerant	Thermophilic
Bacteria	4–15°C	22–37°C		45–60°C
Fungi	4–15°C	15–30°C	20–45°C	45–60°C

Toxin content. Toxins designed to control specific organisms or classes of organisms that might mask organisms of interest can be added to culture medium. Various antibiotics are often added to fungus culture media to avoid overgrowth with bacteria (although use of low pH is probably sufficient in most environments). Rose bengal is a compound that is toxic to bacteria (and to other organisms as well when light activated). It has been used in fungus culture medium to suppress bacteria as well as to limit radial growth of the fungi (Rogerson, 1958). Because of its overwhelming biocidal effects when light activated, it must be used with extreme care. Antifungal agents are also sometimes used (e.g., cycloheximide). The DG-18 formulation includes an antifungal agent to limit spread of fast-growing organisms.

Temperature. Microorganisms have temperature optima as well as ranges at which they will grow and (for the fungi) sporulate (Table 1.4). Temperature can be used to select for organisms with highly specific (or very broad) temperature requirements. For example, *Aspergillus fumigatus* is one of the few common fungi that will grow in culture above 45°C. The thermophilic actinomycetes (filamentous bacteria) require temperatures in excess of 50°C for growth, as do some species of *Bacillus*.

Light. Light can be irrelevant, a stimulant, or a suppressant for microorganisms, depending on the organism and other conditions. Clearly, some wavelengths of ultraviolet light are toxic for most organisms (e.g., Riley et al., 1962), although many fungi have melanized cell walls that provide considerable protection (Leach, 1962). For many bacteria, visible light is probably not important. On the other hand, light across a broad spectrum appears to play a role in fungal morphogenesis. Many fungi require very specific cycles of light and dark, in addition to specific wavelengths of light before sporulation (either sexual or asexual) occurs (see Chapter 5; Leach, 1962).

Aeration. Nearly all fungi and bacteria commonly encountered in indoor air (as far as we know) require oxygen for growth. Fungi usually do best on solid culture media, although most will grow on the surface of liquids. Agitation is required to induce subsurface fungal growth, and sporulation often does not occur within the culture medium. Bacteria, on the other hand, appear readily able to grow submerged in liquid, although surface films can also be formed, and most bacteria also do well on solid media.

Time. The time required to produce a mature microbial colony depends on the nature of the organism, temperature, and other environmental conditions. For many bacteria, well-developed colonies are produced within 24 h at 37°C. However, for some, more than 2 weeks of incubation time is required for visible colony formation (e.g., *Mycobacterium tuberculosis*. At room temperature, up to 5 days may be necessary for most bacteria. Most fungus cultures are incubated at room temperature for at least 7 days. Two or more weeks might be necessary for sporulation in some fungi.

Common errors associated with cultural analysis include:

1. Use of inappropriate culture media: As mentioned above, each culture medium is selective, and one cannot assume the absence of an organism because it fails to appear in culture.
2. Too few or too many colonies on each plate: For bacteria, variance between duplicate culture plates appears to become minimal above about 50 colonies. Depending on colony size, accurate counts of bacteria can be made with numbers in excess of 200 to 300 colonies/plate. For fungi, although variance continues to decrease, recoveries begin to decline at about 10 colonies per plate on malt extract agar, and inhibition becomes severe above 50 colonies. Culture media that limit colony diameter (e.g., DG-18) may allow accurate recoveries at higher colony concentrations.
3. Inaccurate counts: Counting errors increase with the number of colonies on the plate and as the colony size decreases. Counting errors can be avoided by using low-power magnification.
4. Inaccurate identifications: Bacteria can often be identified using standardized "kits" that require little knowledge of bacterial taxonomy. Most of these commercially available methods are designed for clinical specimens, and do not result in identification of many environmental isolates. Identification of bacteria that do not fit these schemes requires extensive experience and effort, often including subculture onto many different kinds of media. Fungal identification has not been standardized, and extensive training is required for accurate identification. This is the major, but unavoidable, drawback of the use of culture for fungal analysis, and is driving the search for more automated methods.

D. Microscopy

Microscopy relies on the existence of characteristics that allow a particle to be recognized visually (or by a computer based on visual characters). Microscopy is especially useful when total counts of some broad category (e.g., asymmetric basidiospores, grass pollen) are desired. This is not adequate information, in most cases, to make close connections between a specific disease process and an agent. Microscopy is an extremely useful probe or monitoring technique, however, that allows one to recognize unusual exposure situations without reliance on culturability.

Light microscopy. Light microscopy is historically and currently the primary method used for evaluation of outdoor aerosols, especially of pollen. Worldwide networks exist that routinely monitor the ambient aerosol using microscopy as an analytical tool (D'Amato, 1993; AAAI, 1994). Data from these analyses is presented as categories, such as tree, grass, weed pollen, or, where possible, with generic identification. For some pollen and large fungal spore types, relatively accurate generic identifications can be made (e.g., *Ambrosia, Plantago, Juglans, Alternaria, Pithomyces*). Some of these groupings are inclusive, meaning that all pollens of the genus are included, but not all pollens counted are necessarily of that genus. For example, *Ambrosia* pollen is consistent across the entire genus but cannot always be distinguished from (for example) *Xanthium* pollen. On the other hand, some categories are exclusive, meaning that all of the particles counted belong to the genus, but not all particles of that genus are counted. A good example here is *Pithomyces*, a common fungus taxon both outdoors and in. Generally, only one species of *Pithomyces* is included in spore counts (*P. chartarum*). It is readily recognized as such and does not closely resemble any other spore type. However, a number of other species exist in the genus for which the spores are not readily recognized.

In general, fungus spores can be counted without staining, although acid fuchsin will enhance visibility of small colorless spores. However, pollen is more readily identified when stained. Commonly used stains include phenosaffranin, basic fuchsin, and cotton blue. Bacteria can be counted using fluorescent stains (Palmgren et al., 1986). In this case, the bacteria are counted as a group, with little, if any, differentiation possible. Some fungus spores can also be counted in this way, although many resist staining, or the fluorescence is masked by dark pigmentation.

Immunospecific fluorescence staining is commonly used to identify *Legionella pneumophila* serotypes (Palmer et al., 1993) and has also been used for pollen (Schumacher et al., 1988). As with other immunoassays, this is a taxon-specific technique that requires preparation of specific antibodies (i.e., you must know what you are looking for) (Drowart et al., 1993).

The primary objection to the use of microscopy for routine monitoring of bioaerosols is that it is time- and skill-intensive. There are relatively few individuals in existence that are able to accurately identify even a small fraction of the ambient fungus aerosol. Efforts are currently under way to develop computerized methods for the identification of bioaerosols, and technology is available that should allow mechanization of the actual counting procedure. Combining image analysis with, for example, immunospecific staining, may allow fast, accurate counting of specific particle types. However, monitoring unknown aerosols may continue to rely on well-trained technicians for many years because of the immense number of different kinds of aerosol particles that can be abundant in the air and the subtle differences that distinguish them.

Common errors associated with microscopic analysis include:

1. Undercounting: This can result from there being too many biological particles on a sample, or from masking of biological particles by nonbiological material. Microscopic analysis is of limited use for samples where extraneous material is present in much higher concentrations than the particles of interest.
2. Overcounting: Overcounting results when microscope fields are counted more than once, when the fraction of a sample counted is inaccurately determined, resulting in inaccurate conversions, or when either biological or nonbiological particles are present that are morphologically similar. For example, small oil droplets are often present on air sample slides that resemble some *Penicillium* or similar small globose spores.
3. Misidentifications: Giving an incorrect name to a biological particle is the most frequent error of this type (e.g., counting oak pollen grains under the category of maple pollen). As mentioned above, nonbiological particles can also be mistaken for fungus spores.
4. Basing inferences on counts of too few particles: This error is particularly a problem if only a fraction of a sample is to be counted, because conversion factors may result in large numbers based on only one or two particles. Our general rule is to require 50 particles in any group if only a fraction of a sample is to be counted.

SEM/TEM. Scanning (SEM) and (to some extent) transmission electron microscopy (TEM) provide much higher resolution than light microscopy. Small particles such as bacteria can be readily seen, and details of pollen and spore surfaces are resolved. As an aerobiological research tool, electron microscopy has some intriguing possibilities, although it has been rarely used (Geisbert et al., 1993; Eduard et al., 1990). For example, carriage of bacteria or nonbiological materials on larger biological particles could be examined, and antigen localization studies using immunostaining techniques are feasible. However, SEM only resolves surface characteristics, so that (for fungus spores) septations and color are lost, and for pollen, internal wall structures cannot be used for identification. Also, the necessary preparative procedures that could easily cause loss of particles and the very limited fraction of any sample that can be counted render electron microscopy of limited use for monitoring studies.

E. PCR

The polymerase chain reaction allows large quantities of specific DNA fractions to be produced, and to be used as probes for very specific biological particles. Such probes could be used in combination with light microscopy (using fluorescent labeling), with SEM or TEM using an electron-detectable label, or with electrophoretic methods such as are used in immunoblotting to evaluate environmental samples for specific organisms, or as monitoring tools for laboratory-generated aerosols. As with the immunoassays, PCR methods are valuable only where a very specific type of particle is of interest. For example, genetic probes have been produced for multiply drug-resistant strains of *Mycobacterium tuberculosis* (Nolte et al., 1993) and for specific serotypes of *Legionella pneumophila* (Palmer et al., 1993).

F. Pyrolysis GCMS

Pyrolysis and gas chromatography/mass spectroscopy (GCMS) take advantage of the fact that biological particles have characteristic profiles of specific chemical constituents, usually fatty acids. The method has been relatively well developed for the identification of some bacteria (White 1983; Elmroth et al., 1993; Fox et al., 1990) and may be adaptable to a wide variety of biological particles. It will remain a research tool until its specificity has been more carefully documented, and until extensive libraries of data on common environmental biological particles have been accumulated.

III. SAMPLE COLLECTION METHODS FOR BIOLOGICAL AGENTS

A. Overview

Sampling for biological agents can include visual observation, collection of bulk or surface samples, or air sampling. Which of these modalities is chosen is determined by the hypothesis under which the investigation is being conducted (see above).

B. Observational Sampling

Observational sampling is simply the process of walking through an environment and using the human senses (sight and smell) to evaluate the biological status of a building. Observation is almost always the first step in a problem-solving building investigation, but can also be used for other kinds of studies. Observational sampling can be qualitative, indicating only the presence or potential presence of bioaerosol sources, or it can be semiquantitative, providing comparative information from one building to another.

Qualitative observation. Qualitative observation focuses on the recognition of indicators of contamination, including visible microbial growth, the presence of animals or arthropods known to produce aerosols, or odors known to be associated with bioaerosol reservoirs (moldy, fishy, or rotting odors, or odors commonly associated with the presence of animals). Noting the presence of appliances (e.g., humidifiers) or conditions (e.g., standing water, flooding, high relative humidity) that are commonly recognized as potential bioaerosol sources is also a part of observational sampling. Often these observations provide sufficient evidence to develop a hypothesis as to the cause of complaints, and recommendations can be made. Presence/absence data can also be used in epidemiological studies. For example, several published studies have correlated health effects with the reported presence of water damage in homes (e.g., Brunekreef et al., 1989).

Quantitative observational techniques. Observations can be made semiquantitative by estimating the extent of contamination and by standardizing observation procedures. Visible areas of contaminated surfaces can be measured, number of potentially contributing appliances counted, and odors categorized on some subjective scale. Checklists can also be developed so that all buildings are evaluated consistently, and descriptors of contamination are the same throughout a study. For example, a checklist might ask the observer to rate the extent of fungus growth on interior wall surfaces on a scale of 1 to 10 according to some preset formula.

C. Bulk and Surface Sampling

The next step up from observational sampling is to collect samples from actual and potential reservoirs (e.g., fungus growth on surfaces, water from potential reservoirs). Qualitatively, these samples provide data on the nature of the contamination, an important addition, since the health risks related to bioaerosols are agent specific. Bulk sampling can also be done in quantitative ways. For example, house dust bulk sampling is the method generally used for mite allergen assessment (see Chapters 7 and 8). Dust samples are usually collected with a vacuum device, analyzed by immunoassay, and results reported as nanograms of mite allergen per gram of dust. Interestingly, although this is the accepted method, and results appear to correlate with health risk (Platts-Mills et al., 1992), such data do not reveal how much dust mite allergen actually exists in an environment. To arrive at that information, one would have to collect the original sample in such a way that estimates could be made of how much dust was actually in the environment. This would involve the collection of a dust sample that is truly representative of the environment. In fact, the difficulty in using bulk sampling in a truly quantitative way lies in the problems faced in collecting a representative sample.

D. Air Sampling

Even if one were to be able to collect a representative bulk sample, human exposure must be extrapolated in order to evaluate health risk. While for some appliances (e.g., humidifiers) this might be relatively simple to measure, in general, reservoir concentrations have not been quantitatively related to exposure for any bioaerosol. Ideally, then, air sampling should be the method most likely to reveal actual exposure information.

Air sampling plans. As for bulk sampling, air sampling is not likely to yield useful data unless a sample that is representative of exposure is collected. Ideally, breathing zone samples collected over the range of aerosol concentrations likely to be encountered are most representative of individual exposure. However, no personal sampling method has yet been proposed that is sensitive enough for any bioaerosol. There is, therefore, dependence on ambient sampling and on sampling plans that are designed so that reasonable estimates of exposures of given populations can be made over representative time periods.

Perhaps the first question to be asked is how many replicate samples are required for each sample site for each time period. It is always best to collect biological air samples at least in duplicate. However, costs of both samplers and analysis mandate, in many cases, that only one device be used so that sequential replicators are the only option. Locations and times for collection of ambient samples are chosen to best represent potential exposures, including release from sporadically active sources. For example, if the ventilation system is considered a potential source, then samples should be taken near supply air vents over periods of time that include both "on" and "off" conditions. Some investigators use aggressive methods to maximize the chances that potential contributions from specific reservoirs are assessed (Morey et al., 1984). Artificial disturbance of reservoirs should mimic natural activities as closely as possible.

For building investigations designed to provide evidence for causality of complaints, as many samples as possible should be taken. We consider ten 1-min Andersen culture plate samples using a single culture medium (replicate samples at five locations or times) to be the minimum for characterizing a home on a single visit. Larger buildings with more people require more intensive sampling. The error most often made by inexperienced field investigators is to collect an inadequate number of samples.

Such intensive sampling is not possible for large epidemiological studies, with the result that exposure measures are often not representative. Continuously operating samplers such as the Burkard indoor recording spore trap (Burkard Manufacturing Company Ltd., Rickmansworth, England) provide better exposure estimates for fungi than grab samples, providing the aerosols of interest are microscopically identifiable and sufficient funding is available for analysis of the samples. To analyze a single Burkard slide (representing 24 h) in 2-h increments takes a minimum of 1 h. On the other hand, significant correlations between fungus levels and health outcomes have been reported using single grab samples (Harrison et al., 1992).

Choice of samplers. Air samplers for biological aerosols are often chosen on the basis of cost or, perhaps, convenience. Actually, the first consideration should always be the method of analysis to be used, since this is what determines the kind of information to be retrieved. The ACGIH Air Sampling Instruments Manual (ACGIH, 1994) provides a review of available bioaerosol samplers that specifies the kinds of analysis that can be performed on resulting samples. Briefly, cultural analysis requires collection of samples in ways such that viability (and hence, culturability) is protected. This usually means that some water-based medium must be used. The most commonly used sampling methods for cultural analysis are the culture-medium impactors (e.g., the sieve plate, slit, and centrifugal impactors) (Muilenberg, 1989). Liquid impingers are also used for bacterial aerosols (Cox, 1987) (fungus spores are generally hydrophobic and difficult to collect in liquid media). Filtration methods are also occasionally used, but losses due to desiccation have yet to be documented for most aerosols. If samples are to be evaluated microscopically, then a sample of good optical quality is needed. The slide or tape impactors are most convenient, although filters can be cleared to reasonable transparency. For immunoassay and chemical assays, the primary considerations with respect to analysis are that the particles must be readily washed from the collection medium and must not be changed by the medium (Milton et al., 1990). Filtration is most often used, although agar gel impaction (Sakaguchi et al., 1990) and liquid impingement are also amenable to such assays.

The next consideration should be the level of aerosol expected, including expected concentrations of extraneous aerosols. The culture-plate impactors overload readily (unless the medium is processed for dilution culture). For a 1-min sample at 28.3 l/min, 10 colonies represent about 360 colony-forming units (CFU)/m^3 air; 50 colonies represent about 1800 CFU/m^3 (remember that 10 to 50 fungal colonies is the optimal range). Thirty-second samples double this upper limit, while longer sampling times (up to 15 min) increase sensitivity. Overall, the effective range for the sieve impactors on 100-mm diameter plates is about 80 to 3500 CFU/m^3. If lower levels are expected, samplers should be chosen that collect larger volumes of air (e.g., the portable culture plate or centrifugal devices). For higher levels, it may be necessary to dilute samples after collection. One method is to collect the sample onto gelatin, then carefully melt the gelatin and use dilution culture (Blomquist et al., 1984).

For slide and tape samples, a primary consideration is the presence of irrelevant aerosols that will use space on the impaction surface (thereby decreasing particle collection efficiency over time) and that will mask the biological particles. In very "dirty" air, a compromise must be achieved, often accepting relatively low biological particle counts in order to prevent losses in collection efficiency.

Immunoassay and chemical assays are usually limited by the lower limit of sensitivity rather than overloading, since dilution is usually possible. Dust mite allergen, for example, has proven extremely difficult to recover from air, at least in part because assays are inadequately sensitive to detect the very low levels of airborne allergen that are apparently necessary for sensitization/ symptom development.

Note that sharp peaks of airborne mite allergen may be the important exposure, and these peaks are not likely to be recovered using any short-term ambient sampling plan. This emphasizes the third factor with respect to sampler choice — time discrimination. Most bioaerosol sources release aerosols sporadically, and short-term grab samples are very likely to miss important peak concentrations. For chemical and immunoassay, long-term filtration sampling is possible. For microscopic identification, the moving tape or slide samplers can provide time-discriminated information. Unfortunately, except for the short-term (1 h) time discrimination offered by the rotating slit samplers (which require relatively low aerosol concentrations), no good method other than a series of closely spaced sequential replicates exists for cultural sampling.

Particle size collection efficiency is another important factor to consider, although all of the commercially available samplers will collect all but the smallest aerosols with an efficiency that is adequate for most studies. The exception is the popular rotating rod impactors that are commonly used to study outdoor aerosols. These devices are efficient for particles above about 15 μm (including nearly all pollens) but are not adequate for characterizing fungal spore aerosols.

Finally, cost and convenience are important considerations. Most good bioaerosol samplers are expensive (with the exception of filtration devices). Convenience, then, becomes the bottom line. Especially for epidemiological studies, small, lightweight, battery-powered devices are more popular than the more efficient bulky line-powered samplers.

Sampler calibration. It cannot be emphasized too strongly that every sampler needs to be carefully calibrated. Most devices are factory calibrated. However, we have found calibration errors exceeding 100% for some devices. Calibration techniques must take into account the capability of the vacuum source to overcome resistance. Several of the culture plate impactors use suction devices that will not maintain flow with any added resistance, so that bubble tubes or other low-resistance calibration methods must be used.

IV. CONCLUSIONS

This chapter has been an introduction to some of the principles of environmental bioaerosol assessment. It should be stressed that each situation or study is unique, and that there is no standard, "off the shelf" method that can be used for bioaerosol investigations. This means that, for each study, good aerobiological expertise is necessary. This is a significant limiting factor at present, because there are so few well-trained aerobiologists available. Aerobiology training programs and funding for such programs are virtually nonexistent.

We consider the primary area of research focus for bioaerosols should be the ecology of sources (the study of factors controlling source strengths) and the relationship of sources to aerosols and human exposures. Such knowledge is essential if adequate sampling plans are to be devised that accurately assess human exposure and lead to the ability to evaluate the risk associated with exposure to specific bioaerosols.

REFERENCES

AAAI, *1993 Pollen and Spore Report,* American Academy of Allergy and Immunology, Milwaukee, WI, 1994.

ACGIH, *Air Sampling Instruments Manual*, 7th ed., American Conference of Governmental Industrial Hygienists, Cincinnati, OH, 1994 (in press).

Blomquist, G., Palmgren, U., Strom, G., Improved techniques for sampling airborne fungal particles in highly contaminated environments, *Scand. J. Work Environ. Health* 10, 253, 1984.

Brunekreef, B., Dockery, D. W., Speizer, F. E., Ware, J. H., Spengler, J. D., Ferris, B. J., Home dampness and respiratory morbidity in children, *Am. Rev. Respir. Dis.* 140, 1363, 1989.

Burge, H. P., Solomon, W. R., Boise, J. R., Comparative merits of eight popular media in aerometric studies of fungi, *J. Allergy Clin. Immunol.* 60(3), 199, 1977.

Chapman, M. D., Heymann, P. W., Wilkins, S. R., Brown, M. J., Platts-Mills, T. A. E., Monoclonal immunoassays for the major dust mite (*Dermatophagoides*) allergens, *Der p* I and *Der f* I, and quantitative analysis of the allergen content of mite and house dust extracts, *J. Allergy Clin. Immunol.* 80, 184, 1987.

Cox, C. S., *The Aerobiological Pathway of Microorganisms,* John Wiley & Sons, New York, 1987.

Croft, W. A., Jarvis, B. B., Yatawara, C. S., Airborne outbreak of trichothecene toxicosis, *Atmos. Environ.* 20, 549, 1986.

D'Amato, G., *Patologia Allergica Respiratoria*, Editrice Kurtis, Milano, 1993.

Dorner, J. W., Blankenship, P. D., Cole, R. J., Performance of 2 immunochemical assays in the analysis of peanuts for aflatoxin at 37 field laboratories, *J. AOAC Int.* 76(3), 637, 1993.

Drowart, A., Cambiaso, C. L., Huygen, K., Serruys, E., Yernault, J. C., Vanvooren, I. P., Detection of mycobacterial antigens present in short-term culture media using particle counting immunoassay, *Am. Rev. Respir. Dis.* 147(6), 1401, 1993.

Eduard, W., Lacey, J., Karlsson, K., Palmgren, U., Strom, G., Blomquist, G., Evaluation of methods for enumerating microorganisms in filter samples from highly contaminated occupational environments, *Am. Ind. Hyg. Assoc. J.* 51(8), 427, 1990.

Elmroth, I., Fox, A., Larsson, L., Determination of bacterial muramic acid by gas chromatography-mass spectrometry with negative-ion detection, *J. Chromatogr.* 628, 93, 1993.

Fox, A., Rogers, J. C., Fox, K. F., Schnitzer, G., Morgan, S. L., Brown, A., and Aono, R., Chemotaxonomic differentiation of Legionellae by detection and characterization of aminodideoxyhexoses and other unique sugars using gas chromatography-mass spectrometry, *J. Clin. Microbiol.* 28(3), 546, 1990.

Geisbert, T. W., Jahrling, P. B., Ezzell, J. W., Use of immunoelectron microscopy to demonstrate *Francisella tularensis*, *J. Clin. Microbiol.* 31(7), 1936, 1993.

Hansen, T. J., Quantitative testing for mycotoxins, *Cereal Foods World* 38(5), 346, 1993.

Harrison, J., Pickering, C. A., Faragher, E. B., Austwick, P. K., Little, S. A., Lawton, L., An investigation of the relationship between microbial and particulate indoor air pollution and the sick building syndrome, *Respir. Med.* 86, 225, 1992.

Hocking, A. D., Pitt, J. I., Dichloran glycerol medium for enumeration of xerophilic fungi from low-moisture foods, *Appl. Environ. Microbiol.* 39, 488, 1980.

Kendrick, B., *The Fifth Kingdom*, 2nd ed., Mycologue Publications, Newburyport, MA, 1992.

Leach, C. M., Sporulation of diverse species of fungi under near-ultraviolet radiation, *Can. J. Bot.* 40, 151, 1962.

Menzies, R., et al., The effect of varying levels of outdoor-air supply on the symptoms of sick building syndrome, *N. Engl. J. Med.* 328, 821, 1993.

Milton, D. K., Feldman, H. A., Neuberg, D. S., Bruckner, R. J., Greaves, I. A., Environmental endotoxin measurement: the kinetic *Limulus* assay with resistant-parallel-line estimation, *Environ. Res.* 57, 212, 1992.

Milton, D. K., Gere, R. J., Feldman, H. A., Greaves, I. A., Endotoxin measurement: aerosol sampling and application of a new *Limulus* method, *Am. Ind. Hyg. Assoc. J.* 51(6), 331, 1990.

Morey, P. R., Hodgson, M. J., Sorenson, W. G., et al., Environmental studies in moldy office buildings: biological agents, sources, and preventive measures, *Ann. Am. Conf. Gov. Ind. Hyg.* 10, 121, 1984.

Muilenberg, M. L., Aeroallergen assessment by microscopy and culture. In: Solomon, W. R. (Ed.), *Airborne Allergens, Immunol. Allergy Clin. North Am.* 9(2), 245, 1989.

Nolte, F. S., Metchock, B., Mcgowan, J. E., Edwards, A., Okwumabua, O., Thurmond, C., Mitchell, P. S., Plikaytis, B., Shinnick, T., Direct detection of *Mycobacterium tuberculosis* in sputum by polymerase chain reaction and DNA hybridization, *J. Clin. Microbiol.* 31(7), 1777, 1993.

Palmer, C. J., Tsai, Y. L., Paszkokolva, C., Mayer C., Sangermano, L. R., Detection of *Legionella* species in sewage and ocean water by polymerase chain reaction, direct fluorescent-antibody, and plate culture methods, *Appl. Environ. Microbiol.* 59(11), 3618, 1993.

Palmgren, L. L., Strom, G., Blomquist, B., Malmberg, P., Collection of airborne microorganisms on Nuclepore filters, estimation and analysis — CAMNEA method, *J. Appl. Bacteriol.* 61(5), 401, 1986.

Peat, J. K., Tovey, E., Mellis, C. M., Leeder, S. R., Woolcock. A. J., Importance of house dust mite and *Alternaria* allergens in childhood asthma, an epidemiological study in two climatic regions of Australia, *Clin. Exp. Allergy* 23(10), 812, 1993.

Platts-Mills, T. A. E., Thomas, W. R., Aalberse, R. C., Vervloet, D., Chapman, M. D., Dust mite allergens and asthma: report of a second international workshop, *J. Allergy Clin. Immunol.* 89, 1046, 1992.

Riley, R. L., Mills, C. C., O'Grady, F., Sulton, L. U., Wittstadt, F., Shivpuri, D. N., Infectiousness of air from a tuberculosis ward: ultraviolet irradiation of infected air; comparative infectiousness of different patients, *Am. Rev. Respir. Dis.* 84, 511, 1962.

Rogerson, C. T., Kansas aeromycology. I. Comparison of media, *Trans. Kansas Acad. Sci.* 61(2), 155, 1958.

Sakaguchi, M., Inouye, S., Yasueda, H., Tatehisa, I., Yoshizawa, S., and Shida, T., Measurement of allergens associated with house dust mite allergy II. Concentrations of airborne mite allergens (*Der* I and *Der* II) in the house, *Int. Arch. Allergy Appl. Immunol.* 90, 190, 1990.

Schumacher, M. J., Griffith, R. D., and O'Rourke, M. K., Recognition of pollen and other particulate aeroantigens by immunoblot microscopy, *J. Allergy Clin. Immunol.* 82(4), 608, 1988.

Strachan, D. P., Flannigan, B., McCabe, E. M., McGarry, F., Quantification of airborne moulds in the homes of children with and without wheeze, *Thorax* 45(5), 382, 1990.

Verhoeff, A. P., van Wijnen, J. H., Boleij, J. S., Brunekreef, B., van Reenen-Hoekstra, E. S., Samson, R. A., Enumeration and identification of airborne viable mould propagules in houses. A field comparison of selected techniques. *Allergy* 45(4), 275, 1990.

White, D. C., Analysis of microorganisms in terms of quantity and activity in natural environments. In: Slater, J. H., Whittenbury, R., and Wimpenny, J. W. T. (Eds.), *Microbes in their Natural Environments*, Symp. 34, Soc. Gen. Microbiol. Ltd., Cambridge University Press, Cambridge, 1983.

Wood, R. A., Laheri, A. N., Eggleston, P. A., The aerodynamic characteristics of cat allergen, *Clin. Exp. Allergy* 23(9), 733, 1993.

2 AIRBORNE CONTAGIOUS DISEASE

Harriet A. Burge

CONTENTS

0-87371-724-4/95/$0.00+$.50

I. INTRODUCTION

A. The Current Problem

Interest in airborne contagion has recently reemerged for several reasons. Tuberculosis, well known to be transmitted by small particle aerosols, has again become a major public health problem (Nardell, 1989). This is due in part to the emergence of antibiotic resistant strains and to the AIDS (acquired immune deficiency syndrome) epidemic, which has provided a large population that is especially susceptible to the disease.

A second impetus for interest in airborne infection is the current concern with the quality of indoor environments where we spend in excess of 90% of our time (Samet and Spengler, 1991). In the interest of energy efficiency, we have gradually restricted the amount of outside air supplied to interiors. While this restriction does not seriously limit the amount of available oxygen, it can allow for the accumulation of indoor-generated air pollutants, including infectious aerosols (Brundage et al., 1988; Moser et al., 1979). Measures taken to reduce energy consumption by limiting outdoor air ventilation rates have also been taken in hospitals, possibly increasing the risk for hospital transmission of contagious disease. In addition, we have taken some of the most dangerous of our infectious disease agents (including the viruses that cause AIDS, Hepatitis B, Lassa fever, etc.) into the laboratory where handling can produce infectious aerosols even when the natural disease process does not.

B. History

Airborne contagion is not new. Pasteur demonstrated bacterial survival in air by showing that sterile solutions could become contaminated when exposed to air (thereby refuting the theory of spontaneous generation). In 1876 Robert Koch proved conclusively that bacteria can cause disease by demonstrating *Bacillus anthracis* in the blood of sick cattle, recovering the bacteria from the blood in artificial culture, reinfecting healthy cattle and producing the disease, and finally, recovering the same bacterium from the blood of newly infected cattle. These steps have become known as Koch's Postulates, and are considered definitive proof that a particular agent causes a particular disease (Evans, 1976).

During these early days when both Pasteur and Koch's work were current, it was recognized that some diseases could be transmitted through the air. Subsequently, it became apparent that some diseases either could not be transmitted from one person to another at all (noncommunicable diseases) or were transmitted from one person to another by other means (e.g., by direct contact, or with the intervention of a vector), and the airborne route for disease transmission was considered less important. This was, at least in part, due to the difficulties involved in conducting transmission experiments in the early days (Cox, 1987). However, airborne transmission of infectious disease has been clearly documented (e.g., Wells, 1955) and continues to occur, and some of the most common human diseases are transmitted in this way (e.g., influenza, some common colds, measles, chicken pox, tuberculosis).

This chapter is a discussion of the factors that allow a disease to be transmitted through the air, a discussion of some of the contagious diseases that are clearly airborne, probably airborne, and perhaps airborne, and a review of approaches to the control of airborne contagious disease.

II. THE NATURE OF THE ORGANISMS CAUSING CONTAGIOUS DISEASE

Contagious diseases can only occur when a living organism overcomes the defenses of the host and multiplies within the host. Organisms that are known to cause contagious disease include viruses and bacteria (including rickettsia, chlamydia, and mycoplasmas, as well as the true bacteria) (Tortora et al., 1992).

A. Viruses

Viruses are minute, simple forms of life that require living cells to reproduce. Viruses are categorized on the basis of the kind of genetic material (RNA or DNA), shape, the presence or absence of a lipid coat, the way the virus acts within the cell (e.g., the retroviruses become a part of the host cell DNA), and other characteristics. Agents causing the common aerosol-transmitted viral diseases fall into most of these groups as outlined in Table 2.1.

B. Bacteria

The bacteria are grouped on the basis of the prokaryotic cell (i.e., a cell with no membrane surrounding the genetic material). The group is multiphyletic and is categorized on the basis of the structure of the cell wall (or its absence) and on biochemical characteristics. The bacteria that are known to produce aerosol-transmitted infectious disease are listed in Table 2.2. Note that some of the diseases listed here are not contagious because they are not transmitted from person to person, but rather from exposure to infected animals, or to contaminated environmental reservoirs. In addition, some bacteria can cause other kinds of airborne diseases (e.g., hypersensitivity diseases, and toxicoses) (see Chapters 4 and 10).

III. THE AEROBIOLOGY OF CONTAGIOUS DISEASE

Aerobiology is the study of airborne organisms and their effluents and the effects of these aerosols on other living organisms (people, animals, plants, fungi, etc). A convenient way to subdivide this complex topic is to consider:

Table 2.1. Characteristics of airborne viruses.

Airborne disease	Morphology	Nucleic acid	Name	Other diseases
Respiratory infection	Naked polyhedral	Double-strand DNA	Adenovirus	Tumors
Chicken pox	Naked polyhedral	Double-strand DNA	Herpesvirus (herpes zoster)	Shingles, fever, blisters, infectious mononucleosis
Smallpox	Complex polyhedral	Double-strand DNA	Poxviruses (variola)	Cowpox
Polio, colds	Naked	Single-strand RNA	Picornavirus	Hepatitis A
Rubella	Enveloped polyhedral	Single-strand RNA	Togavirus	Yellow fever, dengue, St. Louis encephalitis
Influenza	Enveloped helical	Single-strand RNA	Orthomyxovirus	—
Measles, mumps	Enveloped helical	Single-strand RNA	Paramyxovirus	—
Colds	Enveloped helical	Single-strand RNA	Coronavirus	—
Newcastle disease/chickens	Enveloped helical	Single-strand RNA	Rhabdovirus	Rabies
(Not airborne)	Enveloped helical	Double-strand DNA	Retrovirus	AIDS
(Not airborne)	Enveloped polyhedral	Double-strand DNA	Hepadnavirus	Hepatitis B

Table 2.2. Characteristics of bacteria causing airborne disease.

Contagious airborne diseases			
Bordetella pertussis	Gram negative rod	Whooping cough	Explosive epidemics
Yersinia pestis	Gram negative rod	Pneumonic plague	Documented; bubonic not contagious
Neisseria meningitidis	Gram negative coccus	Meningococcal meningitis	Explosive epidemics
Staphylococcus aureus	Gram positive coccus	Wound infections	Nosocomial epidemics
Mycoplasma pneumoniae	Mycoplasma	Pneumonia	Explosive epidemics
Chlamydia pneumoniae	Chlamydia	Chlamydial pneumonia	Explosive epidemics
Corynebacterium diphtheriae	Gram positive rod	Diphtheria	Explosive epidemics
Mycobacterium tuberculosis	Acid-fast rod	Tuberculosis	Documented
Noncontagious airborne diseases			
Pseudomonas aeruginosa	Gram negative rod	Respiratory infections	Contaminated humidifiers
Legionella pneumophila	Gram negative rod	Legionellosis	Contaminated water
Brucella abortus	Gram negative rod	Brucellosis	Animal handling
Francisella tularensis	Gram negative rod	Tularemia	Animal handling
Bacillus anthracis	Gram positive rod	Anthrax	Animal handling
Chlamydia psittaci	Chlamydia	Psittacosis	Bird exposure
Coxiella burnetii	Rickettsia	Q fever	Animal handling, laboratory

Sources and disseminators: the places where the organisms grow, reproduce, from which they are disseminated, and the modes of dissemination.
Aerosol characteristics and transport: the study of the physical and biological nature of the aerosols, and mechanisms and patterns of transport of aerosols through the air.
Deposition and the disease process: the study of particle deposition within the human respiratory tract, how the particle (in this case a living organism) colonizes the new host, and the resulting disease states.
Control: the study of ways to control disease resulting from aerosol transmission.

A. Sources and Disseminators

Humans as reservoirs and disseminators. Infected humans are (by definition) the primary sources (reservoirs) for contagious disease, and the primary disseminators as well. For most of the airborne contagious diseases, the respiratory tract is colonized with disease organisms (Couch et al., 1969; Loosli et al., 1970; Knight, 1980). Most also include respiratory symptoms such as coughing or sneezing, although even talking or singing can spread especially virulent diseases (Houk, 1980). Less common are cases of natural aerosol contagion resulting from explosive diarrhea and vomiting (Sawyer et al., 1988).

Virulent agents can also be released from human skin when the disease produces skin lesions (e.g, as in chickenpox and smallpox). Other modes of release directly from infected humans include sprays of saliva and other respiratory secretions during dental and respiratory therapy procedures (Weintraub and Iftimovici, 1981). Blood sprays that occur during dental and surgical procedures are of potential concern for droplet or even aerosol transmission of the blood-borne diseases, including the AIDS and hepatitis viruses (Aach et al., 1968; Almeida et al., 1971). These activities cause the expulsion of a cloud of relatively large droplets, many containing virulent organisms. Large droplets can transmit infectious particles to those close to the disseminator, and are thought to be the primary mode of infection for some diseases (Wells, 1955). At a rate depending on relative humidity, respiratory and other droplets immediately begin to dry, and become "droplet nuclei" within seconds (or even fractions of a second) of release. These much smaller particles remain airborne, and can spread throughout a room, or even from one room to another, although their concentration becomes more and more dilute with distance from the source (Wells, 1934; Moser et al., 1979).

Secondary exposure from human sources. Fomites (inanimate materials on which infectious agents can survive) are usually considered to be involved in direct contact transmission of disease (i.e., the person handles the contaminated material and then carries the organisms to the mouth or nose). However, organisms that remain viable on substrates that can subsequently be disturbed (vibrated, shaken) present the risk of secondary aerosol exposure. Fabrics can sequester viruses for lengths of time that depend on the kind of fabric and the relative humidity. Poliomyelitis virus persists at low levels for up to 20 weeks on wool fabrics (Dixon et al., 1966). In one study, the smallpox virus (vaccinia) persisted on all fabrics tested long enough to be of epidemiological significance (Sidwell et al., 1966). Poliovirus, adenovirus, herpes simplex virus, human rotavirus, and hepatitis A virus can persist for up to 8 weeks and vaccinia and coxsackie viruses for at least 2 weeks on nonporous surfaces (Mahl and Sadler, 1975; Mbithi et al., 1991) All of these effects are dependent on relative humidity, temperature, and wavelength and intensity of light.

Floors, especially in hospitals, can be reservoirs for organisms that are subsequently re-released into the air (Cox, 1987). While carpeting appears to more firmly trap organisms, conditions within carpets may be more conducive to survival, cleaning is rarely completely effective, and reaerosolization may occur during cleaning procedures.

Reaerosolized disease agents may be borne on particles that differ in size from the originally formed droplet nuclei. A change in particle size can have a profound effect on the site of deposition in the human lung, and therefore on the disease process (see below).

Water is a well-known source for infectious disease agents. The route of infection is usually considered to be ingestion of contaminated drinking water, although water readily produces small particle aerosols containing viable microorganisms. Blanchard and co-workers have demonstrated the scavenging of particles from water by bubbles rising to the surface or being splashed from bodies of water (Baylor et al., 1977a). This group demonstrated that the action of surf aerosolizes experimentally added phage (bacterial virus) particles (Baylor et al., 1977b). They hypothesize that a stretch of surf 25 m wide and 100 km long will emit 7.5×10^{11} bubbles per second (Baylor et al., 1977a). If each bubble contains only a single bacterium, it would take only 12 min for all the bacteria contributed by the Hudson River to this section of surf to be put into the air either as droplets or aerosol particles. They suggest that people near the surf may be at risk from infections (both viral and bacterial) from this source.

Laboratories and hospital accidents as sources. Secondary exposure to human-source infectious agents can also occur in the laboratory. In hospitals, pathology laboratories handle infected blood, respiratory secretions, urine, and, ultimately, cultures of infectious agents. Careful laboratory technique can usually prevent exposure to aerosols of these agents. However, some common procedures do produce aerosols under normal circumstances. For example, blenders and centrifuges are commonly used in labs, and both can produce aerosols and have resulted in unusual cases of infectious disease (Conomy et al., 1977; Pike, 1976). Laboratory accidents are another mode for aerosol exposure to infectious aerosols (Pike, 1976).

Animals as sources. Relatively few agents will infect more than one host, so that most animal diseases are not readily transmitted to humans. In general, those animal diseases that are dangerous for people are not contagious (i.e., cannot be transmitted from one person to another). Among the animal-source infections that are commonly transmitted to humans via aerosols are brucellosis (Huddleson and Munger, 1940), anthrax (Brachman et al., 1966; LaForce et al., 1969), Argentine hemorrhagic fever, and Q-fever (Kenyon et al., 1992).

B. Aerosol Characteristics and Transport

Aerosol physics. Airborne infectious particles behave physically in the same way as any other aerosol-containing particles of similar physical properties (i.e., density, size, electrostatic properties, etc.). Infectious aerosols physically change (decay) over time in response (for example) to gravity, electrostatic forces, impaction, and diffusion, and these changes are dependent on the aerodynamic sizes of the particles in the aerosols, and conditions within the aerosol matrix (Willeke and Baron, 1993). Understanding the physical characteristics of infectious aerosols is essential for understanding how airborne disease transmission occurs. For example, the particle size characteristics of an infectious aerosol will determine how long the aerosol will remain at a concentration sufficient for an infective dose to be inhaled. Models that have been developed to describe the fate of aerosols in indoor environments should apply to infectious aerosols, providing sources can be adequately described. Particle size also determines where in the human respiratory tract the particle will land (Knight, 1973) (see below).

Biological properties of the aerosols. However, it is also essential to consider the biological properties of infectious aerosols. An organism that does not remain virulent in the airborne state cannot cause infection, regardless of how many units of the organism are deposited in a human respiratory tract. Among the factors that affect maintenance of virulence are relative humidity (Akers et al., 1973; Benbough, 1969; de Jong et al., 1974, 1975, 1976, etc.), temperature, oxygen, pollutants such as nitrogen and sulfur oxides (Ehrlich and Miller, 1972; Berendt et al., 1972), ozone, ultraviolet light (Berendt and Dorsey, 1971), and the "open air" factor (Donaldson and Ferris, 1975). All of these interact both individually and synergistically with intrinsic factors within each organism (Berendt et al., 1972). Unfortunately, the situation is so complex that extremely unnatural situations must be created to study the effects of any one environmental factor on a particular organism. For example, the fact that oxygen is toxic to many organisms means that these organisms must be aerosolized in nitrogen or other "inert" atmospheres to study, for example, the effects of humidity. In addition to these complex relationships, the methods of producing and collecting the aerosol, which are necessarily unnatural, affect the responses of the organisms, especially to relative humidity (Cox, 1987; Schaffer et al., 1976).

The effect of any one factor on survival can be different for different organisms. For example, respiratory syncytial virus (RSV) survives essentially unchanged in small particle aerosol for 5 h at humidities above 52% but loses up to 2 logs of infectivity in the same period at 33% relative humidity (RH) (Kingston, 1968). Semliki forest virus, on the other hand, is inactivated most rapidly at high RH with about 40% survival after 4 h at 20% RH, and less than 10% over an RH range of 49 to 90% (Benbough, 1969). Simian virus 40 (an oncogenic virus) and yellow fever virus are stable in aerosols over broad temperature and relative humidity ranges. The mechanism of these relationships between humidity and maintenance of virulence is thought to be due to changes in the protein coat of viruses during either dehydration in the air or rapid rehydration during sampling (Akers and Hatch, 1968; deJong et al., 1973). Rehydration of dry aerosols before sampling changes these virulence/relative humidity relationships (deJong et al., 1975). Note that an organism is essentially rehydrated before sampling (impacting) as it enters the human respiratory tract. Table 2.3 presents a simplified view of some of the published relationships between survival of viruses and bacteria in experimental aerosols within broad ranges of relative humidity.

C. Deposition and the Disease Process

Particle size is a critical determinant of the site of deposition in the human respiratory tract (Task Group on Lung Dynamics, 1966), and the site of deposition controls, in part, the disease process. For example, pneumococcal pneumonia is a droplet-borne disease. Droplets emitted from the source that are inhaled as large particles impact in and colonize the upper airway. These organisms are then aspirated into the lung to cause pneumonia. Droplets that evaporate to droplet nuclei and are inhaled as small particles, penetrating directly to the lower airway, apparently do not cause disease. This difference may be due to a loss in virulence (or viability) of the organism in the dry, small particle state. On the other hand, the upper respiratory tract is apparently resistant to colonization with the tuberculosis organism, so that inhaled large droplets fail to colonize the respiratory tract (Nardell, 1993). On the other hand, the organism retains virulence in droplet nuclei, which are deposited and colonize the lower respiratory tract.

D. Control

The science of aerobiology is also concerned with the control of airborne disease. The primary focus for aerobiological control of disease relates to prevention of exposure to virulent agents. Control methods that have been

Table 2.3. Survival of some viruses and bacteria in aerosol related to relative humidity (highly simplified).

	20% RH	40% RH	60% RH	80% RH
Viruses				
RSV	+	++	+++	—
Lassa	+	++	+++	—
Semliki forest	+++	+	+	—
Japanese B encephalitis	+++	++	+	—
Simian 40	+++	+++	+++	—
Moloney murine sarcoma	++	++	++	—
Yellow fever	++	++	++	—
St. Louis encephalitis	+++	+++	+++	—
Influenza	+++	+	++	—
Pseudorabies	++	+++	++	—
Bacteria				
Escherichia coli				
(Water)[a]	—	+++	++	+
(Buffer)[a]	—	+	++	+++
Franciscella tularensis				
(Wet)[a]	+	+	++	+++
(Dry)[a]	+++	++	+	++

Note: +, minimum; ++, intermediate; +++, maximum; —, no data.

[a] Aerosolizing fluid or state.

studied include filtration, dilution ventilation, local exhaust ventilation, treatment of air with biocides (especially ultraviolet light), and masking (for prevention of both release of agents from infected hosts and exposure to susceptible hosts). These and other approaches for control are discussed below.

IV. APPROACHES TO THE STUDY OF AIRBORNE CONTAGIOUS DISEASE

In order to decide whether or not a disease is spread by aerosols one must decide what will constitute adequate evidence. Such evidence can be obtained from the study of disease outbreaks (either case studies, or larger epidemiological studies), by sampling the air and demonstrating that airborne transport is occurring in the natural environment, or by experimental approaches including artificial transmission to animals or humans, or proving that the infectious agent can remain virulent in an aerosol in the laboratory. All of these methods have been used to document airborne transmission, although no single piece of evidence is adequate to document air transmission, and the various methods are used in combination.

A. Outbreaks and Epidemics

Evidence from outbreaks and epidemics has been used to document the consistent airborne route for transmission of some diseases, and for accidental or unusual airborne transmission for others. Diseases that are transmitted by small particle aerosols often spread rapidly, causing explosive epidemics, and nearly all susceptible people are infected within a short period of time. This kind of evidence indicates, for example, that mumps, measles, and chicken pox are commonly transmitted via aerosols (Davis et al., 1986; Riley, 1980; Remington et al., 1985; Josephson and Gombert, 1988; Ledair et al., 1980). Whooping cough and influenza follow this pattern of small droplet transmission, although both are still considered by many to be transmitted by direct contact (including transmission by large droplets) (Couch, 1981; Knight, 1980; Monto, 1987), and airborne transmission does not preclude direct transmission as an alternate route. A classic airborne outbreak of influenza occurred in an unventilated airplane grounded for several hours in which all susceptible passengers contracted from the disease from a single active case (Moser et al., 1979). Airborne outbreaks of diseases not commonly transmitted by aerosols have also occurred. For example, hepatitis was spread to six patients and a staff member as a result of exposure to a blood aerosol (Almeida et al., 1971). An outbreak of diarrhea and vomiting was traced to exposure to aerosols from explosive diarrhea and vomiting in a hospital emergency room (Sawyer et al., 1988). Chlamydial infections (both *Chlamydia psittaci* and *C. pneumoniae*) have been transmitted nosocomially to staff nurses (Bennett and Brachman, 1986). Lassa fever, an extremely dangerous hemorrhagic fever, has been shown to spread through the air at least nosocomially (Stephenson et al., 1984). Other notable cases include airborne laboratory transmission of rabies (Winkler et al., 1973), hepatitis (Aach et al., 1968), yellow fever, and other arbovirus infections (Berry and Kitchen, 1931; Hanson et al., 1967). Infectious diseases contracted through exposure to laboratory aerosols are summarized by Pike (1976).

B. Sampling During Natural Outbreaks

Air sampling is not often used to document the natural transmission of infectious disease, primarily because it is often difficult to induce infectious agents to grow in culture, and because the numbers of agents necessary to cause disease is often below the detection limit of most sampling methods (Artenstein et al., 1967; Gerone et al., 1966; Bourgueil et al., 1992). Hierholzer (1990) describes a method for ambient viral sampling using a culture-plate impactor, but suggests that routine sampling for viral aerosols is not cost-effective because of the cost of analysis. Natural epidemics of foot and mouth disease (a serious viral disease of cattle) have been studied using the Litton high-volume electrostatic sampler and a cyclone sampler (Donaldson et al., 1982). Natural aerosols of *Bacillus anthracis* associated with an epidemic of inhalation anthrax in industrial workers were sampled using Andersen culture-plate impactors (Dahlgren et al., 1960). Natural aerosols of *Coxiella burnetii* have been sampled in animal environments using liquid impingers (DeLay et al., 1950; Lennette and Welsh, 1951) and cotton filters (Welsh et al., 1958) (see Regnery and McDade, 1990). Note that sampling for viruses usually involves the use of some assay system that allows determination of infectivity. These are usually indicator assays that use cell culture and bypass the immune system of people. For bacteria, culture on artificial media is usually used, with virulence inferred. Air sampling to document airborne transmission of tuberculosis in health care settings has involved the use of sentinel animals — usually guinea pigs — which are as, or more, susceptible to the agent as people (Riley et al., 1957). This method, though complex and expensive, has the advantage of documenting infection rather than only the presence of culturable organisms.

C. Experimental Aerobiology

Much of the evidence for the potential for airborne disease transmission comes from laboratory-based studies. Experimental data on airborne transmission takes two general forms: experimental infection in either humans or laboratory animals, and study of the stability of infectious agents in artificial aerosols. Theoretically, if experimental aerosol transmission is effective, or if organisms can be demonstrated to survive in small particle aerosol, the *possibility* of natural aerosol transmission is demonstrated. However, the failure of experimental aerosol transmission does not prove the negative case because of the many variables involved. On the other hand, aerosol transmission demonstrated in the laboratory does not prove that natural aerosol transmission occurs, only that it might be possible. In the natural environment, death of airborne virus, diluted aerosols, different host responses, etc. could all contribute to prevent aerosol transmission. Especially important in the natural environment may be the number of infectious agents shed from patients. If the disease does not include symptoms that are likely to produce aerosols (e.g., coughing, sneezing, explosive vomiting or diarrhea) the agents are unlikely to become naturally airborne in levels sufficient for transmission to occur. Also, the agent must be infective at relatively low doses, usually via the respiratory tract, for natural airborne infection to be likely.

Experimental infection via the airborne route in susceptible people proves that the airborne route of infection is effective (although epidemiological evidence is necessary to conclude that airborne transmission is the usual mode). Influenza has been transmitted to people via aerosols (Alford et al., 1966). Dick has elegantly demonstrated that rhinoviruses are readily spread to volunteers with the only possible route being small particle aerosols (Dick et al., 1987). Couch et al. (1969) demonstrated that small particle aerosols of Coxsackie virus produced significantly greater antibody rises in exposed people than a larger particle aerosol. Needless to say, experiments that use human volunteers can only be used for relatively innocuous organisms.

Animal models can be used for the more dangerous pathogens, although comparable infectivity for humans must then be inferred. These kinds of experiments have been used (for example) to further document the characteristics of airborne transmission of influenza (Loosli et al., 1970; Shulman, 1967; Sherwood et al., 1988).

Diseases not commonly considered to be airborne have also been experimentally transmitted to laboratory animals via aerosols, including Rift Valley fever (Keefer et al., 1972), Argentine hemorrhagic fever (Kenyon et al., 1992), Rauscher murine leukemia virus (McKissick et al., 1970), Lassa fever (Stephenson et al., 1984), Sendai virus (van der Veen et al., 1972), and Japanese B encephalitis (Larson et al., 1980). It is interesting to note that the agent of Argentine hemorrhagic fever (or a similar hantavirus) is now suspected to be causing airborne disease in the southwest (Koster et al., 1993).

Many infectious agents, some of which are unlikely to cause airborne disease outside of the laboratory, have been demonstrated to survive long enough in aerosols that transmission could occur. These include oncogenic viruses (Akers et al., 1973; Hinshaw et al., 1976), Semliki forest virus (Benbough, 1969; deJong et al., 1976), Venezuelan equine encephalitis virus (Berendt et al., 1972; Berendt and Dorsey, 1971; Ehrlich and Miller, 1972), encephalomyocarditis virus (deJong et al., 1974, 1975), Japanese B encephalitis virus (Larson et al., 1980), yellow fever virus (Mayhew et al., 1968), St. Louis encephalitis virus (Rabey et al., 1969), and pseudorabies virus (Schoenbaum et al., 1990). Most of these experiments were designed to examine the effects of specific factors (aerosolizing fluid, method of aerosolization, humidity, temperature, oxygen, and other factors) on survival of the agents, and used artificial culture rather than infectivity assays. The demonstration that an infectious agent can remain alive in an aerosol is evidence that can support but not prove a hypothesis of airborne transmission. It is also necessary that small particle deposition in the respiratory tract results in colonization by the organism and disease in the host. Both the upper and lower respiratory tract are resistant to colonization by some agents, so that even virulent aerosols of these agents are not infective by the inhalation route.

However, for agents that can produce disease via the respiratory tract and where restriction of airborne transmission depends only on the absence of natural virulent aerosols, or for agents for which this information is unknown, great care must be taken in laboratory situations to prevent the production of accidental aerosols, or exposure to experimental aerosols.

V. CONTROL OF CONTAGIOUS DISEASE

For any health effect endpoint related to an environmental exposure to be reached:

1. A source for the agent must be present. For contagious diseases, the source is always the infected person and the agent enters the air through actions of the person (coughing, sneezing, shedding skin surface material, etc.).
2. The agent must retain characteristics necessary to cause disease. In the case of infectious disease, the agent must be alive and virulent.
3. The agent must enter the air in sufficient quantity and remain for sufficient time so that an infectious dose can be inhaled. Note that this includes conditions where many organisms are present over relatively short periods of time, or fewer organisms over longer periods of time.
4. A susceptible host must be present. For contagious diseases, this means that a person must be present who has no immunity (either natural or artificially acquired) to the specific agent.
5. The susceptible host must come into direct contact (inhale in the case of the airborne contagious diseases) with enough of the agent to produce the disease. For many of the contagious diseases, this amount may be as little as a single agent.

Control can focus on one or more of these factors. The control of most contagious disease today is centered on immunization (preventing susceptibility) and the use of antibiotics (killing the agent within the host, thereby preventing release of virulent organisms). Both immunization and antibiotics have dramatically decreased the incidence of many diseases. In fact, it is now considered that smallpox (a highly contagious airborne disease) has been eliminated from the world population entirely through intensive immunization programs (Henderson, 1976). Likewise, tuberculosis has been considered under control (at least in some parts of the world) because antibiotics effectively render tuberculosis patients noninfective. It is because antibiotics are so effective against most forms of tuberculosis that extensive transmission to health care workers has not occurred in clinics devoted to the care of tuberculosis patients. This is also one of the reasons why there is so much concern over the newly emergent multiply drug-resistant strains of *Mycobacterium tuberculosis* (Edlin et al., 1992).

Antibiotics are largely ineffective against viral agents, and immunization can also be relatively ineffective, often because the disease is caused by a number of different agents (e.g., the common cold) or because the agents change enough over time to render previous immunization ineffective (e.g., influenza; Stuart-Harris, 1981). In addition, we now have a group of people who cannot be immunized, and who are extremely susceptible to infectious agents: patients with AIDS, and immunosuppressed transplant patients. For them, prevention of exposure is the only recourse.

Practically speaking, prevention of exposure involves isolating either susceptible hosts or infected people; physically removing agents either from ambient air or breathing zone air (filtration); diluting aerosols to levels that are not likely to cause infection; and killing agents either at the source (antibiotics), at secondary sources (surface disinfection), or in the air.

A. Isolation

Isolation involves either placing an infected person in an enclosed environment to prevent exposure of anyone else, or isolating a particularly susceptible person to prevent exposure to any agent. Before the development of immunization and antibiotics, control of contagious diseases centered on preventing exposure by isolating infected people. For example, many tuberculosis patients were treated in sanitaria, and quarantine signs were placed on the doors of homes of children with some kinds of contagious disease. Isolation can be an effective method for control, providing the room air is exhausted directly to the outside (i.e., not recirculated), and the room is under negative pressure so that when the door is opened particles cannot spread to adjacent areas (Figure 2.1). Practically speaking, transmission of many contagious diseases happens in public places. Symptoms that allow dissemination can occur before the infected person or the potential host is aware of the nature of the illness, and infected people may continue to occupy public spaces (e.g., work spaces) even when they know they have a contagious disease. Also, the airborne nature of some contagious diseases remains unrecognized.

Particularly susceptible people can be protected by isolation in specially constructed rooms. Isolation rooms in this case must have clean supply air (i.e., air that is free of infectious agents) and must be under positive pressure so that agents from other parts of the building cannot penetrate (Figure 2.1). Anyone entering these rooms must wear masks and protective clothing that reduce dissemination of infectious agents into the room.

Another form of isolation is to use personal protective equipment. Masks are commonly used to reduce the release of respiratory aerosols in (for example) operating rooms or rooms housing extremely susceptible people. Particulate respirators are also used by susceptible people or people at high risk of exposure to virulent agents. In general, the particle sizes on which most infectious agents are borne are large enough so that well-fitted particle respirators will drastically reduce the risk of exposures sufficient to induce infection. However, it is clearly impractical to wear respiratory protection on the subway or in the classroom or office in order to prevent epidemics of influenza, the common cold, and other virulent contagions. Respirators can be worn by health care workers in isolation rooms or rooms where high-risk procedures are performed, but respirators cannot be worn at all times, and undiagnosed disease is often the source of transmission in the institutional setting.

B. Particle Removal from Ambient Air

Infectious aerosols (as for any aerosol) decay in the sense that particle concentrations gradually decrease. Although droplet nuclei are said to have a negligible settling rate, they do impact onto surfaces, and their numbers are greatly reduced with only a few air changes over several hours. The decrease

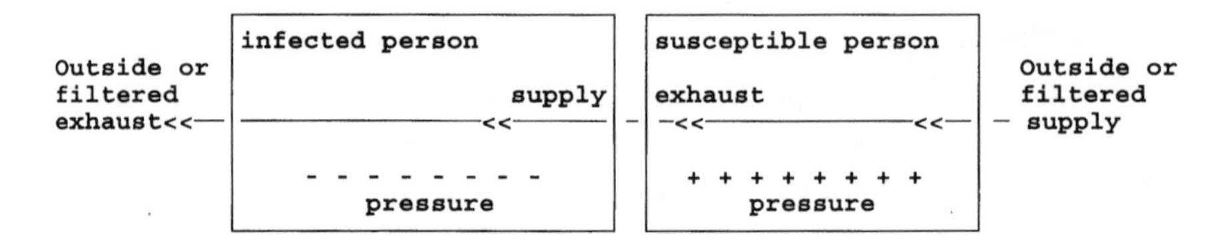

Figure 2.1. Ventilation requirements for two different kinds of isolation facilities.

occurs as the result of impaction onto surfaces, including those in the occupied space and the ventilation system, and by filtration.

Filtration. Room (or building) air filtration can successfully remove many kinds of infectious particles from the air. As for any other aerosol, efficiency of particle removal from the air passing through the filter depends on the pore size of the filter. There are filters that will remove virtually all infectious particles from an airstream (e.g., high-efficiency or HEPA filters). Little work has been done to test filters commonly in use in commercial buildings and homes for retention of natural infectious aerosols. However, some viruses are not effectively removed by the commonly used 34 to 45% dust-spot rated filters (Burmester and Witter, 1972). Under Burmester and Witter's test conditions, 93 to 97% dust-spot rated filters were effective for capturing specific viral aerosols.

The efficiency of filtration for the removal of aerosols does not only depend on the filter type. Assuming all the air in the room eventually passes through the filter, the steady state achieved in the occupied space also depends on the rate of flow through the filter and on the source strength (both instantaneously and with respect to changes over time) in the space. Infectious aerosol sources are usually discontinuous and sporadic, so that the aerosol is characterized by sharp peaks in spatial distribution and in concentration. This means that people closest to an active source will be unlikely to be protected by filtration alone. The ability to filter all air in a given space without stagnant areas is a major concern in the use of high-efficiency filter machines for the control of tuberculosis, for example.

Filtration is, in fact, used to construct clean rooms and biosafety containment facilities. These kinds of environments use very high-efficiency particle filters and very high air flow rates with the air flow patterns (laminar air flow) directed in such a way that infectious particles are not released into ambient air but are pulled immediately into the filter material.

Air ionization. Ionization imparts a charge to airborne particles and causes them to be attracted to surfaces. Prevention of airborne infectious diseases using negative corona discharge has been accomplished for animal diseases (Estola et al., 1979), and has been suggested for other kinds of aerosols, but is not widely used (Lehtimaki and Graeffe, 1976; Makela et al., 1979).

C. Dilution Ventilation

In the past, homes, schools, and some commercial buildings designed for the privileged classes were provided with high ceilings, floor-to-ceiling windows, and large fireplaces that allowed for large amounts of outdoor dilution air, which is essentially free of virulent infectious agents. However, the development of climate control technology has encouraged the sealing of indoor environments to prevent casual penetration of outdoor air. The recognition that heating and cooling are expensive, combined with the increasing cost of energy, resulted in the development of recirculation systems that reduced the amount of outdoor air entering buildings, and has led to an overall increase in levels of indoor-source pollutants including the agents of airborne contagious disease.

With some common diseases infected people are able to continue to appear in public during the transmissible stage (e.g., influenza, measles, tuberculosis) and the trend toward reducing ventilation and increasing occupancy in some buildings has probably to some degree increased the risk of exposure to an infectious dose. Brundage et al. (1988) present data that are suggestive of this effect, although they did not actually measure either ventilation rates or levels of airborne organisms. According to Riley and Nardell (1993), at very low ventilation rates, the rate of infection of susceptible people in the presence of a virulent aerosol of the agent causing tuberculosis can be close to 100%. It is relatively straightforward to reduce this risk to about 20% chance using dilution ventilation (i.e., ventilation with outside or other germ-free air). However, reduction below 20% is neither cost-effective nor acceptable from a comfort standpoint. These kinds of predictions have not been made for other contagious diseases, but are likely to be similar for any disease for which a single agent-containing particle is sufficient to initiate infection.

D. Killing the Organisms

Infectious agents must be alive to cause disease, so that killing the agent can be an effective means of control. This is the principle of hand washing and cleaning surfaces with disinfectants (although neither of these is particularly effective in the control of airborne disease). As mentioned earlier, air itself is a hostile environment, and making air even more hostile is an approach that can be used to control airborne infectious disease. One simple, but probably relatively ineffective method is to maintain humidity in a range in which many infectious agents are damaged (Arundel et al., 1986). In practice, as has been discussed above and as listed in Table 2.3, no single relative humidity range is useful for all organisms. More aggressive techniques are clearly effective. Wells, Riley, and co-workers have demonstrated that ultraviolet light can be an effective means for tuberculosis control under some circumstances (Riley et al., 1962; Wells et al., 1942). Ozone has also been proposed, although preventing human exposure to toxic levels of ozone is more difficult than for ultraviolet light, and exposure to ozone may actually increase the susceptibility of exposed people to infection (Fairchild, 1974; Purvis et al., 1961).

VI. CONCLUSIONS

Airborne infectious disease remains a public health concern, and probably always will, considering the many variables involved in its control and the continuing emergence of "new" diseases or new modes of transmission for old diseases (Plouffe, 1990). The recent hantavirus epidemic mentioned earlier is a good example. It is difficult to predict the emergence of new agents or changes in the epidemiology of recognized agents, so that immunization and other agent-specific control measures must be implemented after the fact. Environmental control, however, is not agent specific. We have the technology available to maintain indoor environments in such a way that the risk of transmission of infectious disease can be minimized. What remains is to educate people (architects, engineers, managers, homeowners, etc.) regarding the means and the cost-effectiveness of such control.

REFERENCES

Aach, R. D., Evans, J., and Losee, J., An epidemic of infectious hepatitis possibly due to airborne transmission, *Am. J. Epidemiol.*, 87, 99, 1968.

Akers, T. G. and Hatch, M. T., Survival of a picornavirus and its infectious ribonucleic acid after aerosolization, *Appl. Microbiol.*, 16, 1811, 1968.

Akers, T. G., Prato, C. M., and Dubovi, E. J., Airborne stability of simian virus 40, *Appl. Microbiol.*, 26, 146, 1973.

Alford, R. H., Kasel, J. A., Gerone, P. J., and Knight, V., Human influenza resulting from aerosol inhalation, *Proc. Soc. Exp. Biol. Med.*, 122, 800, 1966.

Almeida, J. D., Chisholm, G. D., Kulatilake, A. E., MacGregor, A. B., MacKay, D. H., O'Donogue, E. P. N., Shackman, R., and Waterson, A. P., Possible airborne spread of serum-hepatitis virus within a haemodialysis unit, *Lancet,* 2, 849, 1971.

Artenstein, M. S., Miller, W. S., Rust, J. H., and Lamson, T. H., Large-volume air sampling of human respiratory disease pathogens, *Am. J. Epidemiol.*, 85, 479, 1967.

Arundel, A. V. et al., Indirect health effects of relative humidity in indoor environments, *Environ. Health Perspect.*, 65, 351, 1986.

Baylor, E. R., Baylor, M. B., Blanchard, D. C., Syzdek, L. D., and Appel, C., Virus transfer from surf to wind, *Science,* 198, 575, 1977a.

Baylor, E. R., Peters, V., and Baylor, M. B., Water-to-air transfer of virus, *Science,* 197, 763, 1977b.

Benbough, J. E., The effect of relative humidity on the survival of airborne Semliki Forest virus, *J. Gen. Virol.,* 4, 473, 1969.

Bennett, J. V. and Brachman, R. S., *Hospital Infections*, 2nd ed., Little, Brown, Boston, 1986.

Berendt, R. F., Dorsey, E. L., and Hearn, H. J., Virucidal properties of light and SO_2 I. Effect on aerosolized Venezuelan equine encephalomyelitis virus, *Proc. Soc. Exp. Biol. Med.,* 139, 1, 1972.

Berendt, R. F. and Dorsey, E. L., Effect of simulated solar radiation and sodium fluorescein on the recovery of Venezuelan equine encephalomyelitis virus from aerosols, *Appl. Microbiol.*, 21, 447, 1971.

Berry, G. P. and Kitchen, S. F., Yellow fever accidentally contracted in the laboratory, *Am. J. Trop. Med.*, 11, 365, 1931.

Bourgueil, E., Hutet, E., Cariolet, R., and Vannier, P., Air sampling procedure for evaluation of viral excretion level by vaccinated pigs infected with Aujeszky's disease (pseudorabies) virus, *Res. Veterinary Sci.*, 52, 182, 1992.

Brachman, P. S., Kauffman, A. F., and Dalldorf, F. G., Industrial inhalation anthrax, *Bacteriol. Rev.*, 30, 646, 1966.

Brundage, J. F., Scott, R. M., Lednar, W. M., Smith, D. W., and Miller, R. N., Building-associated risk of febrile acute respiratory diseases in army trainees, *JAMA,* 259(14), 2108, 1988.

Burmester, B. R. and Witter, R. L., Efficiency of commercial air filters against Marek's disease virus, *Appl. Microbiol.*, 23(3), 505, 1972.

Conomy, J. P., Leibovitz, A., McCombs, W., and Stinson, J., Airborne rabies encephalitis: demonstration of rabies virus in the human central nervous system, *Neurology,* 27, 67, 1977.

Couch, R. B., Knight, V., Gerone, P. J., Cate, T. R., and Douglas, R. G., Factors influencing response of volunteers to inoculation with coxsackie virus A type 21, *Am. Rev. Respir. Dis.*, 99, 24, 1969.

Couch, R. B., Viruses and indoor air pollution, *Bull. N.Y. Acad. Med.*, 57, 907, 1981.

Cox, C. S., *The Aerobiological Pathway of Microorganisms*, John Wiley & Sons, New York, 1987.

Dahlgren, C. M., Buchanan, L. M., Decker, H. M. et al., *Bacillus anthracis* aerosols in goat hair processing mills, *Am. J. Hyg.,* 72, 24, 1960.

Davis, R. M., Orenstein, W. A., Frank, J. A. et al., Transmission of measles in medical settings, *JAMA,* 255(6), 1295, 1986.

deJong, J. C., Harmsen, M., Plantinga, A. D., and Trouwborst T., Inactivation of Semliki forest virus in aerosols, *Appl. Environ. Microbiol.*, 32, 315, 1976.

deJong, J. C., Harmsen, M., Trouwborst, T., and Winkler, K. C., Inactivation of encephalomyocarditis virus in aerosols: fate of virus protein and ribonucleic acid, *Appl. Microbiol.*, 27, 59, 1974.

deJong, J. G., Harmsen, M., and Trouwborst, T., The infectivity of the nucleic acid of aerosol-inactivated poliovirus, *J. Gen. Virol.*, 18, 83, 1973.

deJong J. C., Harmsen, M., and Trouwborst, T., Factors in the inactivation of encephalomyocarditis virus in aerosols, *Infect. Immun.,* 12, 29, 1975.

DeLay, P. D., Lennette, E. H., and DeOme, K. B., Q fever in California. II. Recovery of *Coxiella burneti* from naturally-infected air-borne dust, *J. Immunol.*, 65, 211, 1950.

Dick, E. C., Jennings, L. C., Mink, K. A., Wartgow, C. P., and Inborn, S. L., Aerosol transmission of rhinovirus colds, *J. Infect. Dis.*, 156, 442, 1987.

Dixon, G. J., Sidwell, R. W., and McNeil, E., Quantitative studies on fabrics as disseminators of viruses. II. Persistence of poliomyelitis virus on cotton and wool fabrics, *Appl. Microbiol.*, 14(2), 183, 1966.

Donaldson, A. I., Ferris, N. P., and Gloster, J., Air sampling of pigs infected with foot-and-mouth disease virus: comparison of Litton and cyclone samplers, *Res. Vet. Sci.*, 33, 384, 1982.

Donaldson, A. I. and Ferris, N. P., The survival of foot and mouth disease virus in open air conditions, *J. Hyg. Camb.*, 74, 409, 1975.

Edlin, B. R., Tokars, J. I., Grieco, M. H., Crawford, J. T., Williams, J., Sordillo, E. M., Ong, K. R., Kilburn, J. O., Dooley, S. M., Castro, K. G., Jarvis, W. R., and Holmberg, S. D., An outbreak of multidrug-resistant tuberculosis among hospitalized patients with the acquired immunodeficiency syndrome, *N. Engl. J. Med.*, 326, 1514, 1992.

Ehrlich, R. and Miller S., Effect of NO_2 on airborne Venezuelan equine encephalomyelitis virus, *Appl. Microbiol.*, 23, 481, 1972.

Estola, T., Makela, P., and Hovi, T., The effect of air ionization on the air-borne transmission of experimental Newcastle disease virus infections in chickens, *J. Hyg. Camb.*, 83, 59, 1979.

Evans, A. S., Causation and disease: the Henle-Koch postulates revisited, *Yale J. Biol. Med.*, 49, 175, 1976.

Fairchild, G. A., Ozone effect on respiratory deposition of vesicular stomatitis virus aerosols, *Am. Rev. Respir. Dis.*, 109, 446, 1974.

Gerone, P. J., Couch, R. B., Keefer, G. V., Douglas, R. G., Derrenbacher, E. B., and Knight, V., Assessment of experimental and natural viral aerosols, *Bacteriol. Rev.*, 30, 576, 1966.

Hanson, R. P., Sulkin, S. E., Buescher, E. L., Hammon, W. M., McKinnery, R. W., and Work, T. H., Arbovirus infections of laboratory workers, *Science,* 158, 1283, 1967.

Henderson, D. A., The eradication of smallpox, *Sci. Am.,* 235, 25, 1976.

Hierholzer, J. C., Viruses, mycoplasmas as pathogenic contaminants in indoor environments, in *Biological Contaminants in Indoor Environments*, Morey, P. H., Feeley, J., and Otten, J., Eds., ASTM STP 1071, Philadelphia, 1990.

Hinshaw, V. S., Schaffer, F. L., and Chatigny, M. A., Evaluation of Moloney murine sarcoma and leukemia virus complex as a model for airborne oncogenic virus biohazards: survival of airborne virus and exposure of mice, *J. Natl. Cancer Inst.* 57(4), 775, 1976.

Houk, V., Spread of tuberculosis via recirculated air in a naval vessel: the Byrd study, *Ann. N.Y. Acad. Sci.*, 353, 10, 1980.

Huddleson, I. F. and Munger, M., A study of an epidemic of brucellosis due to *Brucella melitensus*, *Am. J. Public Health*, 30, 944, 1940.

Josephson, A. and Gombert, M., Airborne transmission of nosocomial varicella from localized zoster, *J. Infect. Dis.*, 158, 238, 1988.

Keefer, G. V., Zebarth, G. L., and Allen, W. P., Susceptibility of dogs and cats to Rift Valley fever by inhalation or ingestion of virus, *J. Infect. Dis.*, 125, 307, 1972.

Kenyon, R. H., McKee, K. T., Jr., Zack, P. M., Rippy, M. K., Vogel, A. P., York, C., Meegan, J., Crabbs, C., and Peters, C. J., Aerosol infection of Rhesus macaques with junin virus, *Intervirology,* 33, 23, 1992.

Kingston, D., Towards the isolation of respiratory syncytial virus from the environment, *J. Appl. Bacteriol.*, 31, 498, 1968.

Knight, V., Airborne transmission and pulmonary deposition of respiratory viruses, in *Viral and Mycoplasmal Infections of the Respiratory Tract*, Knight, V., Ed., Lea & Febiger, Philadelphia, 1973.

Knight, V., Viruses as agents of airborne contagion, *Ann. N.Y. Acad. Sci.*, 353, 147, 1980.

Koster, F., Levy, H., Mertz, G., Cushing, A., Young, S., Foucar, K., McLaughlin, J. et al., Infectious diseases update — outbreak, Hantavirus infection — Southwestern United States, *Morbid. Mortal. Weekly Rep.,* 42, 441, 1993.

LaForce, F. M. et al., Epidemiologic study of a fatal case of inhalation anthrax, *Arch. Environ. Health,* 18, 798, 1969.

Larson, E. W., Dominik, J. W., and Slone, T. W., Aerosol stability and respiratory infectivity of Japanese B encephalitis virus, *Infect. Immun.,* 30, 397, 1980.

Ledair, J., Zaia, J., Levin, M. et al., Airborne transmission of chickenpox in a hospital, *N. Engl. J. Med.*, 302, 450, 1980.

Lehtimaki, M. and Graeffe, G., The effect of the ionization of air on aerosols in closed spaces, in *Proceedings of the Third Symposium on Contamination Control,* Copenhagen, 2, 370, 1976.

Lennette, E. H. and Welsh, H. H., Q fever in California X: Recovery of *Coxiella burneti* from the air of premises harboring infected goats, *Am. J. Hyg.*, 54, 44, 1951.

Loosli, C. G., Hertweck, M. S., and Hockwald, R. S., Airborne influenza PR8 — a virus infections in actively immunized mice, *Arch. Environ. Health,* 21, 332, 1970.

Mahl, M. C. and Sadler, C., Virus survival on inanimate surfaces, *Can. J. Microbiol.*, 21, 819, 1975.

Makela, P., Ojajarvi, J., Lehtimaki, M., and Graeffe, G., Studies on the effects of ionization on bacterial aerosols in a buns-and-plastic surgery unit, *J. Hyg.,* 83, 199, 1979.

Mayhew, C. J., Zimmerman, W. D., and Hahon, N., Assessment of aerosol stability of yellow fever virus by fluorescent-cell counting, *Appl. Microbiol.*, 16, 263, 1968.

Mbithi, J. N., Springthorpe, V. S., and Sattar, S. A., Effect of relative humidity and air temperature on survival of hepatitis A virus on environmental surfaces, *Appl. Environ. Microbiol.*, 57(5), 1394, 1991.

McKissick, G. E., Griesemer, R. A., and Farrell, R. L., Aerosol transmission of Rauscher murine leukemia virus, *J. Natl. Cancer Inst.*, 45, 625, 1970.

Monto, A. S., Influenza: quantifying morbidity and mortality, *Am. J. Med.,* 82(Suppl. 6A), 20, 1987.

Moser, M. R., Bender, T. R., Margolis, H. S. et al., An outbreak of influenza aboard a commercial airliner, *Am. J. Epidemiol.*, 110, 1, 1979.

Nardell, E. A., Tuberculosis in homeless, residential care facilities, prisons, nursing homes, and other close communities, *Semin. Respir. Infect.*, 4, 206, 1989.

Nardell, E., Pathogenesis of tuberculosis, in *Tuberculosis*, Reichman, L. B. and Hershfield, E., Eds., Marcel Dekker, New York, 1993, chap. 5.

Pike, R. M., Laboratory-associated infections: summary and analysis of 3,921 cases, *Health Lab. Sci.,* 13, 105, 1976.

Plouffe, J. F., New microorganisms and their health risk, in *Biological Contaminants in Indoor Environments*, Morey, P. H., Feeley, J., and Otten, J., Eds., ASTM STP 1071, Philadelphia, 1990.

Purvis, M. R., Miller, S., and Ehrlich, R., Effect of atmospheric pollutants on susceptibility to respiratory infection. I. Effect of ozone, *J. Infect. Dis.*, 109, 238, 1961.

Rabey, F., Janssen, R. J., and Kelley, L. M., Stability of St. Louis encephalitis virus in the airborne state, *Appl. Microbiol.*, 18, 880, 1969.

Regnery, R. L. and McDade, J. E., *Coxiella burnetii* (Q fever), a potential microbial contaminant of the environment, in *Biological Contaminants in Indoor Environments*, Morey, P., Feeley, J., and Otten, J., Eds., American Society for Testing Materials STP 1071, Philadelphia, 1990.

Remington, P. L., Hall, W. N., Davis, I. H. et al., Airborne transmission of measles in a physician's office, *JAMA,* 253, 1574, 1985.

Riley, E. C., The role of ventilation in the spread of measles in an elementary school, *Ann. N.Y. Acad. Sci.*, 353, 25, 1980.

Riley, R. L., Wells, W. F., Mills, C. C., Nyka, W., and McLean, R. L., Air hygiene in tuberculosis: Quantitative studies of infectivity and control in a pilot ward, *Am. Rev. Tuber. Pulmon. Dis.*, 75, 420, 1957.

Riley, R. L., Mills, C. C., O'Grady, F., Sulton, L. U., Wittstadt, F., and Shivpuri, D. N., Infectiousness of air from a tuberculosis ward: ultraviolet irradiation of infected air; comparative infectiousness of different patients, *Am. Rev. Respir. Dis.*, 84, 511, 1962.

Riley, R. L. and Nardell, E. A., Controlling transmission of tuberculosis in health care facilities: ventilation, filtration, and ultraviolet air disinfection, *PTSM,* Series No. 1, 25, 1993.

Samet, J. M. and Spengler, J. D., *Indoor Air Pollution — A Health Perspective*, Johns Hopkins University Press, Baltimore, 1991.

Sawyer, L. A., Murphy, J. J., Kaplan, J. E., Pinsky, P. F., Chacon, D., Walmsley, S., Schonberger, L. B., Phillips, A., Forward, K., Goldman, C., Brunton, J., Fralick, R. A., Cater, A. O., Gary, W. G., Jr., Glass, R. I., and Low, D. E., 25 to 30-nm Virus particle associated with a hospital outbreak of acute gastroenteritis with evidence for airborne transmission, *Am. J. Epidemiol.*, 127, 1261, 1988.

Schaffer, F. L., Soergel, M. E., and Straube, D. C., Survival of airborne Influenza virus: effects of propagating host, relative humidity, and composition of spray fluids, *Arch. Virol.*, 51, 263, 1976.

Schoenbaum, M. A., Zimmerman, J., Beran, G. W., and Murphy, D. P., Survival of pseudorabies virus in aerosol, *Am. J. Vet. Res.*, 51, 331, 1990.

Schulman, J. L., Experimental transmission of influenza virus infection in mice: IV. Relationship of transmissibility of different strains of virus and recovery of airborne virus in the environment of infector mice, *J. Exp. Med.*, 125, 479, 1967.

Sherwood, R. L., Thomas, P. T., Kawanishi, C. Y., and Fenters, J. D., Comparison of *Streptococcus zooepidemicus* and influenza virus pathogenicity in mice by three pulmonary exposure routes, *Appl. Environ. Microbiol.*, 54, 1744, 1988.

Sidwell, R. W., Dixon, G. J., and McNeil, E., Quantitative studies on fabrics as disseminators of viruses. I. Persistence of Vaccinia virus on cotton and wool fabrics, *Appl. Microbiol.*, 14(1), 55, 1966.

Stephenson, E. H., Larson, E. W., and Dominik, J. W., Effect of environmental factors on aerosol-induced Lassa virus infection, *J. Med. Virol.*, 14, 295, 1984.

Stuart-Harris, C., The epidemiology and prevention of influenza, *Am. Sci.,* 69, 166, 1981.

Task Group on Lung Dynamics, Deposition and retention models for internal dosimetry of the human respiratory tract, *Health Phys.*, 12, 173, 1966.

Tortora, G. J., Funke, B. R., and Case, C.L., *Microbiology*, Benjamin/Cummings, Redwood City, CA, 1992.

van der Veen J., Poort, Y., and Birchfield, D. J., Effect of relative humidity on experimental transmission of Sendai virus in mice, *Proc. Soc. Exp. Biol. Med.,* 140, 1437, 1972.

Weintraub, T. L. and Iftimovici, R., Investigations on the risk of virus infection in dental surgery. Preliminary data, *Virologie,* 32(2), 155, 1981.

Wells, W. F., Wells, M. W., and Wilder, T. S., The environmental control of epidemic contagion I. An epidemiologic study of radiant disinfection of air in day schools, *Am. J. Hyg.*, 35, 97, 1942.

Wells, W. F., On air-borne infection: II. Droplets and droplet nuclei, *Am. J. Hyg.,* 20, 611, 1934.

Wells, W. F., *Airborne Contagion and Air Hygiene: An Ecological Study of Droplet Infections*, Harvard University Press, Cambridge, MA, 1955.

Welsh, H. H., Lennette, E. H., Abinanti, F. R., and Winn, J. F., Airborne transmission of Q fever: the role of parturition in the generation of infective aerosols, *Ann. N.Y. Acad. Sci.*, 70, 528, 1958.

Willeke, K. and Baron, P. A., *Aerosol Measurement: Principles, Techniques and Applications*, Van Nostrand Reinhold, New York, 1993.

Winkler, W. G., Fashinell, T. R., Leffingwell, L. et al., Airborne rabies transmission in a laboratory worker, *JAMA,* 266, 1219, 1973.

3 *LEGIONELLA* ECOLOGY

Carl B. Fliermans

CONTENTS

I. INTRODUCTION

Legionella is a fascinating organism that has told us many things about the environments in which we live and the relationships between natural environments and infectious diseases. The scope of this chapter is to provide insight into the ecological niche of *Legionella*, establish its relationships in the natural world, and provide information as to its ecology so that its effect on human health can be better understood and treated.

0-87371-724-4/95/$0.00+$.50

The American Legion Convention of 1976 in Philadelphia was America's salute to her war heroes during the bicentennial: a tribute to those who provided the ultimate sacrifice for freedom if required. During the summer of 1976, 221 of 4400 individuals associated with the Legionnaires' Convention in Philadelphia became ill, 34 of whom died (Fraser et al., 1977). Through the overlapping of human and bacteriological niches (i.e., cooling towers), the courses of life for hundreds of families, as well as the disciplines of microbial ecology and medical microbiology, were forever changed. That single event sent the epidemiological and medical communities on a biological sleuth pilgrimage that is one of the most intriguing of medical and microbiological history. The pursuit sought to find, identify, treat, and subsequently control a pneumonic killer. Out of the tragedy came insights into the ecology of pathogenic microorganisms, and this led to the discovery of a previously unrecognized human pathogen, *Legionella*, named for the convention from which it claimed its first recognized victims.

Not until Legionnaires' Disease and the discovery of its etiologic agent, *Legionella*, had such a successful modern-day marriage between an agent of human disease and its environmental niche been recognized and understood. The format for such a marriage between microbial ecology and medical microbiology was established with the work of Robert Koch in 1886 on anthrax, a disease of cattle caused by *Bacillus anthracis*. Koch's Postulates, which were developed to study medical microbiological principles, are applicable to the study of microbial ecology and, adapted for microbial ecology, provide a framework from which we can expand the study of *Legionella* in the following ways:

Koch's Postulates	Ecological analogs
A specific organism can always be found in association with a given disease.	The selected microorganism or consortia of organisms should always be found in a habit that demonstrates a given characteristic and should neither be found in high densities nor demonstrate high activity in habitats that do not respond in that characteristic manner.
The organism can be isolated and grown in pure culture in the laboratory.	The microorganisms or microbial consortia must be isolated and cultivated outside the habitat.
The pure culture will produce the disease when inoculated into a susceptible animal.	When the isolated consortia or microorganism are inoculated back into the environment or a subsample of the environment, the characteristic response of the habitat is again observe
It is possible to recover the organism in pure culture from the experimentally infected animal.	The microorganisms or microbial consortia ca be isolated from the inoculated habitat, cultured in the laboratory, and be the same microorganism or consortia as the original.

Koch's contributions have provided the stage for subsequent ecological investigations over the past century, and it is from this heritage that the ecological and epidemiology research on *Legionella* has subsequently drawn its breath.

II. *LEGIONELLA,* THE BACTERIUM

A large amount of information has been obtained about the bacterium *Legionella pneumophila* since the first recognized outbreak of Legionnaires' Disease in Philadelphia in 1976. The cause of this epidemic was unknown until the winter of 1976 when investigators from the Centers for Disease Control (CDC) isolated the responsible bacterium, subsequently named *Legionella pneumophila* (lung-loving) (McDade et al., 1977).

Legionella pneumophila was first recognized in yolk sacs of embryonated eggs by McDade et al. (1979) and was shown to be the etiologic agent of Legionnaires' Disease. Weaver (1978) at CDC isolated *L. pneumophila* on nonliving media and characterized the bacterium culturally and biochemically. Legionellosis, the term used for both Legionnaires' Disease and Pontiac fever, is caused by a group of rod-shaped bacteria that are usually very small ($1 \times 3\ \mu$) but can form long filaments when grown on particular types of media. The bacterium is Gram negative, non-acid-fast, motile, and does not produce endospores. *Legionella* species do not grow on ordinary laboratory media, but require special media supplemented with L-cysteine, soluble iron, and a pH adjusted to 6.9 (Weaver and Feeley, 1978). When cultured on charcoal yeast-extract agar, the colonies are circular, gray to white in color, and present a characteristic cut-glass appearance with green to purple iridescence. Only a few cultural and biochemical tests are currently used for the routine identification and biochemical testing of *Legionella.* Current bacteriological media, BCYE (buffered charcoal yeast extract, the medium of choice with a variety of modifications) included, do not allow *Legionella* to exhibit all the physiological capacities it displays in nature.

Studies of the biochemical characteristics, molecular weight of the genome, guanine-cytosine content, and DNA homology have not demonstrated relatedness between these and any other known family of bacteria (Band and Fraser, 1986). The legionellae are not a single species, but they are a collection of at least 34 species grouped (based on DNA homology) under the genus *Legionella* and the family legionellae. To date, only 11 of these species have been isolated in culture from both clinical and environmental sources. Further characterization of the organisms has been done using monoclonal antibodies or isoenzyme analyses. This technique is commonly called "fingerprinting" and can be most helpful in associating environmental sources with clinical cases.

III. *LEGIONELLA* ECOLOGY

A. Methods for Studying the Ecology of *Legionella*

Synecology vs. autecology. There are two basic approaches to the study of microbial systems in nature. They are the synecological and the autecological approaches. Both have their merits and shortcomings. The synecological approach is one in which the relationship between the entire microbial community and a given habitat is investigated. In this approach there is little concern about the types of microorganisms involved in the functioning of the habitat, but rather the extent of microbial processing that takes place. This approach provides the investigator with process-oriented data and information regarding the overall picture of the functioning of the habitat. Synecological approaches include the study of cycling of nutrients through an ecosystem and the transformation of chemical wastes by microbiological components of the ecosystem without regard to the microorganisms involved. These investigations often rely on the mineralization or transformation of isotopically labeled compounds that provide a marker for transformation processes. The marker can be followed during various stages of incubation, degradation, and mineralization, enabling one to measure breakdown products and intermediates.

The autecological approach focuses on a particular microorganism and asks questions concerning that organism and how it functions in the studied habitat. The data and information gained from this perspective are especially valuable when one is dealing with an organism that is potentially pathogenic. Such an approach allows one to pursue questions about the organism, its particular functioning in a given ecological niche, and potential approaches for control of the organism that could not otherwise be made. The vast majority of ecological studies on *Legionella* have been from an autecological perspective.

The disadvantage of the autecological approach is that it requires specialized techniques for the identification and assessment of the bacterium. Few bacteria can be characterized by their morphological structure and then only to the genus level, in part because many are pleomorphic depending on their nutritional status. Specific serological and genetic probes allow identification of bacteria to species and subgroup without culture. Under appropriate conditions the bacterium can be assessed as to its viability, state of physiological health, and ability to transform selected isotopically labeled compounds using microautoradiographic techniques (Fliermans and Schmidt, 1975), cytochrome activity (Fliermans et al., 1981b), and fluorescent probe techniques (Rodriguez et al., 1992). Specific polyvalent and/or monoclonal fluorescent antibody probes have been developed to allow visualization of *Legionella* and evaluation of the role of *Legionella* in ecological environments that are microbiologically complex and difficult to sample.

Immunofluorescent techniques. Immunofluorescent techniques were originally used in diagnostic medical microbiology, and only in the last few decades have they been employed in investigations of agricultural and natural terrestrial and aquatic microbial ecosystems. Most recently, they have been used to study the ecology of microorganisms associated with the degradation of toxic and hazardous wastes. The direct fluorescent antibody (DFA) technique allows the visualization of specific kinds of bacterium in samples from very complex habitats, such as soils, metal piping, sediments, vascular plant material, sewage sludge, and virtually any aquatic habitat. If one employs the DFA technique for *Legionella*, a single cell per milliliter of water can be visualized after concentrating the sample 500-fold by continuous-flow centrifugation or positive-pressure filtration. This detection limit could be pushed even further given a larger sample size and a greater concentration factor.

Fluorescent-labeled serospecific polyvalent antibodies against whole cells or cell walls of *Legionella* are the most useful tools to date for quantification and visualization of the bacterium in its natural habitats. Polyvalent fluorescent antibodies can be made specific enough to react with single serogroups of a *Legionella* species without cross-reacting with organisms outside the serogroup or with other species or genera of bacteria. The DFA approach can also be used with monoclonal antibodies. However, the high specificity of monoclonal antibodies allows detection of only a very limited number of organisms from a single serotype, and, although *Legionella* may be present in the habitat, if it will not react with the specific monoclonal, it will not be detected, and the number of *Legionella* will be greatly underestimated. Monoclonal applications are valuable, however, in relating *Legionella* from a selected environment to a clinical infection.

One limitation of the DFA technique is that the conjugate stains both living and dead organisms. If the *Legionella* are being stressed and lysed, either by natural events in their habitat or by man-made events such as the addition of biocides, then cell wall debris resulting from such events will stain and be observed microscopically. However, once dead, bacteria are no longer capable of keeping the "wolves" from the door and are readily lysed and serve as food for other members of the microbial population. Thus, in aquatic or terrestrial habitats, dead bacteria do not remain intact for an extended time. This phenomenon has been observed in both natural and sterile soils and in aquatic systems where known DFA-positive organisms were killed and added back to the system. In sterile systems, the DFA-positive bacteria remained intact for extended periods of time (days), while in nonsterile systems DFA-positive organisms were converted to debris within a few hours. Thus, it is likely that most of the intact bacteria that one observes in natural habitats are viable, although they may be nonculturable, as Hussong et al. (1987) have demonstrated for *Legionella*.

Tetrazolium dyes, which are reduced by the cytochrome system of bacteria during metabolic activity (Zimmerman et al., 1978; Fliermans et al., 1981b; Fliermans and Tyndall, 1992), can be used in conjunction with DFA to indicate actively metabolizing bacteria. Once the oxidized dye enters the cell, it is reduced and forms a formazan crystal inside the cell that can be viewed by bright-field microscopy. One can view directly the DFA staining of the cell for identification, distribution and enumeration as well as the formazan crystals for metabolic activity. Thus, the autecological activity of a particular bacterium or group of bacteria can be measured in the natural habitat under a variety of environmental conditions. This technique has been very useful for investigations of the ecology of *Legionella*.

DFA has also been used in conjunction with microautoradiography (Fliermans and Schmidt, 1975) to determine whether specifically stained organisms are active with regard to the incorporation of selected radioactive substrates. Although the technique is often as much art as science, it provides information that can be gained by no other technique.

Isolation of Legionellae. Recovery of *Legionella* from environmental samples continues to be difficult. This is particularly true because *Legionella* generally comprises less than 1% of the total bacterial population in environmental samples and is easily out-competed by other organisms in the environment. As a result, the levels of *Legionella* observed on selective media greatly underestimate the density of the bacterium in the natural environment. Attempts to overcome this problem have included the use of "selective" media consisting of a buffered charcoal yeast-extract base supplemented with iron, cysteine, and a variety of bacterial inhibitors. Wadowsky and Yee (1981) and Vickers et al. (1981) described a glycine-vancomycin-polymyxin B (DGVP) agar, which incorporated glycine, the antibiotics vancomycin and polymyxin B as selective agents, and the dyes bromothymol blue and bromocresol purple. Bopp et al. (1981) produced a medium containing cephalothin, colistin, cycloheximide, and vancomycin. Edelstein (1982) described a medium containing glycine, polymyxin B, vancomycin, and anisomycin, as well as bromothymol blue and bromocresol purple. Although these media are designed to inhibit other bacteria and select for the recovery of legionellae, many aquatic and terrestrial bacteria will grow on all of the above media. There is at present no defined or complex medium that allows the full expression of *Legionella*'s capability as demonstrated by DFA, and, currently, there is still no selective medium that is specific for legionellae.

Pretreatment methodology, used in conjunction with the previously mentioned media, does improve isolation efficiency. Exposure of water samples to acid (pH 2.2) for 5 min (Bopp et al., 1981), heating to 50°C for 30 min or 60°C for 2 min (Dennis et al., 1984b), along with the addition of sterile photosynthates (Fliermans, unpublished results) before plating has enhanced recovery. A portion of the legionellae in any sample is adversely affected by the pretreatment, and studies on the sensitivity of *Legionella* spp. to such procedures indicate a need for constraint when interpreting quantitative culture results from environmental samples (Calderon and Dufour, 1984; Roberts et al., 1987). Finally, environmental samples clearly contain legionellae that are viable but nonculturable so that any culturing technique employed for environmental samples will underestimate the total densities of legionellae present (Hussong et al., 1987).

B. Ecology of *Legionella*

Surveys of lakes, ponds, and streams indicate that *Legionella* is a common inhabitant of natural waters (Fliermans et al., 1979; 1981a; Politi et al., 1979). Recent work has expanded these habitats to include deep terrestrial subsurface environments (Fliermans and Tyndall, 1992). A microbiologically controlled bore hole was drilled into the southeastern coastal plain sediments of the United States to a depth of 538 m. Samples were processed for a variety of microbial components including *Legionella pneumophila* serogroups 1 to 6. Only serogroup 1 was present in any significant density and was present in isolated geological strata, rather than uniformly distributed through the sediments. These data are important because subsurface environments that house *Legionella* may be brought to the surface through excavation, bioremedial cleanup of subsurface wastes, well drilling, or spray irrigation, and have the potential to encounter a susceptible host. The data suggest that it may be important to insure that irrigation wells and excavation activities are not located in geological formations that house high concentrations of *Legionella* that are capable of being aerosolized to susceptible hosts.

There are man-made environments that tend to select for *Legionella* by providing appropriate temperature and consortia of other microorganisms. Of particular concern are those which people come into close contact with, including cooling towers and evaporative condensers (Glick et al., 1978; Cordes et al., 1980; Dondero et al., 1980; Gorman et al., 1985); the plumbing systems of hospitals, hotels, gymnasiums, and homes (Cordes et al., 1981; Tison and Seidler, 1983; Wadowsky et al., 1982; Stout et al., 1987); air wash systems; whirlpools; process water; and respiratory therapy equipment.

The number of studies that have documented the occurrence of *Legionella* spp. within the internal plumbing systems of buildings (Tobin et al., 1981a; Cordes et al., 1981; Wadowsky et al., 1982; Arnow et al., 1985), suggests that one of the habitats of legionellae is municipal drinking water supplies, and that these supplies serve as a pathway for the contamination of buildings (Fliermans and Harvey, 1984; Hsu et al., 1984; Stout et al., 1982; Wadosky et al., 1982). Tison and Seidler (1983) examined the incidence of *Legionella* spp. in raw water at various stages of treatment, and in the distribution system of several water supplies. While they obtained positive direct fluorescent antibody results, they were unable to isolate legionellae by animal inoculation or culture procedures.

States et al. (1987a), over a several-year period, surveyed the drinking water system of Pittsburgh. *Legionella* spp. could not be isolated from any of the samples, other than the river. These investigations indicate that, while legionellae are widely available in raw source waters and have been detected in numerous internal plumbing systems, the organism probably occurs in public supplies only in low numbers or sporadically.

Thermal characteristics of *Legionella* populations. Microbiologists recognize that populations of aquatic microorganisms generally follow the thermal cycle of their habitat. This means that a seasonal temperature change is reflected in the activity and density of the microbial populations in the habitat. Under extreme habitat conditions, the diversity of microbial populations (i.e., the number of species) is usually reduced while the densities of the populations (i.e., the number of individuals of the remaining species) may be quite high. This increase in population density is probably due to the lack of competition and the successful exploitation of a particular ecologic niche by selected organisms. In both natural and man-made thermal habitats such as those in Yellowstone National Park and those created by heat rejection from electrical and nuclear generation facilities, microbial communities are adapted to habitat temperatures ranging from 30 to 90°C. Although populations can be dense in these habitats, only a few kinds of organisms (e.g., purple and green photosynthetic bacteria, *Thiobacillus, Thermoplasma*, Gram negative anaerobes, and sulfur-dependent Archaebacteria) are usually dominant, especially at the upper temperature extremes.

Fliermans et al. (1981a, 1981b) sampled 67 rivers and lakes in the United States and using guinea pig inoculation, recovered *Legionella* in waters at temperatures ranging from 5.7 to 63°C. While these studies did not determine whether or not *Legionella* multiplies throughout this temperature range, they demonstrated that the bacterium can survive and remain viable. However, on an equal-effort sampling basis, *Legionella* is more readily isolated from warm or thermally altered habitats than from ambient ones. *Legionella* spp. survive high temperatures, as demonstrated by their occurrence in natural hot springs as well as domestic hot water tanks and hot water plumbing systems. Ecological investigations in the hot spring environments of Yellowstone National Park indicated that *L. pneumophila* is part of this natural environment that has remained unchanged since the early 1900s (Fliermans, 1983; Allen and Day, 1935). *Legionella* has also been isolated from lakes and thermal habitats formed by the eruption of Mt. St. Helen's (Fliermans, 1983, 1985).

Since 1947, the consumption of electricity has increased from 0.033×10^{15} Btu to a peak in 1979 of 78.9×10^{15} Btu (U.S. Department of Energy, 1983), an increase of more than 2000-fold in just over 30 years. This represents a tremendous increase in the amount of energy used to perform useful work, as well as a tremendous amounted of wasted heat dissipated back into the environment. The ubiquity of *Legionella* ensures that when this heat dissipation creates a suitable habitat, a ready inoculum is available. Thus, through the advancement of human technology, habitats have become established that have promoted the growth and distribution of *Legionella*.

The bacteria associated with these thermally elevated habitats have a characteristic in common with the clinical isolates of *Legionella*, that is, a large number of branched-chain fatty acids. The relationship between *L. pneumophila* and thermal environments is also indicated by the cellular fatty acid composition of the bacterium, which is similar to that of known thermophilic bacteria (Heinen, 1970; Moss et al., 1977). Initially, this character was used as a diagnostic tool for the identification of bacterial isolates that were suspected of being *Legionella.* Such information provided an ecological as well as an epidemiological tool for the autecological investigations of *Legionella* that followed.

As demonstrated in the field (Fliermans et al., 1981b), *Legionella* shows maximum cytochrome activity at a temperature optimum of 45°C with cardinal temperature activities at 4 and 63°C. These data agree with the biochemical analysis of the fatty acid content of *Legionella* and its relationship to thermophilic bacteria (Heinen, 1970). The observed resistance of *Legionella* to higher temperatures and low pH may be part of its genetic makeup. This resistance may also be enhanced by the associations that *Legionella* forms with correspondingly heat-resistant cyanobacteria and amebae. In the case of amebae, cysts of some pathogenic *Naegleria* strains have been shown to survive 51°C for 145 min and 58°C for 45 min (Chang, 1978). Similarly, preheating samples at 55°C for 30 min has seen successfully used for the selective isolation of *Naegleria* from sediments (R. L. Tyndall, personal communication). The sequestering of *Legionella* within an amoeba may in fact confer some resistance to heat (see Chapter 6).

In the laboratory, Wadowsky et al. (1985) investigated the effects of temperature on multiplication of naturally occurring legionellae in membrane-filtered tap water. They observed that *L. pneumophila* multiplied between 20 and 37°C, with a maximum increase occurring in laboratory test systems adjusted to 32 and 35°C. Dennis et al. (1984a) compared the effects of elevated temperatures on several types of bacteria and found that *L. pneumophila* exhibited little loss in viability at 50°C relative to a *Pseudomonas* species, a *Micrococcus* species, and coliforms.

Although temperature is an important parameter in the distribution of *Legionella,* it is not an overriding paradigm, since no single environmental parameter has been shown to be an effective predictor of the density of *Legionella* (Fliermans, 1985).

pH. Survival and growth of *Legionella* have also been shown to be substantially affected by pH levels. Laboratory and field studies differ in that naturally occurring *L. pneumophila* survives and grows in habitats with pH levels of 5.5 to 9.2 (Fliermans et al., 1981a; Wadowsky et al., 1985), while growth in laboratory cultures has been restricted to near neutrality. While developing one of the original culture media for legionellae, Feeley et al. (1978) reported a narrow pH limit of 6.5 to 7.5 for growth of agar-maintained *L. pneumophila* stock cultures on artificial media. Work with cooling-tower water suggested that elevated pH values are inhibitory to *Legionella* multiplication (States et al., 1987b). Several in vitro ameba studies have demonstrated that *Naegleria* spp. tolerate a wide pH range of 2.1 to 9.5 (Sykora et al., 1986; Carter, 1970; Kyle and Noblett, 1986). If *Legionella* spp. are associated with amebae in the natural environment, then the effect of pH levels on *L. pneumophila* in laboratory stock cultures, natural waters, and cooling-tower waters may be related to the susceptibility of amebae to those pH values (see Chapter 6).

Associations with algae. All the carbon and energy needs of *Legionella* spp. can be met with nine amino acids (Tesh and Miller, 1982). *Legionella pneumophila* can apparently synthesize all other necessary constituents de novo and seems to have no vitamin requirements (Ristroph et al., 1981). In aquatic environments, *Legionella* must obtain these amino acids either from other living organisms that produce them in excess or from the decomposition of organic matter, or both. Since legionellae typically exist in association with other microorganisms, this suggests that these microorganisms may indeed support *Legionella* growth.

The microbial group initially considered for its influence on legionellae were the algae. A survey of aquatic habitats at the Savannah River Site near Aiken, South Carolina, was made for the presence of *L. pneumophila* using serospecific polyvalent fluorescent antibodies as well as culture. Bacteria morphologically similar to *L. pneumophila*, reactive with a specific serogroup conjugate, and subsequently identified as *L. pneumophila* were observed in the naturally occurring algal-bacterial mat communities in thermal effluents ranging in temperature from ambient to 55°C. This algal mat community was composed of cyanobacteria of the genera *Fisherella*, *Phormidium,* and *Oscillatoria,* based on the taxonomic scheme of Rippka et al. (1979). Each cyanobacterium was isolated in unialgal culture, and bacteria associated with the algae were cultured and identified. Initially only *Fisherella* cultures contained a bacterium antigenically and morphologically similar to *L. pneumophila* serogroup 1, but subsequently the other cyanobacteria were shown to harbor *L. pneumophila* serogroup 1 as determined by fatty acid composition, antigenicity homology, and DNA homology. (Fliermans et al., 1979; Pope et al., 1982). Tison et al. (1980) isolated *L. pneumophila* from an algal-bacterial mat community growing in a man-made thermal effluent. When *Fisherella*, a thermophilic blue-green alga, was subsequently inoculated with *L. pneumophila* cultures, bacterial multiplication was observed. Multiplication occurred, however, only when the test system was exposed to light, and only minimal *Legionella* growth was observed under conditions of darkness or in the absence of *Fisherella*. Pope et al. (1982) expanded these findings to include additional species of green algae. They inoculated environmental and clinical isolates of *L. pneumophila* into minimal salts medium with various algal strains. The medium, containing no organic nutrients other than those provided by algae, supported legionellae replication. Growth reoccurred with successive transfers to new green algal or cyanobacteria cultures.

Additional experiments initiated with *Legionella* growing in the presence of the cyanobacterium *Fisherella* indicated that at 45°C in minimal salts medium *L. pneumophila* serogroup 1 had a doubling time of 2.7 h (Fliermans, 1985). This growth rate is twice as fast as that reported for the growth of *Legionella* in complex or defined media. These data suggest that *Legionella* uses the complex organic material that is produced by the *Fisherella* culture as a sole carbon and energy source. Thus, it is apparent that the presence of certain photosynthetic products in the environment may be critical in maintaining *Legionella* populations. During photosynthesis, oxygen is produced that is readily used by the bacterial community in general and *Legionella* in particular during aerobic respiration. Such respiration produces CO_2, which in turn is available for fixation by the algal community for photosynthates. This mutual association between algae, cyanobacteria, and heterotrophic bacteria occurs in a variety of natural habitats where light is the driving energy source.

Besides supporting growth, algae may promote the aerosol transmission of legionellae. Berendt (1981) demonstrated that survival of *L. pneumophila* in aerosols was improved when the bacterium was associated with *Fisherella* spp. This enhancement of survival may result from physical protection from desiccation provided by the algal mucilaginous matrix.

The importance of associations between *Legionella* and other microorganisms for its survival, growth, and pathogenicity is still not completely understood. One aspect of this association is the possible role of algal and amebic hosts' interactions in explaining observed differences between agar-passed and non-agar-passed *Legionella* strains. Data over the years have indicated that the virulence of *Legionella* is rapidly lost during agar passage, but can be regained or enhanced through non-agar passage techniques. Fliermans (1983, 1985) indicates that, in the presence of *Fisherella*, the virulence of *L. pneumophila* serogroup 1 is enhanced with respect to guinea pig infection. This virulence trigger may play a role in nature and provide conditions that enhance the virulence of the organism as well as its viability and density. We are still unaware of all the niches that house *Legionella* and the various aspects of those niches that control the survival, density, viability, and virulence of *Legionella*.

Associations with amebae. Evidence indicates that amebae and other protozoa act as natural hosts and amplifiers for *Legionella* in the environment (Rowbotham, 1980a, 1983; Anand et al., 1983; Tyndall and Dominque, 1982; Tyndall et al., 1983; Skinner et al., 1983; Fields et al., 1984; Holden et al., 1984; Barbaree et al., 1986). It has also been suggested that the host relationship affects the virulence of *Legionella* spp. (Fliermans, 1983, 1985; Rowbotham, 1980a; Fields et al., 1986) (Chapter 6). It is believed that *Legionella* are phagocytized (ingested) by amebal trophozoites, multiply within vesicles, and are either released when the vesicles and amebae rupture or, under certain conditions, remain encapsulated when the amebae encyst (Rowbotham, 1986). This ability to remain viable after ingestion is apparently important in *Legionella's* pathogenic mechanism (Eisenstein, 1987) and resembles the human disease process where *Legionella* multiplies within host macrophages.

The *Legionella*–amebae relationships may be an important factor in the ecology of *Legionella* as well as the epidemiology of legionellosis. A review of previous studies suggests that an interaction with amebae may also explain some of the behavior of *Legionella* in the laboratory. Of particular importance is the probable protection from the effects of disinfectants, commercial biocides in the field, low pH, desiccation, and heat granted by amebic hosts to *Legionella*. Recent studies provide evidence that free-living amebae are important determinants for survival and multiplication of *Legionella* in drinking water systems. Fields et al. (1989) isolated hartmannellid amebae from tap water stock cultures used in the laboratory to study *L. pneumophila* multiplication. Subsequent experiments indicated that a growth factor, removed from the culture by 1-μm filtration, could be reintroduced by inoculating hartmannellid amebae into the culture (Wadowsky et al., 1988). Since these amebae are widely present in potable water systems, they may contribute to the amplification of *L. pneumophila*. A study conducted on the city of Pittsburgh's public water supply indicated the presence of amebae in stagnant areas including the bottom and corners of reservoirs and the surface of treatment plant filters (States et al., 1988). These sites are similar to those previously determined to support *Legionella* growth (States et al., 1987a), and suggested the hypothesis that amebae could be an important parameter in *Legionella* contamination of public drinking water systems.

While there are still many questions concerning the extent of the association between amebae, algae, and *Legionella*, the *Legionella*–amebae and *Legionella*–algae models are consistent with many of the earlier environmental findings concerning the organism. This suggests that the best way to understand the ecology of environmental *Legionella* spp. may be through an improved understanding of the ecology of amebae and specific algae.

C. Dissemination of *Legionella*

It is not well understood how legionellae are transmitted to humans. Person-to-person spread has not been documented. Rather, the major mechanism of infection appears to be direct transmission from the environment by inhalation of aerosolized water droplets or particles containing viable *Legionella*, and subsequent deposition in the alveoli of the lungs. One of the critical flow paths to infection is the aerosolization and distribution of the bacterium in air in such a way that allows *Legionella* to maintain its viability, virulence, and infectivity. Thus, any parameter that promotes or enchances the sequestering, survival, and growth of *Legionella* will contribute to infection. Cooling towers were initially implicated in outbreaks of legionellosis from the standpoint that they were amplifiers and disseminators of the organism in a form that was conducive to establishing infection.

Several investigators have documented the occurrence of *Legionella* spp. at sites within domestic plumbing systems that readily allow the production of

Legionella-containing aerosols. *Legionella* has been isolated from faucet nozzles, aerators, and showerheads, as well as in domestic water piping and water heaters (Wadowsky et al., 1982; Witherall et al., 1988; Joly et al., 1986; Lee et al., 1988). These few surveys of domestic plumbing systems suggest the possibility of contaminated domestic plumbing systems acting as disseminators of the pathogens, although evidence for disease transmission from these sources remains limited.

Legionella pneumophila has been isolated from water samples collected from hot and cold taps, showers, and storage tanks in a number of hotels (Tobin et al., 1981a,1981b). Some of the hotels had previously been associated with cases of Legionnaires' Disease. Dennis et al. (1982) examined water samples from the plumbing systems of 52 hotels, ranging in size from small Victorian buildings to large, modern multistory structures. Although none of these hotels had been associated with cases of legionellosis, *L. pneumophila* was recovered in 33% of the buildings.

IV. *LEGIONELLA* AND DISEASE

A. Legionnaires' Disease

Legionella pneumophila causes a severe respiratory illness, and while the illness is treatable with antibiotics, the overall case-fatality rate is still high. Among previously healthy individuals, 7 to 9% still die when treated with erythromycin, while 25% die when hospitalized but not treated with appropriate antibiotics. Among individuals with seriously impaired immune systems, the mortality rate is 24% for the adequately treated and 80% for those without treatment.

Reported attack rates for Legionnaires' Disease after exposure have ranged from 0.1 to 4.0% with an incubation time of 2 to 10 days. The risk of contracting Legionnaires' Disease is significantly enhanced for individuals with underlying disease, such as diabetes, malignancy, chronic cardiopulmonary disease, surgery, renal transplantation, use of immunosuppressive medication, radiation therapy, or other conditions causing general impairment of the immune system. Additional risk factors include being a male over 50 years of age, cigarette smoking, and excessive use of alcohol.

Legionnaires' Disease occurs as epidemic clusters and sporadic cases. Many cases are hospital-acquired while others have been community-acquired. While Legionnaires' Disease has been documented nearly worldwide, the actual incidence is unknown. To date, all reported outbreaks (i.e., clusters of cases) of Legionnaires' Disease have been caused by *L. pneumophila*. Sporadic cases have been caused most frequently by *L. pneumophila*, particularly serogroup 1, followed by *L. micdadei* and less commonly by *L. bozemanii*, *L. dumoffii*, and other *Legionella* spp.

Retrospective examination of preserved sera and bacterial specimens from earlier explosive outbreaks of pneumonia indicates that Legionnaires' Disease is not a new syndrome but has occurred undetected for decades (McDade et al., 1979). An important retrospective study done by Foy et al. (1979) estimated the community incidence of Legionnaires' Disease using stored paired sera from 500 patients treated for pneumonia from 1963 to 1975 in Seattle, Washington. Based on 1% of the patients showing a fourfold rise in antibody titer to the Legionnaires' Disease antigen, the incidence of Legionnaires' Disease was estimated to be 0.4 to 2.8 cases per 10,000 persons in the population per year. These findings led to the often used estimate of 25,000 cases of *Legionella* pneumonia occurring annually in the United States. According to a 1991 community-based pneumonia incidence study conducted by the U.S. Public Health Service, Centers for Disease Control, the estimated incidence of pneumonia among adults due to *Legionella* is 6.1/100,000 or 11,000 cases a year in the United States.

B. Pontiac Fever

Legionella is also responsible for another form of legionellosis termed Pontiac fever after a 1968 epidemic of respiratory illness occurring in Pontiac, Michigan. Pontiac fever is a nonpneumonic, self-limiting, nonfatal influenzalike disease with an attack rate of nearly 100% of those exposed and an incubation period of 5 to 66 h. The illness has only been recognized in epidemic form and has been caused by *L. pneumophila* serogroups 1 and 6, and in one outbreak by *L. feelii.* No specific drug therapy has been established for Pontiac fever because of its benign nature (Band and Fraser, 1986).

The basis for the two different clinical and epidemiological presentations of legionellosis is not yet understood. No characteristics of either the organisms or the mode of transmission have been identified that account for the difference between Legionnaires' Disease and Pontiac fever. Additionally, *Legionella* have never been recovered directly from Pontiac fever patients, and diagnosis has been established through seroconversion in the context of the characteristic clinical symptoms (Muder et al., 1986). Several hypotheses to explain Pontiac fever include a change in virulence factors (Broome, 1984); toxic or hypersensitivity reaction to *Legionella* (Kaufmann et al., 1981); and hypersensitivity to amebae containing *Legionella* (Rowbotham, 1980b).

Several reports have been published concerning the occurrence of Pontiac fever in the work environment. The first, occurring in Pontiac, Michigan, in 1968, is the outbreak after which the illness was named (Glick et al., 1978). In this outbreak, 144 people who had worked in or visited a health department building contracted Pontiac fever. Aerosols from a contaminated evaporative condenser in a defective air-conditioning system were the apparent source of infection. Other Pontiac fever outbreaks have occurred among workers performing maintenance in a steam turbine condenser (Fraser et al., 1979); among employees in an auto assembly plant from an oil–water industrial coolant system; and among workers in a New York City office building from a contaminated cooling tower.

C. Epidemiology of Legionellosis

A question immediately arises concerning the observation that *Legionella* species are often found in aquatic environments, but, in most instances, no disease is reported. If, in fact, legionellae are common in nature, why are there so few cases of legionellosis? Ahtough the reason for the discrepancy is unknown, a number of possible factors have been proposed, including (1) underdiagnosis of the disease; (2) a critical number of legionellae required for infection; (3) relationships between *Legionella* and other microorganisms triggering survival and dispersion; (4) variations in *Legionella* virulence associated with species and subspecies differences; (5) presence of adequate and effective disseminators; and (6) variations in host susceptibility for infection.

Underdiagnosis. A particular problem associated with determining the incidence of Legionnaires' Disease is the difficulty of diagnosing the disease. If specialized laboratory techniques are not routinely available, including the use of highly selective culture media, direct immunofluorescent probes, DNA probes, and serological detection of antibodies, Legionnaires' Disease may remain undetected. Broome (1984) has speculated that some of the 30 to 40% of pneumonia cases of undetermined etiology may be due to legionellae. Other investigators have estimated that 1 to 5% of the pneumonias observed in a community are *Legionella* related (Marrie et al., 1989; Research Committee, 1987; Woodhead and McFarlane, 1987). Bartlett et al. (1986) suggest that where Legionnaires' Disease is endemic or common the incidence of *Legionella* isolation from reservoirs is very high, while in other areas such recovery is low, indicating that there may truly be a geographical variation in the incidence of the disease.

Hotel outbreaks. Since the initial documented outbreak of Legionnaires' Disease, which created the impetus for isolation of the bacterium, other hotel outbreaks and case clusters have been documented in virtually every country of the world (Broadbent et al., 1991; Reid et al., 1978; Tobin et al., 1981a,1981b; Jorgensen et al., 1981, Bartlett et al., 1984, 1986). *Legionella pneumophila* has been isolated from water samples collected from hot and cold water taps, showers, and storage tanks in a number of hotels (Tobin et al., 1981a,1981b), some of which had previously been associated with cases of Legionnaires' Disease.

Although institutional plumbing systems have been implicated in outbreaks, little work has been done on the role of residential hot water systems as a source for *Legionella* infection. Arnow et al., (1985) recovered

L. pneumophila from 51% of 37 Chicago apartment buildings. These buildings contained large hot water tanks having water temperatures lower than 55°C. However, health surveys and blood serum analyses indicated that this bacterial contamination did not represent a significant health risk to the residents.

Stout and co-workers (1987) and Phillips et al. (1987) have annotated cases of community acquired legionellosis associated with contamination of domestic plumbing systems. Because of medical problems, these patients had been confined to their homes prior to the onset of symptoms. Isolates of *L. pneumophila* recovered from hot water tanks and faucets of the residential water system were paired with serogroups obtained from clinical samples. Additionally, *Legionella* was isolated from the faucet of an infected cardiac transplant patient's kitchen sink as well as an ultrasonic humidifier used prior to transplantation. The humidifier was heavily colonized with *L. pneumophila* serogroup 1, the same serogroup cultured from the clinical specimens. These data suggest that the household water supply was the source for the organism.

The question remains, however, of what role potable water, both in municipal and residential systems, plays in the dissemination of *Legionella*. Variations in the occurrence of contaminated source waters, differences in *Legionella* virulence and density, suitable means of aerosol transmission, and the proximity of susceptible populations lead to wide differences in the risk of acquiring Legionnaires' Disease or Pontiac fever.

Although *Legionella* spp. have been implicated in numerous epidemic, endemic, and sporadic cases of community and hospital acquired pneumonias, it is important to note that since *Legionella* is ubiquitous, the mere presence of the bacterium does not mean that a particular habitat is the cause of an outbreak or has caused disease. Both virulence and density of the *Legionella* are key parameters in assessing the data, and although *Legionella* is not uncommon in large buildings, the occurrence of the bacterium is generally not associated with illness in a healthy human population. If Legionnaires' Disease occurs in these structures, it is likely to occur in the form of sporadic cases, rather than the more easily detected outbreaks or epidemics.

Sporadic cases of Legionnaires' Disease have originated in the work place. Bornstein et al. (1986) described a fatal case of Legionnaires' Disease acquired in an administrative building in France. Serotyping with monoclonal antibody provided a link between the infection and contamination of the building's hot water supply. Muraca et al. (1988) documented a fatal case of Legionnaires' Disease contracted in a small plastics factory, and identified the industrial cooling water of the plastics factory, through monoclonal techniques, as the source of the pathogen. Studies by Muraca et al. (1988) indicate that it is often difficult if not impossible to determined the source of infection, since the ubiquity of the organism allows an individual to be associated with environments that are readily positive for *L. pneumophila*.

Although numerous cases of work place acquired Legionnaires' Disease have not been reported, the risk of contracting the disease from the work place is not high. Other workers within the same locations were not affected, underscoring the role played by host susceptibility factors. Most of the work-related legionellosis cases have resulted in Pontiac fever rather than Legionnaires' Disease presentations, indicating those host risk factors among typically healthy workers are less crucial for this manifestation.

V. CONTROL

Barbaree (1991) has described seven events that are necessary for the contraction of legionellosis. These events are summarized in Figure 3.1 and represent a flow of the organism from its natural habitat to a full-blown case of Legionnaires' Disease. The major area of concern for control is the amplification and dissemination of the organism. The ability to control the organism at this interactive point is critical. From an ecological perspective it is not realistic to consider that *Legionella* can be controlled in its natural habitat. Thus, effort needs to be concentrated on the control of the organism in those habitats associated with human occupation that are most likely to serve as amplifiers and disseminators for *Legionella*. This discussion centers primarily on the control of *Legionella* from cooling-tower systems, since amplification and dissemination are readily facilitated in these habitats.

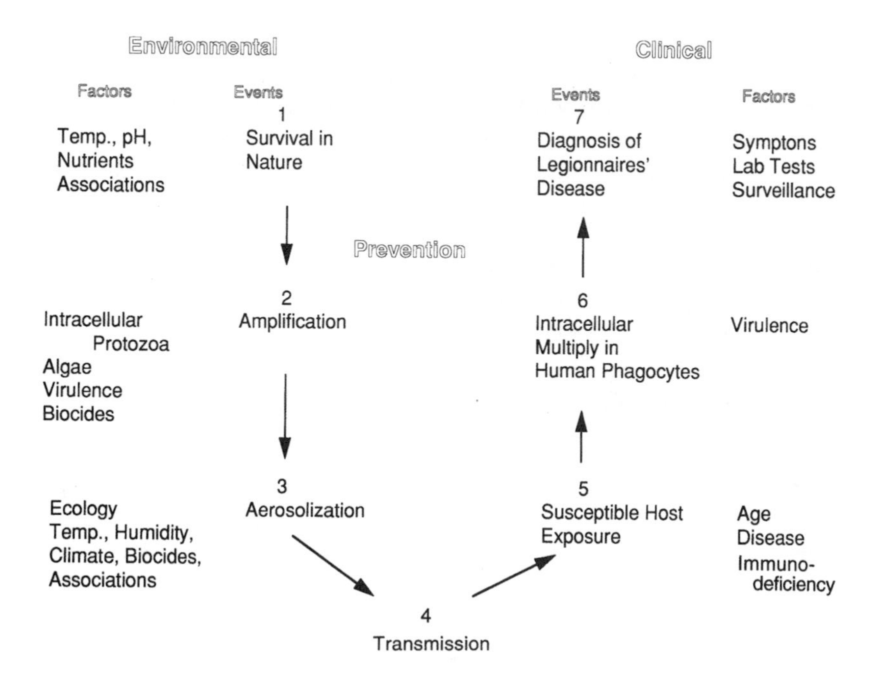

Figure 3.1. Routes for *Legionella* transmission.

A. Chlorine

Laboratory studies. Since chlorine is the most widely used water disinfectant, it has received the most attention. In the laboratory, Kuchta et al. (1983, 1985), using chlorine residuals characteristically found in public water supplies, demonstrated that agar-passed *Legionella* were more resistant than coliform bacteria to chlorine, regardless of how the coliforms were maintained (i.e., on artificial media or directly isolated from a natural river population). At 21°C, pH 7.6, and 0.1 mg free chlorine residual per liter, a 99% kill of *L. pneumophila* required up to 40 min, compared with less than 1 min for coliforms. Further testing indicated that water stock cultures of *L. pneumophila* (which are assumed to more closely resemble the natural state) have even greater resistance than their agar medium-passed counterparts. Roberts et al. (1987), studying the effects of selective isolation techniques on *Legionella*, found that natural water cultures of *Legionella* were significantly more resistant to the combinations of acid treatment, antibiotics, and heat treatment than were BCYE-passed cultures.

Legionella isolated from a hyperchlorinated (2 ppm) hospital sink faucet were even more chlorine resistant. Other laboratory disinfection experiments showed that *Legionella* isolated from populations having survived elevated chlorine concentrations in a hospital plumbing system were more resistant to the action of chlorine than isolates from natural sites (Kuchta et al., 1983, 1985).

These findings indicate that *Legionella* can survive for relatively long periods of time at low concentrations of chlorine, and imply that small numbers of *Legionella* may survive at chlorine concentrations that are assumed to make waters microbiologically acceptable. The mechanism by which *Legionella* develops resistance to chlorine is unclear. Protection may be associated with variations in growth rate, alterations of the competing biota, degree of nutrient limitation, changes in the cellular envelope, and/or interactions that shelter *Legionella* in algal or protozoan consortia. Investigation by King and coworkers (1988) indicated that *Escherichia coli, Enterobacter* spp., *Klebsiella* spp., *Salmonella typhimurium, Yersinia entercolitica, Shigella sonnei*, and *Campylobacter jejuni*, as well as *Legionella* spp. form protozoa–bacteria host relationships. Consequently, these organisms may tolerate levels of chlorine that kill free-living bacteria. These data make the point that natural systems and naturally occurring rather than agar-passed cultures of legionellae need to be evaluated for response to control measures.

The effectiveness of chlorine dioxide for inactivating *L. pneumophila* grown in a chemostat containing a liquid growth medium was examined by Berg et al. (1984, 1985). Under these conditions, both *L. pneumophila* and *E. coli* were inactivated at a comparable rate by 0.75 mg/l chlorine dioxide. In another experiment, conducted to compare the effectiveness of chlorine dioxide and free chlorine in the disinfection of batch-grown *Legionella* cultures, chlorine dioxide was superior to chlorine when both were applied at equal mass doses.

Field studies. Fliermans et al. (1982) demonstrated the effectiveness of chlorine against *Legionella* in cooling towers in the southeastern United States. The data indicated that chlorine at a free residual concentration of 1.5 ppm was necessary to reduce the densities of *Legionella* below those in the makeup water coming into the cooling tower. Fliermans and Nygren (1987) have used this information to demonstrate the effectiveness of chlorination in maintaining industrial cooling and process systems "free" of *Legionella*.

Although the test results have been mixed, a number of chlorine field studies have demonstrated the effectiveness of continuous hyperchlorination. Fliermans et al. (1982) concluded that elevated *L. pneumophila* densities in cooling systems could be reduced and maintained near levels found in the source water by either continuous or shock chlorination. Band et al. (1981) confirmed that continuous chlorination of a contaminated cooling tower at 3 mg/l eliminated *L. pneumophila*, although a 7-day time lag was observed between the initiation of treatment and elimination of the organism.

When DeJonckheere and van de Voorde (1976) examined inactivation of amebic cysts by free chlorine, they found that cysts of *Acanthamoeba* spp. survived a free chlorine concentration of 4 ppm for more than 3 h of contact time. Chang (1978) observed that a 25-min exposure was required to inactivate 99.9% of pathogenic *Naegleria* cysts at a free Cl_2 concentration of 1.4 mg/l. Additionally, Skinner et al. (1983) suggested that an amebic cyst wall may provide protection to *Legionella* enabling it to survive in chlorinated waters. Rowbotham (1986) suggested that the failure of periodic dosing with biocide to control *L. pneumophila* in a cooling tower may provide a selective advantage for chlorine-resistant amebae infected with *Legionella*. Thus, it is possible that the chlorine resistance observed in *Legionella* can be credited in part to the presence of chlorine-resistant protozoa.

B. Other Biocides

Numerous other biocides have been tested in the laboratory and in the field. R. J. Soracco (Ph.D. thesis, Rensselear Polytechnic Institute, Troy, NY, 1981) surveyed the reaction of environmental and clinical isolates of *Legionella* to 12 selected biocides licensed by the Environmental Protection Agency for use in cooling towers. These studies indicated that many of the biocides were not biocidal in the laboratory at the recommended "slug" or maintenance doses against *Legionella*. Three of the most effective biocides (potassium *n*-methyldithiocarbamate; disodium cyanodithioimidocarbonate; *bis*-tributyltin oxide with dimethylbenzyl ammonium chlorides) in the laboratory analyses were employed in industrial-sized cooling towers (E. B. Braun, Ph.D. thesis, Rensselear Polytechnic Institute, Troy, NY, 1982). None of the biocides demonstrated the effectiveness seen with chlorine.

Fliermans and Harvey (1984) evaluated the use of bromocide in reducing the activity and density of *Legionella* in an industrial cooling tower. The data indicated that even at concentrations of over 2 ppm of free halogen, the levels of *Legionella* were not reduced to below the densities of *Legionella* in the makeup water to the cooling tower. Additionally, the activity of *Legionella* as measured by INT activity was not reduced significantly by the various concentrations of bromocide tested in the field.

Experiments of Cargill et al. (1992) demonstrated that biofilm-associated *Legionella* were 100 times more resistant to iodine disinfection than unattached *Legionella* and 100,000 times more resistant than agar-grown *Legionella*. These data demonstrate that the environmental conditions under which *Legionella* exist and grow are very important in assessing the response of the organism. Thus, it is imperative that well-conducted laboratory results be expanded into field studies to observe whether the laboratory findings are paralleled in the field.

An additional disinfection technique has been proposed for cooling towers based upon an investigation by States et al. (1987b) that indicated that *Legionella* growth is inversely correlated with water pH and alkalinity. These laboratory findings suggest that increasing pH and alkalinity above the cardinal range for *Legionella* spp. may be a useful approach for controlling this organism. The operation of cooling towers at high alkalinity and high pH values has been recommended for corrosion control and to enhance chlorine persistence (Kemmer and McCallion, 1979; Characklis et al., 1980). However, full-scale field investigations have not been conducted that document the efficacy of this approach.

The laboratory investigations of Dominque and Tyndall et al. (1988) compared the effects of ozone, hydrogen peroxide, and free chlorine on agar-maintained *L. pneumophila* and found that ozone was the most efficacious, with 99% kill occurring during a 5-min exposure to 0.10 to 0.30 mg/l of O_3. Hydrogen peroxide was the least effective, with a 30-min exposure to 1,000 mg/l of H_2O_2 being required for a 99% reduction.

Several investigations have dealt with the response of *Legionella* to ultraviolet (UV) radiation. Antopol and Ellner (1979) examined the response of agar-grown *L. pneumophila* to UV disinfection. *Legionella* suspensions were sensitive to low doses of germicidal UV radiation, with 99% kill occurring at a dosage of 1840 μW–s/cm^2. Subsequently, Gilpin (1984) found that *Legionella* suspended in phosphate buffer required moderate-range UV dosages for 99% kill when compared to those reported for other bacteria. Dutka (1984) evaluated the sensitivity of agar-grown *L. pneumophila* to sunlight in water and demonstrated that both UV and visible light play a role in killing the microorganisms. Knudson (1985) exposed lawns of seven *Legionella* species grown on charcoal yeast-extract agar to low doses (240 μW/cm^2) of UV light. While *Legionella* were susceptible to short exposures, all species repaired the germicidal damage of the UV light when subsequently exposed to photoreactivating light. Such data suggest caution should be used when assessing the effectiveness of UV radiation, particularly if the water will later be exposed to visible light.

C. Guidelines for Cooling-Tower Maintenance

Fliermans and Nygren (1987) have proposed guidelines for the maintenance of industrial cooling towers "free" from *Legionella.* The guidelines are based upon the field evaluation of hundreds of industrial cooling towers and process water since 1978. The guidelines recognize that *Legionella* is part of the environment and is likely to seed many industrial systems on a continuous basis. The guidelines not only address the presence of the organism, but place limits on the density of the organism above which disinfection techniques are used to eliminate *Legionella* from the system. The guidelines recognize that any disinfection techniques for *Legionella* are not long-lasting because reseeding of the system due to the ubiquity of the organism will occur.

While variable chlorine sensitivity makes the occurrence of high levels of *Legionella* spp. less likely, it does not eliminate the possibility that low numbers of legionellae pass from municipal supplies into cooling towers and internal plumbing systems. This emphasizes the need for sound conventional public water treatment practices, including maintenance of a disinfectant residual, and, where indicated, the application of additional control measures at the point of use. Variations in the occurrence of contaminated source waters, the density and virulence of the legionellae, suitable means for transmission of aerosols, and the proximity of susceptible populations lead to wide differences in the risk of infection in hospitals, other large buildings, private homes, the work place, and the community.

Australia, New Zealand, and United Kingdom governments have established guidelines for the control and prevention of legionellosis in their countries related to cooling towers (Standards Australia, 1989; NHMRC, 1989; Health Department, Victoria, 1989; CIBSE, 1991). These governments have established risk evaluations from the aspect that those cooling systems located close to susceptible hosts, that is, nursing-home residents, hospital areas with immunocompromized patients, etc., are by law to receive a greater scrutiny than other conventionally located industrial cooling towers. Additionally, each cooling tower in Australia is registered with the government and the owner is required to keep maintenance records. If towers are found to contain *Legionella* spp. at a predetermined level, then disinfection protocols are implemented depending on the density of the *Legionella* contamination. These guidelines are the most aggressive in the world.

D. Litigation

A number of legal cases have been brought into the United States court systems for compensation claims. These have ranged from workmen's compensation cases to civil trials where manufacturers of chemicals, owners of cooling towers, and municipalities have all been sued by individuals who deemed they were infected by the negligence of the defendant. In each of these cases the burden of proof has rested on the following: demonstrating effectively to the jury that (1) legionellosis was contracted from a particular system and (2) that the party responsible for that system was negligent with respect to applying all available knowledge about *Legionella* to the eradication of the problem. Neither of these tasks has been particularly easy.

REFERENCES

Allen, E.T. and Day, A.L., *Hot Springs of the Yellowstone National Park.* Carnegie Institution of Washington, D.C., p. 208, 1935.

Anand, C. M., Skinner, A. R., Malic, A., and Kurta, J. B., Interaction of *L. pneumophila* and a free-living amoeba (*Acanthamoeba palestinenis*), *J. Hyg.,* 91, 167, 1983.

Antopol, S. C. and Ellner, P. D., Susceptibility of *Legionella pneumophila* to ultraviolet radiation, *Appl. Environ. Microbiol.,* 38, 347, 1979.

Arnow, P. M., Weil, D., and Para, M. F., Prevalence and significance of *Legionella pneumophila* contamination of residential hot-tap water systems, *J. Infect. Dis.,* 152, 145, 1985.

Band, J. D. and Fraser, D. W., Legionellosis (Legionnaires' Disease and Pontiac fever), pp. 831-841. *In*: Braude, A. I., Davis, C. E., and Fierer, J. (Eds.), *Infectious Diseases and Medical Microbiology*, W. B. Saunders, Philadelphia, 1986.

Band, J. D., LaVenture, M., David, J. P., Mallison, G. F., Skaliy, P., Hayes, P. S., Schell, W. L., Weiss, H., Greenberg, D. J., and Fraser, D. W., Endemic Legionnaires disease: Airborne transmission down a chimney, *J. Am. Med. Assoc.,* 245, 2404, 1981.

Barbaree, J. M., Controlling *Legionella* in cooling towers, *ASHRAE*, 38, June 1991.

Barbaree, J. M., Fields, B. S., Feeley, J. C., Gorman, G. W., and Martin, W. T., Isolation of protozoa from water associated with a Legionellosis outbreak and demonstration of intracellular multiplication of *Legionella pneumophila*, *Appl. Environ. Microbiol.,* 51, 422, 1986.

Bartlett, C. L. R., Swann, R. A., Casal, J., Canada, L., and Taylor, A. G., Recurrent Legionnaires disease from a hotel water system, pp. 237-239. *In*: Thornsberry, C., Balows, A., Feeley, J. C., and Jakubowski, W. (Eds.), *Legionella: Proceedings of the Second International Symposium*, American Society for Microbiology, Washington, D.C., 1984.

Bartlett, C. L. R., Macrae, A. D., and MacFarlane, J. T., *Legionella Infections*, Edward Arnold, Ltd., London, UK, 1986.

Berendt, R. F., Influence of blue-green algae (cyanobacteria) on survival of *Legionella pneumophila* in aerosols, *Infect. Immun.,* 32, 690, 1981.

Berg, J. D., Hoff, J. C., Roberts, P. V., and Matin, A., Growth of *Legionella pneumophila* in continuous culture and its sensitivity to inactivation by chlorine dioxide, pp. 68-70. *In*: Thornsberry, C., Balows, A., Feeley, J. C., and Jakubowski, W. (Eds.), *Legionella: Proceedings of the Second International Symposium,* American Society for Microbiology, Washington, D.C., 1984.

Berg, J. D., Hoff, J. C., Roberts, P. V., and Matin, A., Disinfection resistance of *Legionella pneumophila* and *Escherichia coli* grown in continuous and batch culture, pp. 603-613. *In:* Jolly, R. L., Bull, R. L., Davis, W. P., Katz, S., Roberts, M. H., and Jacobs, V. A. (Eds.), *Water Chlorination: Environmental Impact and Health Effects*, Vol. 5, Lewis Publishers, Chelsea, MI, 1985.

Bopp, C. A, Sumner, J. W., Morris, G. K., and Wells, J. G., Isolation of *Legionella* spp. from environmental water samples by low pH treatment and use of a selective medium, *J. Clin. Microbiol.,* 13, 714, 1981.

Bornstein, N., Veilly, C., Nowicki, M., Paucod, J. C., and Fleurette, J., Epidemiological evidence of Legionellosis transmission through domestic hot water supply systems and possibilities of control, *Isr. J. Med. Sci.,* 22, 655, 1986.

Broadbent, C. R., Marwood, L. N., and Bentham, R. H., *Legionella* in cooling towers: report of a field study in South Australia, *ASHRAE*, Far East Conference on Environmental Quality, Hong Kong, FE91-10, 55, 1991.

Broome, C. V., Current issues in epidemiology of Legionellosis, pp. 205-209. *In*: Thornsberry, C., Balows, A., Feeley, J. C., and Jakubowski, W. (Eds.), *Legionella. Proceedings of The Second International Symposium*, American Society for Microbiology, Washington, D.C., 1984.

Calderon, R. L. and Dufour, A. P., Media for detection of *Legionella* spp. in environmental water supplies, pp. 290-292. *In*: Thornsberry, C., Balows, A., Feeley, J. C., and Jakubowski, W. (Eds.), *Legionella*: *Proceedings of the Second International Symposium*, American Society for Microbiology, Washington, D.C., 1984.

Cargill, K. L., Pyle, B. H., Sauer, R. L., and McFeters, G. A., Effects of culture conditions and biofilm formation on the iodine susceptibility of *Legionella pneumophila, Can. J. Microbiol.*, 38, 423, 1992.

Carter, R. F., Description of a *Naegleria* species isolated from two cases of primary amebic meningoencephalitis and of the experimental pathological changes induced by it, *J. Pathol.,* 100, 217, 1970.

Chang, S. L., Resistance of pathogenic *Naegleria* to some common physical and chemical agents, *Appl. Environmental Microbiol.,* 35, 368, 1978.

Characklis, W. G., Trulear, M. A., Stathopoulos, N., and Chang, L. C., Oxidation and destruction of microbial films, pp. 349-368. *In*: Jolly, R. L. (Ed.), *Water Chlorination: Environmental Impact and Health Effects*, Butterworth Publishers, Stoneham, MA, 1980.

Chartered Institution of Building Services Engineers, CIBSE, Minimizing the Risk of Legionnaires' Disease, Technical Memorandum No. 13, London, 1991.

Cordes, L. G., Fraser, D. W., Skaliy, P., Pernio, C. A., Else, W. R., Malllison, G. F., and Hayes, P. S., Legionnaires disease outbreak at an Atlanta, Georgia, country club: evidence for spread from an evaporate condenser, *Am. J. Epidemiol.,* 111, 425, 1980.

Cordes, L. G., Wisenthal, A. M., Gorman, G W., Phair, J. P., Sommers, H. M., Brown, A., Yu, V. L., Magnussen, M. H., Meyer, R. D., Wolf, J. S., Shands, K. N., and Fraser, D. W., Isolation of *Legionella pneumophila* from shower heads, *Ann. Intern. Med.,* 94, 195, 1981.

DeJonckheere, J. and van de Voorde, H., Differences in destruction of cysts of pathogenic and nonpathogenic *Naegleria* and *Acanthamoeba* by chlorine, *Appl. Environ. Microbiol.,* 31, 294, 1976.

Dennis, P. J., Barlett, C. L. R., and Wright, A. E., Comparison of isolation methods for *Legionella* spp., pp. 294-296. *In*: Thornsberry, C., Balows, A., Feeley, J. C., and Jakubowski, W. (Eds.), *Legionella: Proceedings of the Second International Symposium*, American Society for Microbiology, Washington, D.C., 1984a.

Dennis, P. J., Fitzgeorge, R. B., Taylor, J. A., Barlett, C. L. R., and Barrow, G. I., *Legionella pneumophila* in water plumbing systems, *Lancet,* i, 949, 1982.

Dennis, P. J., Green, D., and Jones, B. P., A note on the temperature tolerance of *Legionella, J. Appl. Bacteriol.,* 56, 349, 1984b.

Dominque, E. L., Tyndall, E. L., Mayberry, W. R., and Pancorbo, O. C., Effects of three oxidizing biocides on *Legionella pneumophila* serogroup 1, *Appl. Environ. Microbiol.,* 54, 741, 1988.

Dondero, T. J., Rendtorff, R. C., Mallison, G. F., Weeks, R. M., Levy, J. S., Wong, E. W., and Schaffner, W., An outbreak of Legionnaires disease associated with a contaminated air conditioning cooling tower, *N. Engl. J. Med.,* 302, 362, 1980.

Dutka, B. J., Sensitivity of *Legionella pneumophila* to sunlight in fresh and marine waters. *Appl. Environ. Microbiol.,* 48, 970, 1984.

Edelstein, P. H., Comparative study of selective media for isolation of *Legionella pneumophila* from potable water, *J. Clin. Microbiol.,* 16, 697, 1982.

Eisenstein, B. J., Pathogenic mechanisms of *Legionella pneumophila* and *Escherichia coli, Am. Soci. Microbiol. News,* 53, 621, 1987.

Feeley, J. C., Gorman, G. W., Weaver, R. E., Mackel, D. C., and Smith, H. W., Primary isolation media for Legionnaires' Disease bacterium, *J. Clin. Microbiol.,* 8, 320, 1978.

Fields, B. S., Barbaree, J. M., Shotts, E. B., Feeley, J. C., Morill, W. E., Sanden, G. N., and Dykstra, M. J., Comparison of guinea pig and protozoan models for determining virulence of *Legionella* species, *Infect. Immun.,* 53, 553, 1986.

Fields, B. S., Sanden, G. N., Barbaree, J. M., Morill, W. E., Wadowsky, R. M., White, E. H., and Feeley, J. C., Intracellular multiplication of *Legionella pneumophila* in amebae isolated from hospital hot water tanks, *Current Microbiol.,* 18, 191, 1989.

Fields, B. S., Shotts, E. B., Feeley, J. C., Gorman, G. W., and Martin, W. J., Proliferation of *Legionella pneumophila* as an intracellular parasite of the ciliated protozoan *Tetrahymena pyriformis, Appl. Environ. Microbiol.,* 47, 467, 1984.

Fliermans, C. B., Autecological of *Legionella pneumophila. In*: Workshop Conference on Legionnaires' Disease, Fehrenbach, F. J. (Ed.) *Legionellose-Arbeitstagung des Bundesgesundheitsamt,* Dietrich Reimer Verlag, West Berlin, 255, 58, 1983.

Fliermans, C. B., Ecological niche of *Legionella pneumophila, In*: Katz, R. S., (Ed.), *Critical Reviews of Microbiology,* p. 75-116, 1985.

Fliermans, C. B., Bettinger, G. E., and Fynsk, A. W., Treatment of cooling systems containing high levels of *Legionella pneumophila, Water Res.,* 16, 903, 1982.

Fliermans, C. B., Cherry, W. B., Orrison, L. H., Smith, S. J., Tison, D. L., and Pope, D. H., Ecological distribution of *Legionella pneumophila, Appl. Environ. Microbiol.,* 41, 9, 1981a.

Fliermans, C. B., Cherry, W. B., Orrison, L. H., and Thacker, L. Isolation of *Legionella pneumophila* from non-epidemic related aquatic habitats, *Appl. Environ. Microbiol.,* 37, 1239, 1979.

Fliermans, C. B. and Harvey, R. S., Effectiveness of bromicide against *Legionella pneumophila* in a cooling tower, *Appl. Environ. Microbiol.,* 47, 1307, 1984.

Fliermans, C. B. and Nygren, J. A., Maintaining industrial cooling systems "free" of *Legionella pneumophila*, Transactions of American Society of Heating, Refrigerating, and Air-Conditioning Engineers, 93, NT-87-09-4, 1987.

Fliermans, C. B. and Schmidt, E. L., Autoradiography and immunofluorescence combined for autecological study of single cell activity with *Nitrobacter* as a model system, *Appl. Microbiol.,* 30, 676, 1975.

Fliermans, C. B., Soracco, R. J., and Pope, D. H., Measurement of *Legionella pneumophila* activity in situ, *Current Microbiol.,* 6, 89, 1981b.

Fliermans, C. B. and Tyndall, R. L., Associations of *Legionella pneumophila* with natural ecosystems, 1992 International Symposium on *Legionella*, Orlando, FL, p. 30, 1992.

Foy, H. M., Hayes, P. S., Cooney, M. K., Broome, C. V., Allan, I., and Tobe, R., Legionnaires' Disease in a prepaid medical-care group in Seattle, 1963-75, *Lancet,* i, 767, 1979.

Fraser, D. W., Deubner, D. C., Hill, D. L., and Gilliam, D. K., Nonpneumonic, short-incubation-period legionellosis (Pontiac fever) in men who cleaned a steam turbine condenser, *Science,* 206, 690, 1979.

Fraser, D. W., Tsai, T. F., Orenstein, W., Parkin, W. E., Beecham, H. J., Sharrar, R. C., Harris, J., Mallison, G. F., Martin, S. M., McDade, J. E., Shepard, C. C., Brachman, P. S., and The Field Investigation Team, Legionnaire's disease: A description of an epidemic of pneumonia, *N. Engl. J. Med.,* 297, 1189, 1977.

Gilpin, R. W., Laboratory and field applications of U.V. light disinfection on six species of *Legionella* and other bacteria in water, pp. 337-339. *In*: Thornsberry, C., Balows, A., Feeley, J. C., and Jakubowski, W. (Eds.), *Legionella: Proceedings of the Second International Symposium*, American Society for Microbiology, Washington, D.C., 1984.

Glick, T. H., Gregg, M. B., Berman, B., Mallison, G. W., Rhodes, W. W., and Kassnnoff, I., Pontiac fever: An epidemic of unknown etiology in a Health Department. I. Clinical and epidemiological aspects, *Am. J. Epidemiol.,* 107, 149, 1978.

Gorman, G. W., Feeley, J. C., Steigerwalt, A., Edelstein, P. H., Moss, C. W., and Brenner, D. J., *Legionella anisa*: A new species of *Legionella* isolated from potable waters and a cooling tower, *Appl. Environ. Microbiol.,* 49, 305, 1985.

Health Department, Victoria, Guidelines for the Control of Legionnaires' Disease. Environmental Health Standards. Health Department, Victoria, Australia, 1989.

Heinen, W., Extreme thermophilic bacteria fatty acids and pigments, *Antonie van Leeuwenhoek J. Microbial. Serol.,* 36, 582, 1970.

Holden, E. P., Winkler, H. H., Wood, D. O., and Leinbach, E. D., Intracellular growth of *Legionella pneumophila* within *Acanthamoeba castellanii* Neff, *Infect. Immun.* 45, 18, 1984.

Hsu, S. C., Martin, R., and Wentworth, B. B., Isolation of *Legionella* species from drinking water, *Appl. Environ. Microbiol.,* 48, 830, 1984.

Hussong, D., Colwell, R. R., O'Brien, M., Weiss, E., Pearson, A. D., Weiner, R. M., and Burge, W. D., Viable *Legionella pneumophila* not detectable by culture on agar media, *Bio/Technology,* 5, 947, 1987.

Joly, J. R., Dewaily, E., Bernard, L., Ramsey, D., and Brisson, J., *Legionella* and domestic water heaters in Quebec City area, *Can. Med. Assoc. J.,* 132, 160, 1986.

Jorgensen, K. A., Korsager, B., Johannsen, G., Freund, L. G., and Wilkinson, H. W., Legionnaires disease imported to Denmark from Italy, *Scand. J. Infect. Dis.,* 13, 133, 1981.

Kaufmann, A., McDade, J., and Patton, C., Pontiac fever: Isolation of the etiologic agent (*Legionella pneumophila*) and demonstration of its mode of transmission, *Am. J. Epidemiol.,* 114, 337, 1981.

Kemmer, F. N. and McCallion, J. (Eds.), *The Nalco Water Handbook*, McGraw-Hill, New York, 1979.

King, C.H., Shotts, E. B., Wooley, R., Porter, E., and Porter, K. G., Survival of coliforms and bacterial pathogens within protozoa during chlorination, *Appl. Environ. Microbiol.,* 54, 3023, 1988.

Knudson, G. B., Photoreactivation of U.V. irradiated *Legionella pneumophila* and other *Legionella* species, *Appl. Environ. Microbiol.,* 49, 975, 1985.

Kuchta, J. M., States, S. J., McGaughlin, J. E., Overmeyer, J. H., Wadowsky, R. M., McNamara, A. M., Wolford, R. S., and Yee, R. B., Enhanced chlorine resistance of tap water adapted *Legionella pneumophila* as compared with agar-medium passaged strains, *Appl. Environ. Microbiol.,* 50, 21, 1985.

Kuchta, J. M., States, S. J., McNamara, A. M., Wadowsky, R. M., and Yee, R. B., Susceptibility of *Legionella pneumophila* to chlorine in tap water, *Appl. Environ. Microbiol.,* 46, 1134, 1983.

Kyle, D. E. and Noblett, G. P., Vertical distribution of potentially pathogenic free-living amebae in freshwater lakes, *J. Protozool.,* 32, 99, 1986.

Lee, T. C., Stout, J. E., and Yu, L. V., Factors predisposing to *Legionella pneumophila* colonization in residential water systems, *Arch. Environ. Health,* 43, 59, 1988.

McDade, J. E., Brenner, D. J., and Bozeman, F. M., Legionnaires' Disease bacterium isolated in 1947, *Ann. Int. Med.*, 91, 659, 1979.

McDade, J. E., Shepard, C. C., Fraser, D. W., Tsai, T. F., Redus, M. A., Dowdle, W. R., and The Laboratory Investigation Team, Legionnaire's disease: Isolation of a bacterium and demonstration of its role in other respiratory disease, *N. Engl. J. Med.,* 297, 1197, 1977.

Marrie, T. J., Durant, H., and Yates, L., Community-acquired penumonia requiring hospitalization: a five year prospective study, *Rev. Infect. Dis.*, 11, 586, 1989.

Moss, C. W., Weaver, R. E., Dees, S. B., and Cherry, W., Cellular fatty acid composition of isolates from Legionnaires disease, *J. Clin. Microbiol.,* 6, 140, 1977.

Muder, R. R., Yu, V. L., and Woo, A. H., Mode of transmission of *Legionella pneumophila*, a critical review, *Arch. Int. Med.,* 146, 1607, 1986.

Muraca, P. W., Stout, J. E., Yu, V. L., and Lee, Y. C., Legionnaires' Disease in the work environment: Implications for environmental health, *Am. Ind. Hyg. Assoc. J.,* 49, 584, 1988.

National Health and Medical Research Council, Australian guidelines for the control of Legionella and Legionnaires' Disease, Canberra: National Health and Medical Research Council, Australian Government Publishing Service, 1989.

Phillips, S. J., Zeff, R. H., and Gervich, D., Legionnaires' Disease, *Ann Thorac. Surg.,* 44, 564, 1987.

Politi, B. D., Fraser, D. W., Mallison, G. F., Mohatt, J. V., Morris, G. K., Patton, C. M., Feeley, J. C., Telle, R. D., and Bennett, J. V., A major focus of Legionnaires' Disease in Bloomington, Indiana, *Ann. Intern. Med.*, 90, 587, 1979.

Pope, D. H., Soracco, R. J., Gill, H. K., and Fliermans, C. B., Growth of *Legionella pneumophila* in two-membered cultures with green algae and cyanobacteria, *Current Microbiol.,* 7, 319, 1982.

Reid, D., Grist, N.R., and Najera, R., Illness associated with "package tours": a combined Spanish-Scottish study, *Bull. WHO,* 56, 117, 1978.

Research Committee of the British Thoracic Society and the Public Health Laboratory Service, Community-acquired pneumonia in adults in British hospitals in 1982-1983: a survey of aetiology, mortality, prognostic factors and outcome, *Q. J. Med.*, 62, 195, 1987.

Rippka, R., Deruelles, J., Waterbury, J. B., Herdman, M., and Stanier, R. Y., Generic assignments, strain histories and properties of pure cultures of cyanobacteria, *J. Gen. Microbiol.*, 111, 1, 1979.

Ristroph, J. D., Hedlund, K. W., and Gowda, S., Chemically defined medium for *Legionella pneumophila* growth, *J. Clin. Microbiol.,* 13, 116, 1981.

Roberts, K. P., August, C. M., and Nelson, J. D., Relative sensitivities of environmental legionellae to selective isolation procedures, *Appl. Environ. Microbiol.,* 12, 2704, 1987.

Rodriguez, G. G., Phipps, D., Ishiguro K., and Ridgeay, H. F., Use of fluorescent redox probe for direct visualization of actively respiring bacteria, *Appl. Environ. Microbiol.*, 58, 1801, 1992.

Rowbotham, T. J., Preliminary report on the pathogenicity of *Legionella pneumophila* for freshwater and soil amebae, *J. Clin. Pathol.,* 33, 1179, 1980a.

Rowbotham, T. J., Pontiac fever explained?, *Lancet,* ii, 969, 1980b.

Rowbotham, T. J., Isolation of *Legionella pneumophila* from clinical specimens via amebae and the interaction of those and other isolates with amebae, *J. Clin. Pathol.,* 36, 978, 1983.

Rowbotham, T. J., Current views on the relationship between amoeba, *Legionella*, and man, *Isr. J. Med. Sci.*, 22, 679, 1986.

Skinner, A. R., Anand, C. M., Malic, A., and Kurtz, J. B., Acanthamebae and environmental spread of *Legionella pneumophila*, *Lancet,* ii, 289, 1983.

Standards Australia, AS3666, Airhandling and water systems of buildings, microbial control. Sydney: Standards Australia, 1989.

States, S. J., Conley, L. F., Knezivich, C. R., Keleti, G., Sykora, J. L., Wadowsky, R. M., and Yee, R. B., Free-living amebae in public water supplies: Implications for *Legionella, Giardia*, and *Cryptosporidia* spp. pp. 109-126. *In*: *Proceedings of the American Water Works Association Water Quality Technology Conference*, St. Louis, MO, 1988.

States, S. J., Conley, L. F., Kuchta, J. M., Oleck, B. M., Lipovich, M. J., Wolford, R. S., Wadowsky, R. M., McNamara, A. M., Sykora, J. L., Keleti, G., and Yee, R. B., Survival and multiplication of *Legionella pneumophila* in municipal drinking water systems, *Appl. Environ. Microbiol.,* 53, 979, 1987a.

States, S. J., Conley, L. F., Towner, S. G., Wolford, R. S., Stephenson, T. E., McNamara, A. M., Wadowsky, R. M., and Yee, R. B., An alkaline approach to treating cooling towers for control of *Legionella pneumophila*, *Appl. Environ. Microbiol.,* 63, 1776, 1987b.

Stout, J. E., Yu, V. L., and Muraca, P., Legionnaires' Disease acquired within the homes of two patients, *J. Am. Med. Soc.,* 257, 1215, 1987.

Stout, J., Yu, V. L., Vickers, R. M., Zuravleff, J., Best, M., Brown, A., Yee, R. B., and Wadowsky, R., Ubiquitousness of *Legionella pneumophila* in the water supply of a hospital with endemic Legionnaires' Disease, *N. Engl. J. Med.,* 806, 466, 1982.

Sykora, J. L., Keleti, G., and Martinez, J., Occurrence and pathogenicity of *Naegleria fowleri* in artificially heated waters, *Appl. Environ. Microbiol.,* 45, 974, 1986.

Tesh, M. J. and Miller, R. D., Growth of *Legionella pneumophila* in defined media: requirement for magnesium and potassium, *Can. J. Microbiol.,* 28, 1055, 1982.

Tison, D. L., Pope, D. H., Cherry, W. B., and Fliermans, C. B., Growth of *Legionella pneumophila* in association with blue-green algae (cyanobacteria), *Appl. Environ. Microbiol.,* 39, 456, 1980.

Tison, D. L. and Seidler, R. J., *Legionella* incidence and density in potable drinking water supplies, *Appl. Environ. Microbiol.,* 45, 337, 1983.

Tobin, J. O'H., Swann, R. A., and Bartlett, C. L. R., Isolation of *Legionella pneumophila* from water systems: Methods and preliminary results, *Br. Med. J.,* 282, 515, 1981a.

Tobin, J. O'H., Bartlett, C. L. R., and Waitkins, S. A., Legionnaires' Disease: Further evidence to implicate water storage and distribution systems as sources, *Br. Med. J.,* 282, 573, 1981b.

Tyndall, R. L. and Domingue, E. L., Cocultivation of *Legionella pneumophila* and free-living amebae, *Appl. Environ. Microbiol.,* 44, 954, 1982.

Tyndall, R. L., Gough, S. B., Fliermans, C. B., Domingue, E. L., and Duncan, C. B., Isolation of a new *Legionella* species from thermally altered waters, *Current Microbiol.,* 9, 77, 1983.

Vickers, R. M., Brown, A., and Garrity, C. L. R., Dye-containing buffered charcoal-yeast extract medium for differentiation of the family Legionellaceae, *J. Clin. Microbiol.,* 13, 380, 1981.

Wadowsky, R. M., Butler, L. J., Cook, M. K., Verma, S. M., Paul, M. A., Fields, B. S., Keleti, G., Sykora, J. L., and Yee, R. B., Growth-supporting activity for *Legionella pneumophila* in tap water cultures and implication of harmannellid amebae as growth factors, *Appl. Environ. Microbiol.,* 54, 2677, 1988.

Wadowsky, R. M., Wolford, R., McNamara, A. M., and Yee, R. B., Effect of temperature, pH, and oxygen level on the multiplication of naturally occurring *Legionella pneumophila* in potable water, *Appl. Environ. Microbiol.*, 49, 1197, 1985.

Wadowsky, R. M. and Yee, R. B., Glycine-containing selective medium for isolation of Legionellaceae from environmental specimens, *Appl. Environ. Microbiol.,* 42, 768, 1981.

Wadowsky, R. M., Yee, R. B., Mezmar, L., Wing, E. J., and Dowling, J. N., Hot water systems as sources of *Legionella pneumophila* in hospital and nonhospital plumbing fixtures, *Appl. Environ. Microbiol.,* 43, 1104, 1982.

Weaver, R. E., Cultures and staining characteristics, pp. 39-44. *In*: Jones, G. L. and Herbert, G. A. (Eds.), *"Legionnaires'" the Disease, the Bacterium and Methodology,* Centers for Disease Control, Atlanta, GA, 1978.

Weaver, R. E. and Feeley, J. C., Cultural and biochemical characterization of the Legionnaires' Disease bacterium, pp. 20-25 *In*: Jones, G. L. and Herbert, G. A. (Eds.), *"Legionnaires'" the Disease, the Bacterium and Methodology,* Centers for Disease Control, Atlanta, GA, 1978.

Witherall, L. E., Duncan, R. W., Stone, K. M., Stralton, L. J., Orciari, L., Kappel, S., and Jillson, D. A., Investigation of *Legionella pneumophila* in drinking water, *Am. Water Works Assoc. J.,* 80, 87, 1988.

Woodhead, M. A. and McFarlane, J. T., Prospective study of the aetiology and outcome of pneumonia in the community, *Lancet,* 146, 671, 1987.

Zimmerman, R., Iturriaga R., and Becker-Birck, J., Simultaneous determination of the total number of aquatic bacteria and the number thereof involved in respiration, *Appl. Environ. Microbiol.*, 36, 926, 1978.

4 ENDOTOXIN

Donald K. Milton

CONTENTS

I. NATURE AND DISTRIBUTION

Endotoxin is the name given to a class of biological molecules that have certain characteristic toxic effects. Endotoxin is now recognized to be lipopolysaccharide (LPS) in the outer membrane of Gram negative bacteria. The lipid portion of LPS (lipid A) is chemically distinct from all other lipids in biological membranes, and is responsible for the molecule's characteristic toxicity. Lipid A contains 3-hydroxy fatty acids attached to hydroxyl and amino groups on a disaccharide back bone. Because 3-hydroxy fatty acids do not accumulate elsewhere in nature, they may be used as chemical markers for endotoxin. We should consider LPS as a class of molecules because bacterial species differ widely in the polysaccharide portion of the molecule and to a lesser extent in the composition of their lipid A. Even within one organism,

0-87371-724-4/95/$0.00+$.50

molecules of LPS are somewhat heterogeneous (Nowotny, 1984; Rietschel and Brade, 1992).

Endotoxin is present in the environment as whole cells, large membrane fragments, or macromolecular aggregates of about 1 million Da (free endotoxin). Environmental endotoxin is associated with variable amounts of membrane protein. The aggregation of individual LPS molecules is essential for toxicity. Thus, detergents that break LPS aggregates into individual molecular units (2500 to 5000 Da) also block toxicity. However, LPS is relatively heat stable (110°C) and thus is not eliminated by autoclaving.

Airborne endotoxin is ubiquitous, as would be expected considering the ubiquitous nature of the Gram negative bacteria in nature (Andrews and Hirano, 1992). High levels have been reported from a variety of environments, where Gram negative bacteria flourish in specific reservoirs. The highest reported airborne endotoxin levels (2 to 7 $\mu g/m^3$) arise from processing vegetable fibers (e.g., in cotton mills), fecal materials in agriculture or wastewater treatment, industrial washwater mists, and contaminated room humidifiers (Rylander and Morey, 1982; Tyndall et al., 1991; Walters et al., 1993). The most important kind of reservoir is recirculated water-based fluids that produce aerosols. Background levels are usually in the subnanogram range (Olenchock et al., 1990; Rose et al., 1993).

II. HEALTH EFFECTS

Endotoxin causes fever and malaise, changes in white blood cell counts, respiratory distress, shock, and even death when it gains access to the bloodstream in microgram quantities. Early experimental inhalation studies demonstrated effects that included fever, cough, diffuse aches, nausea, shortness of breath, and acute air flow obstruction (Cavagna et al., 1969; Pernis et al., 1961). More recent studies have demonstrated that endotoxin in cotton dust accounted for the acute bronchoconstriction observed in experimental subjects (Castellan et al., 1987; Rylander and Haglind, 1985). Also, asthmatics were found to respond to endotoxin inhalation with acute bronchoconstriction, and even normal subjects show airway inflammation after endotoxin inhalation (Michel et al., 1989, 1992; Rylander et al., 1989; Sandstrom et al., 1992). Until recently, however, epidemiologic studies were unable to show correlations between personal breathing zone endotoxin concentrations and acute changes in lung function (Donham et al., 1989; Kennedy et al., 1987; Rylander et al., 1983) although a correlation between decreased baseline forced expiratory volume at 1 atm (FEV1) and cumulative endotoxin exposure has been observed (Kennedy et al., 1987; Smid et al., 1992). Using new techniques, however, a correlation between personal endotoxin exposure and decrements in serial measurements of peak expiratory flow over the ensuing 24 h was found in fiberglass manufacturing workers (Milton et al., 1993). In addition, an index of the duration of employment in high endotoxin exposure areas was associated both with increased asthmalike variability of lung function and with low baseline lung function.

In contrast to its deleterious effects on lung function, endotoxin's ability to stimulate the immune system (i.e., its adjuvant properties) can have beneficial effects. The adjuvant effects of endotoxin are primarily due to specific binding of lipid A. Binding of lipid A to a cell surface receptor on macrophages results in production and secretion of a variety of important cytokines and arachidonic acid metabolites. Tumor necrosis factor (TNFα), the cytokine most closely associated with the toxic effects of endotoxin, is produced in large quantities by macrophages stimulated with picogram per milliliter quantities of endotoxin. However, a number of other cytokines, including interferons, and granulocyte-macrophage colony-stimulating factor, are also produced. These cytokines play important roles in regulating T-cell as well as phagocytic function. Lipid A has been shown to reverse T-cell tolerance to polysaccharide antigens, a property that is being exploited in the design of new antimalarial vaccines (Baker et al., 1992; Richards et al., 1989). Unfortunately, the adjuvant properties of endotoxin may also play a role in allergic sensitization, as suggested by at least one recent study (Rose et al., 1993).

Endotoxin has long been recognized to possess antitumor properties (Engelhardt et al., 1991). In addition, it has been repeatedly observed that populations with high-level exposure to airborne endotoxin have low rates of lung cancer (Enterline et al., 1985; Hodgson and Jones, 1990; Rylander, 1990). An animal model of lung metastasis and endotoxin inhalation supports a protective role for endotoxin in lung cancer (Lange, 1992). The adjuvant effects of endotoxin together with the antitumor effects of TNFα are likely to account for the antitumor and anticarcinogenic properties of endotoxin.

III. SAMPLING AND ANALYSIS

A. General Considerations

Endotoxin is ubiquitous in the laboratory. Therefore, all glassware, forceps, and metal cassettes must be baked at 210°C for 1 h (or 270°C for 30 min) to destroy background endotoxin. Sterile plasticware is often, but not always, endotoxin free. Where possible, certified endotoxin-free plasticware should be used. Polystyrene is the preferred polymer. Polypropylene should be avoided due to the tendency of some lots of this polymer to irreversibly adsorb large amounts of LPS (Novitsky et al., 1986).

B. Sample Collection

Most of the early studies of airborne endotoxin used cellulose or polyvinyl chloride (PVC) filters for sample collection. Extraction efficiencies for these media were undocumented. Because endotoxin is a surface-active material and binds to glass and plastic surfaces (Novitsky et al., 1986), this was an important omission. A wide range of filter media have been shown to bind endotoxin (Milton et al., 1990). The most avid binding occurs with PVC (Milton et al., 1990; Sonesson et al., 1990). A study examining the effect of sampling media on airborne endotoxin measurement revealed that extraction efficiency depended not only on the filter medium, but also on the type of aerosol (cotton dust, machining oil, or saline mist) (Gordon et al., 1992). Gordon and colleagues did not confirm the earlier findings (Milton et al., 1990; Sonesson et al., 1990) that PVC filters avidly bound endotoxin in agricultural dust extractions. However, this study did not examine the effect of different extraction media or of sonication on yields from each of the filter media and did not use a chemical method of endotoxin analysis to confirm the results. A new study, however, using a different extraction method and validation with a chemical marker, found that polycarbonate capillary pore membranes are optimal collection media (Walters et al., 1994).

Collection of endotoxin using glass impingers may be an alternative to filter cassettes. However, impingers are not efficient for collection of submicron particles; thus, they may underestimate endotoxin aerosols in some environments. Fluid from aerosol reservoirs such as humidifiers may be sampled directly. These present little problem with analysis when appropriate techniques are used as described below. Bulk samples should be collected in endotoxin-free, preferably glass, containers and analyzed promptly to prevent bacterial growth.

C. Sampling Handling

In general, samples for endotoxin analysis should not be frozen. While endotoxin may be stable in some media with freezing (Olenchock et al., 1989), it is not stable to freezing in the presence of oil/surfactant mixtures. Appropriate preservation techniques need to be investigated for each type of sample to be analyzed.

D. Extraction Protocols

Extraction protocols can have significant impact on the results of endotoxin analyses (Milton et al., 1988; Olenchock et al., 1989). Important considerations are (1) the fluid used to solubilize the endotoxin, and (2) the method of applying the fluid. In tests examining endotoxin yields from filters spiked with endotoxin in water or endotoxin in an oil–surfactant emulsion (metal working fluid), Milton (1992) demonstrated that a buffer (0.05 *M* potassium phosphate, 0.01% triethylamine, pH 7.5) formulated to optimize LPS solubility while controlling pH and ionic strength is necessary to ensure efficient endotoxin extraction from the sample medium. In addition, sonication in the buffer is necessary to efficiently remove particles from the filter media. Validation of this extraction method was achieved by comparison with methanolysis of samples followed by 3-hydroxy fatty acid analysis by gas chromatography-mass spectrometry (GC-MS) (Walters et al., 1994). Simultaneously collected filter samples were either directly subjected to methanolysis in 3.6 *M* methanolic HCl for 18 h at 100°C or were first extracted by sonication in buffer followed by methanolysis of extracts. The two extraction methods yielded similar 3-hydroxy fatty acid content ($p = 0.72$); the average difference was 2.1%. Thus, extraction of endotoxin from the filters by the buffer-bath sonication method was essentially complete as compared with exhaustive methanolysis.

E. Analytical Methods

Endotoxin measurement can be achieved through chemical or biological means. There are several chemical markers for endotoxin that have great appeal because of their specificity. However, one must establish that the moiety detected arises from intact, toxic LPS rather than remnants of degraded non-toxic material (e.g., 3-hydroxymyristate in blood) (Maitra et al., 1978). However, chemical methods are not sufficiently sensitive for use in analysis of personal air samples (Walters et al., 1994).

Biological methods include the classic rabbit pyrogen test, in vitro production of tumor necrosis factor from monocytes and macrophages, and the *Limulus* amebocyte lysate (LAL) test (Levin, 1987). Airborne endotoxin is usually measured with an LAL method because the test is simple, fast, and extremely sensitive (subpicogram per milliliter quantities) (Jacobs, 1989). *Limulus* tests are in vitro biological assays that respond to the heterogeneous group of molecules comprising LPS. The potency of a given LPS depends on its aggregation state, polysaccharide content, and the arrangement of fatty acids in the lipid A portion of the LPS (Baker et al., 1992; Grabarek et al., 1990; Helander et al., 1980; Morrison et al., 1987; Pearson, 1985). The potency of LPS in the LAL test corresponds in a general way to variations in the toxic potency of the LPS for mammals. Thus, the LAL test is a comparative method providing an estimate of relative toxicity rather than an analytic method providing a quantitative measure of a physical substance. Results of a comparative method may not be valid under conditions other than those of a given assay because factors other than a change in concentration of LPS can affect the results (Finney, 1978b).

Because LAL tests are comparative dilution assays they must be rigorously standardized if any generalization or comparisons with other observations are to be made. In pharmaceutical applications, for which the LAL test was developed, repeated tests of sample matrix can account for interference with the assay. However, in environmental analysis this is not feasible. The wide variety of aerosol and reservoir constituents in the environment, such as bactericides and corrosion inhibitors added to humidifiers, detergents in metal working fluids, or acid aerosols in ambient air, necessitates that pH and ionic strength must be controlled throughout the extraction, dilution, and assay phases. When conditions are not strictly controlled, nonspecific interference with the LAL reaction can produce either over- or underestimates of endotoxin concentration. To guard against residual interference, even in the presence of a buffer, both the endotoxin standard and environmental samples must be tested over a range of concentrations and tested for the presence of parallel dose-response relationships. Once parallel dose responses have been established, valid measurements for the environmental sample can be computed. Statistical methods for analysis of parallel line biological assays (Finney, 1978a) allow calculation of endotoxin content (relative potency) and an estimate of intra-assay variance.

The forgoing considerations have been incorporated in a *Limulus* assay specifically designed to measure environmental endotoxin: the kinetic *Limulus* assay with resistant-parallel-line estimation (KLARE) method (Milton et al., 1992). The method uses the buffer described above with parallel-line bioassay experimental design and resistant regression analysis. The validity of the method was tested by analyzing buffer extracts by both the KLARE method and by GC-MS for 3-OHFA content. Endotoxin estimates were similar ($p = 0.23$); the average difference between the methods was 0.88% (Walters et al., 1994). Thus, the method of extract preparation and *Limulus* analysis together were able to accurately detect all of the endotoxin present in the sample. Earlier reports (Reynolds and Milton, 1993; Rylander et al., 1989; Sonesson et al., 1990) suggested that LAL tests may underreport endotoxin content due to inability to detect cell-bound endotoxin. Walters et al. (1994) show that sonication, and possibly the buffer containing triethylamine, appears to allow complete release of endotoxin from cell membranes. Thus, the accuracy of the method is not due only to the statistical methods employed. Replicate sampling in an environment with high levels of endotoxin-contaminated mist showed that the standard sampling, extraction, and *Limulus* assay by the KLARE method was highly reproducible; the 95% confidence interval for individual endotoxin measurements was ±0.28 log 10.

F. Reporting Results

Appropriate units for reporting endotoxin results have been discussed in several publications (Olenchock et al., 1989; Pearson, 1985; Pearson et al., 1982; Reynolds and Milton, 1993). Because endotoxin potency rather than an absolute concentration is measured by the *Limulus* assay, standard potency units are often preferred. The endotoxin unit (EU) is defined as the potency of 0.10 ng of a reference standard endotoxin (EC5, U.S. Pharmacopoeia). A recent study showed that failure to convert to standard units can result in erroneous interlaboratory comparisons (Reynolds and Milton, 1993). However, because endotoxin units were standardized based on potency for fever and *Limulus* amebocyte lysate activation, their relevancy for other biological effects remains unclear. Thus, some LPS molecules are "non-toxic" but retain potent immunostimulatory activities. Such immuno-adjuvant activity may be important when inhaled, resulting in increased likelihood of allergic sensitization to antigens present in the aerosol. Thus, some authors may prefer to report endotoxin in terms of nanograms of reference standard to avoid the inference that endotoxin potency for all biological effects is summarized in the reported value.

IV. PROPOSED STANDARD METHOD FOR SAMPLING AND ANALYSIS OF AIRBORNE ENDOTOXIN

The development of standard protocols is essential if a base of bioaerosol data is to be accumulated that will allow standard setting and for routine, straightforward environmental investigations. The methods currently being used for the assessment of airborne endotoxin are not comparable, and misleading and erroneous conclusions are likely. We propose that the KLARE method for endotoxin assessment combined with samples collected on 0.4-μm polycarbonate filters be considered the standard. Although the method requires care and precision, it will provide consistent, reliable, and, above all, comparable results regardless of the environment or reservoir from which the samples are drawn. Briefly, the method can be outlined as follows:

1. Collect samples on endotoxin-free 0.4-mm capillary-pore membrane filters.
2. Do not freeze samples; store desiccated and at 4°C where possible.
3. Extract samples in 0.05 *M* potassium phosphate, 0.01% triethylamine, pH 7.5, using bath sonication.
4. Analyze samples serially diluted in extraction buffer in parallel with standard or reference endotoxin serially diluted in the same buffer. Compute sample potency using the parallel-line bioassay method. Resistant regression methods are preferred.
5. Report results as endotoxin units (EU of reference standard) or in nanograms of reference standard.

REFERENCES

Andrews, J. H., and Hirano, S. S., *Microbial Ecology of Leaves*, Springer-Verlag, New York, 1992.

Baker, P. J., Hraba, T., Taylor, C. E., Myers, K. R., Takayama, K., Qureshi, N., Stuetz, P., Kusumoto, S., and Hasegawa, A., Structural features that influence the ability of lipid A and its analogs to abolish expression of suppressor T cell activity, *Infect. Immunol.,* 60(7), 2694, 1992.

Castellan, R. M., Olenchock, S. A., Kinsley, K. B., and Hankinson, J. L., Inhaled endotoxin and decreased spirometric values, an exposure-response relation for cotton dust, *N. Engl. J. Med.,* 317, 605, 1987.

Cavagna, G., Foa, V., and Vigliani, E. C., Effects in man and rabbits of inhalation of cotton dust or extracts and purified endotoxins, *Br. J. Ind. Med.*, 26, 314, 1969.

Donham, K., Haglind, P., Peterson, Y., Rylander, R., and Belin, L., Environmental and health studies of farm workers in Swedish swine confinement buildings, *Br. J. Ind. Med.,* 46(1), 31, 1989.

Engelhardt, R., Mackensen, A., and Galanos, C., Phase I trial of intravenously administered endotoxin (*Salmonella abortus equi*) in cancer patients, *Can. Res.*, 51(10), 2524, 1991.

Enterline, P. E., Sykora, J. L., Keleti, G., and Lange, J. H., Endotoxins, cotton dust, and cancer, *Lancet,* 2(8461), 934, 1985.

Finney, D. J., *Statistical Method in Biological Assay*, Charles Griffin and Company, Ltd., London, 1978a, 69.

Finney, D. J., *Statistical Method in Biological Assay*, Charles Griffin and Company, Ltd., London, 1978b, 41.

Gordon, T., Galdanes, K., and Brosseau, L., Comparison of sampling media for endotoxin-containing aerosols, *Appl. Occup. Environ. Hyg.*, 7(7), 472, 1992.

Grabarek, J., Her, G. R., Reinhold, V. N., and Hawiger, J., Endotoxic lipid A interaction with human platelets. Structure-function analysis of lipid A homologs obtained from *Salmonella minnesota* Re595 lipopolysaccharide, *J. Biol. Chem.*, 265(14), 8117, 1990.

Helander, I., Salkinoja-Salonen, M., and Rylander, R., Chemical structure and inhalation toxicity of lipopolysaccharides from bacteria on cotton, *Infect. Immun.,* 29, 859, 1980.

Hodgson, J. T., and Jones, R. D., Mortality of workers in the British cotton industry in 1968-1984, *Scand. J. Work Environ. Health*, 16(2), 113, 1990.

Jacobs, R. R., Airborne endotoxins: an association with occupational lung disease, *Appl. Ind. Hyg.*, 4, 50, 1989.

Kennedy, S. M., Christiani, D. C., Eisen, E. A., Wegman, D. H., Greaves, I. A., Olenchock, S. A., Ye, T., and Lu, P., Cotton dust and endotoxin exposure-response relationships in cotton textile workers, *Am. Rev. Respir. Dis.*, 135, 194, 1987.

Lange, J. H., Anticancer properties of inhaled cotton dust: a pilot experimental investigation, *J. Environ. Sci. Health*, A27(2), 505, 1992.

Levin, J., The *Limulus* amebocyte lysate test: perspectives and problems, *Prog. Clin. Biol. Res.*, 231, 1, 1987.

Maitra, S. K., Schotz, M. C., Yoshikawa, T. T., and Guze, L. B., Determination of lipid A and endotoxin in serum by mass spectroscopy, *Proc. Natl. Acad. Sci.*, 75, 3993, 1978.

Michel, O., Duchateau, J., and Sergysels, R., Effect of inhaled endotoxin on bronchial reactivity in asthmatic and normal subjects. *J. Appl. Physiol.*, 66(3), 1059, 1989.

Michel, O., Ginanni, R., Le Bon, B., Content, J., Duchateau, J., and Sergysels, R., Inflammatory response to acute inhalation of endotoxin in asthmatic patients, *Am. Rev. Respir. Dis.*, 146(2), 352, 1992.

Milton, D. K., Endotoxin in Metal Working Fluids. Report to United Auto Workers — General Motors Joint National Committee on Occupational Health and Safety, 1992.

Milton, D. K., Feldman, H. A., Neuberg, D. S., Bruckner, R. J., and Greaves, I. A., Environmental endotoxin measurement: the kinetic *Limulus* assay with resistant-parallel-line estimation, *Environ. Res.,* 57, 212, 1992.

Milton, D. K., Gere, R. J., Feldman, H. A., and Greaves, I. A., Endotoxin measurement: aerosol sampling and application of a new *Limulus* method, *Am. Ind. Hyg. Assoc. J.*, 51(6), 331, 1990.

Milton, D. K., Gould, M. C., Roslansky, P.F., and Novitsky, T. J., New estimates of endotoxin content of cotton dust: SDS suspension and kinetic LAL test, in *Proceedings of the Twelfth Cotton Dust Research Conference*, 1988, 163.

Milton, D. K., Kriebel, D., Wypij, D., Walters, M., Hammond, K., and Evans, J. S., Acute and chronic airflow obstruction and endotoxin exposure, *Am. J. Respir. Crit. Care Med.,* 149, A399, 1994.

Morrison, D. C., Vukajlovich, S. W., Ryan, J. L., and Levin, J., Structural requirements for gelation of the *Limulus* amebocyte lysate by endotoxin, *Prog. Clin. Biol. Res.*, 231, 55, 1987.

Novitsky, T. J., Schmidt-Gengenbach, J., and Remillard, J. F., Factors affecting recovery of endotoxin adsorbed to container surfaces, *J. Parenter. Sci. Technol.*, 40, 284, 1986.

Nowotny, A., Heterogeneity of endotoxin, in *Handbook of Endotoxin Volume I: Chemistry of Endotoxin*, Reitschel, E. T., Ed., Elsevier, New York, 1984, 308.

Olenchock, S. A., Christiani, D. C., Mull, J. C., Ting-ting, T., and Pei-lian, L., Airborne endotoxin concentrations in various work areas within two cotton textile mills in the People's Republic of China, *Biomed. Environ. Sci.*, 3, 443, 1990.

Olenchock, S. A., Lewis, D. M., and Mull, J. C., Effects of different extraction protocols on endotoxin analysis of airborne grain dusts, *Scand. J. Work Environ. Health,* 15, 430, 1989.

Pearson, F. C., A comparison of the pyrogenicity of environmental endotoxins and lipopolysaccharides, in *Bacterial Endotoxins: Structure, Biomedical Significance, and Detection with the Limulus Amebocyte Lysate Test*, ten Cate, J. W., Buller, H. R., Sturk, A., and Levin, J., Eds., Alan R. Liss, New York, 1985, 251.

Pearson, F. C., Weary, M. E., Bohon, J., and Dabbah, R., Relative potency of "environmental" endotoxin as measured by the *Limulus* amebocyte lysate test and the USP rabbit pyrogen test, in *Endotoxins and Their Detection with the Limulus Amebocyte Lysate Test,* Alan R. Liss, New York, 1982, 65.

Pernis, B., Vigliani, E. C., Cavagna, C., and Finulli, M., The role of bacterial endotoxins in occupational diseases caused by inhaling vegetable dusts, *Br. J. Ind. Med.*, 18, 120, 1961.

Reynolds, S., and Milton, D. K., Comparison of methods for analysis of airborne endotoxin, *Appl. Occup. Environ. Hyg*., 8(9), 761, 1993.

Richards, R. L., Swartz, G. J., Schultz, C., Hayre, M. D., Ward, G. S., Ballou, W. R., Chulay, J. D., Hockmeyer, W. T., Berman, S. L., and Alving, C. R., Immunogenicity of liposomal malaria sporozoite antigen in monkeys: adjuvant effects of aluminium hydroxide and non-pyrogenic liposomal lipid A, *Vaccine,* 7(6), 506, 1989.

Rietschel, E. T., and Brade, H., Bacterial endotoxins, *Sci. Am.,* 55, 1992.

Rose, C. S., Newman, L. S., Martyny, J. W., Weiner, D., Kreiss, K., Milton, D. K., and King, T. E., Outbreak of hpersensitivity pneumonitis in an indoor swimming pool: clinical, pathophysiologic, radiographic, pathologic, lavage and environmental findings, *Am. Rev. Respir. Dis.*, 141, A315, 1994.

Rylander, R., Bake, B., Fischer, J. J., and Helander, I. M., Pulmonary function and symptoms after inhalation of endotoxin, *Am. Rev. Respir. Dis*., 140(4), 981, 1989.

Rylander, R., Haglind, P., and Butcher, B. T., Reactions during work shift among cotton mill workers, *Chest,* 84, 403, 1983.

Rylander, R., Environmental exposures with decreased risks for lung cancer?, *Int. J. Epidemiol.*, 19(1), S67, 1990.

Rylander, R., and Haglind, P. M. L., Endotoxin in cotton dust and respiratory function decrement among cotton workers in an experimental cardroom, *Am. Rev. Respir. Dis.,* 131, 209, 1985.

Rylander, R., and Morey, P., Airborne endotoxin in industries processing vegetable fibers, *Am. Ind. Hyg. Assoc. J.*, 43(11), 811, 1982.

Sandstrom, T., Bjermer, L., and Rylander, R., Lipopolysaccharide (LPS) inhalation in healthy subjects increases neutrophils, lymphocytes and fibronectin levels in bronchoalveolar lavage fluid, *Eur. Respir. J.*, 5(8), 992, 1992.

Smid, T., Heederik, D., Houba, R., and Quanjer, P. H., Dust- and endotoxin-related respiratory effects in the animal feed industry, *Am. Rev. Respir. Dis.*, 146, 1474, 1992.

Sonesson, A., Larsson, L., Schutz, A., Hagmar, L., and Hallberg, T. Comparison of the *Limulus* amebocyte lysate test and gas chromatography-mass spectromety for measuring lipopolysaccharides (endotoxins) in airborne dust from poultry processing industries, *Appl. Environ. Microbiol.*, 56, 1271, 1990.

Tyndall, R. L., Bowman, E. K., Ironside, K. S., Milton, D. K., Barbaree, J., and Lehman, E., Aerosolization of microorganisms and endotoxin from home humidifiers, *Am. Soc. Microbiol.*, abstract, 1991.

Walters, M., Milton, D. K., Larsson, L., and Ford, T., Airborne environmental endtoxin: a crossvalidation of sampling and analysis techniques, *Appl. Environ. Microbiol.*, in press.

Walters, M. D., Evans, J. S., Hammond, S. K., and Milton, D. K., Worker exposure to endotoxins and other air contaminants in a fiberglass insulation manufacturing facility, *Am. Ind. Hyg. Assoc. J.*, submitted.

5 FUNGI

Estelle Levetin

CONTENTS

0-87371-724-4/95/$0.00+$.50

I. INTRODUCTION

The presence of fungal spores in both the atmosphere and the indoor environment is well established. Although these spores are normal components of the atmosphere, indoors they are usually considered contaminants. The control of fungal contamination requires a thorough understanding of the fungi and the indoor ecosystem, as well as knowledge of the outdoor air spora and the factors that influence fungal survival.

II. THE NATURE OF THE FUNGI

A. Morphology

Fungi are eukaryotic organisms that belong to a kingdom distinct from plants and animals. Fungi include inconspicuous yeasts, molds, and mildews, as well as large mushrooms, puffballs, and bracket fungi. Structurally, fungi exist as single cells such as yeast or, far more commonly, as threadlike hyphae. Hyphae usually branch extensively, and the collective mass of interwoven hyphal filaments is referred to as a mycelium. Depending on the species, each hypha may have many short cells, or it may be nonseptate with multiple nuclei existing in a common cytoplasm. While individual hyphae are microscopic, the mycelium is often visible to the naked eye. Highly specialized reproductive structures such as mushrooms and brackets are actually compact masses of tightly interwoven hyphae and are the visible portion of an extensive mycelium within the substrate.

One feature the fungi share with plants is the presence of cell walls. The fungal wall, consisting of fibrils embedded in a matrix, is largely composed of polysaccharides (often over 90%) but also contains significant amounts of protein and lipid (Deacon, 1984; Ruiz-Herrera, 1991). In most fungi, the major fibrillar component of the cell wall is chitin (a straight-chain β-1,4-polymer of *N*-acetyl glucosamine); however, some fungi possess cellulose fibrils. The matrix, on the other hand, contains a variety of carbohydrates and proteins. Although chitin is considered the characteristic wall material, in many fungi the matrix polysaccharides are far more abundant. The most thoroughly studied wall polysaccharides are the β-glucans (glucose polymers). The β-1,3-glucans with some β-1,6-branching can be considered the most important, with many different compounds included in this category. Some β-glucans form part of the amorphous matrix while others are part of the fibrillar wall structure (Ruiz-Herrera, 1991). In addition to its obvious structural and osmotic functions, the wall serves as a binding site for some extracellular enzymes and has antigenic properties as well (Deacon, 1984).

Fungi normally reproduce through the formation of spores that may result from sexual or asexual processes. Many fungi undergo both asexual and sexual reproduction at different stages in their life cycle. The sexual phase of the life cycle is known as the teleomorphic (or perfect) state or stage, while the asexual phase is called the anamorphic (or imperfect) stage. Sexual spores are the result of genetic recombination and normally follow karyogamy (fusion of nuclei) and meiosis. These processes occur in specialized cells, and in some groups the sexual spores form in specialized fruiting bodies. Asexual spores result from mitosis and may also be characteristic of certain groups. Asexual spores may form enclosed within a sporangium (sporangiospores) or be produced directly by the hyphae without any enclosing wall (conidia).

Spores contain one to many cells and differ greatly in size, shape, color, and method of formation. However, they are always microscopic, ranging from less than 2 μm to more than 100 μm. Spores may be formed from the fragmentation of undifferentiated hyphal elements or on highly specialized hyphal branches that may be contained within sporocarps or fruiting bodies. Mushrooms, puffballs, brackets, and morels are well-known examples of these specialized sporocarps. In most groups of fungi, spore shape and method of formation are important elements for identification and classification. Most fungal spores are adapted for airborne dispersal, although some are specialized for dispersal, by insects or water. Hyphal fragments may also serve as propagules and can readily found in atmospheric samples.

Under adverse environmental conditions some fungi are able to form chlamydospores, thick-walled dormant spores that develop from transformed vegetative hyphae. These are not usually dispersed but become free only when the adjacent hyphae decay. Another type of resistant or resting body is a sclerotium, which is a hardened mycelial mass that also enables the fungus to survive harsh conditions.

B. Physiology

All fungi depend on external sources of organic material for both energy requirements and carbon skeletons. Fungi produce extracellular enzymes that digest complex organic compounds into smaller molecules, which can then be absorbed. They can exist as parasites (obtaining nutrients from a living host) or mutualistic symbionts (obtaining nutrients from a living host while providing some benefit to that host). However, the majority of species, including those most abundant in the environment, are saprobes, which obtain nutrients from nonliving organic material. The line between saprobes and parasites is not always clear, and some fungi are able to utilize both lifestyles under certain conditions. Facultative parasites or opportunistic pathogens can live as saprobes but will invade living tissues when suitable hosts are available. Many saprobic species are found in the soil or on leaf surfaces. In the outdoor environment, fungal saprobes play an essential role as aerobic decomposers, degrading organic materials and recycling nutrients.

Fungi produce a vast number of secondary metabolites that have no recognized role in the maintenance of life in the producing organisms (Moss,

1984). Secondary metabolites are often species dependent (Deacon, 1984) and are normally produced by older parts of the colony, frequently by cells that have stopped growing or whose growth has become restricted (Deacon, 1984; Moss, 1984).

Thousands of secondary products from fungi have been chemically characterized. Included in this group are many antibiotics, the mycotoxins, and alkaloids. In many respects these secondary metabolites have made a significant impact on society. The value of the antibiotics (and even some of the fungal alkaloids) to medicine and society is clearly recognized, as are some of the harmful effects caused by toxic compounds. Although alkaloids and antibiotics may be toxic to some organisms, a compound is normally referred to as a fungal toxin when it is primarily known for its adverse affects on humans or other animals (Sharma and Salunkhe, 1991). Fungal toxins broadly fall into two groupings: mushroom toxins (or poisons) formed in the fleshy fruiting bodies of some fungi, and mycotoxins formed by hyphae of common molds growing under a variety of conditions (Powell, 1990). Mushroom toxins are well known and have been broadly categorized by the physiological effects of the toxin: (1) protoplasmic poisons that attach to cells in the liver and kidneys and are often lethal; (2) neurotoxins that affect either the central or autonomic nervous system and often lead to hallucinations; and (3) gastrointestinal irritants, which generally lead to nausea, vomiting, cramps, and diarrhea but seldom cause fatalities.

The most familiar mycotoxins are produced by fungi growing on grain or nuts in the field or, more commonly, in storage (Wogan, 1965; Deacon, 1984; Sharma and Salunkhe, 1991). These toxins have been shown to have profound acute and chronic effects on both humans and livestock. Mycotoxins are believed to be among the most potent known carcinogens (Burge, 1990; Autrup and Autrup, 1992). Most research on mycotoxins has focused on health effects following ingestion of contaminated food. Many species of fungi that are common indoor contaminants produce toxins, raising questions about possible health effects due to airborne exposure to these toxins. There is only one report in the literature of well-documented acute illness following airborne exposure. However, significant levels of airborne mycotoxins have been found in occupational settings, and this is clearly an area that needs further research (Croft et al., 1986; Burge, 1990; Lacey, 1991a).

Some fungi produce volatile metabolites that cause unpleasant odors, including the characteristic moldy smell associated with damp basements. Although many different types of compounds have been identified, prominent among these volatile organic compounds (VOCs) are short-chained alcohols and aldehydes (Samson, 1985) (see Chapter 13). The health effects of exposure to VOCs have not been well studied, but they may be responsible for symptoms of headache, dizziness, and eye and mucous membrane irritation among individuals in fungus-contaminated buildings (Samson, 1985; Burge, 1990).

Antigens are compounds that are recognized by antibodies of any class, while allergens more specifically trigger specific IgE or IgG (immunoglobin class E or G) or cell-mediated immune (allergic) responses (see Chapter 10). Components of any fungus can be antigenic, and fungal spores are well-known carriers of aeroallergens. Unlike toxins, which are frequently small molecules, most fungal allergens are large molecules (molecular weight normally between 10,000 and 80,000 Da). Fungal allergens identified to date are glycoproteins that may have structural or metabolic roles in the fungus. Using immuno-electron microscopy, antigenic molecules have been localized in the cell wall, plasma membrane, and even the cytoplasm of various fungi (Latge and Paris, 1988; Reijula et al., 1991). Some fungal allergens may naturally function as extracellular enzymes rather than structural wall components (Deacon, 1984).

Any one fungal species may produce dozens of allergens (Drouhet, 1988). Studies have shown that allergen content of a particular species may vary with the age of the culture, with the substrate, and even with each strain within a species. Burge et al. (1989) found a total of 32 allergens in spore extracts of 9 isolates of *Alternaria.* The majority of allergens were unique to one or more isolates with only four allergens present in every extract from the nine strains. Crenshaw et al. (1992) compared the specific allergen content of 11 strains of *Alternaria.* They found that four strains lacked measurable quantities of a specific 70,000-Da allergen in both the culture filtrate and cellular extracts; three other strains contained measurable amounts of this allergen only in the culture filtrate. In a study of multiple isolates of *Epicoccum nigrum*, Fentress (1991) found that individual strains had as many as 15 unique proteins identified by sodium dodecyl sulfate polyacrylamide gel electrophoresis (SDS-PAGE) gels. When strains were grown on different media, 18 unique proteins were identified in extracts from one of the isolates.

III. FUNGAL ECOLOGY

Fungal spores germinate to produce hyphae, which grow and branch within the substrate, typically producing a colony that eventually forms a new generation of spores. The types of fungi and their abundance in an area depend on the availability of nutrients, water, and temperature. Saprobic fungi are especially versatile and able to utilize many different substrates for growth; therefore, for many saprobes, temperature and especially water availability are the most important limiting factors.

Like all organisms, fungi have an absolute requirement for water, yet exhibit a broad range of tolerance with respect to water availability. Scientists use various terms to describe water relations in various organisms. Many mycologists use the term "water activity" (a_w), which is a measure of the moistness of the substrate (Griffin, 1981; Kendrick, 1992). Water activity is expressed as a decimal that is directly related to substrate relative humidity (i.e., if the substrate relative humidity is 95%, the a_w is 0.95). Many fungi are exceptional in their ability to grow under conditions of low a_w. Theoretical limits to growth lie between a_w values of 1.0 and 0.55, with the low limit defined by the point where DNA is denatured (Griffin, 1981). Most multicellular plants and animals have adaptations to conserve water so that their cells are not exposed to low a_w. Animals function at a_w of 0.99, many plants wilt at 0.98 and below, and most bacteria will only grow at a_w 0.95 or higher. By contrast, many fungi and yeasts can grow at a_w down to 0.7, with some remarkably xerophilic or xerotolerant species, such as *Aspergillus echinulatus*, able to grow at a_w as low as 0.62 (Kendrick, 1992). Fungi able to grow in substrates with these low water activities also have low internal a_w. Yeasts control their internal a_w by converting sugars to polyhedric alcohols, especially glycerol. While this has not been extensively studied in mycelial fungi, the same mechanisms are believed to operate, since polyhedric alcohols are abundant in the fungi (Griffin, 1981; Kendrick, 1992). Many species can also withstand extended periods of desiccation through the formation of spores, sclerotia, chlamydospores, or other resistant structures (Griffin, 1981; Deacon, 1984; Pritchard and Bradt, 1984).

Many fungi are able to survive over a broad temperature range with growth occurring from 10 to 40°C; these mesophiles generally show optimum growth from 20 to 30°C. Some human pathogens have optima in the 35 to 40°C range, while true thermophilic species are adapted to higher temperatures (45 to 60°C). At the opposite end of the range, psychrophilic species are adapted to low temperatures (below 10°C). While many psychrophilic species occur in cold climates, they are also important in temperate areas and can cause serious problems in cold storage facilities (Deacon, 1984).

In general the physical and nutritional requirements for spore formation are more strict than for vegetative growth. Often particular nutrients or environmental triggers are needed to initiate spore development or maturation. Daily rhythms in light, temperature, and humidity are well known to affect spore formation as well as growth and release (Moore-Landecker, 1990; Lysek, 1984). In natural environments, these diurnal rhythms are evident in the fungal taxa identified from the air spora at different times of the day (Gregory, 1973).

IV. CLASSIFICATION SYSTEMS

Approximately 69,000 species of fungi have been described, and estimates for the total number of species exceed 1.5 million (Hawksworth, 1991). There is virtual agreement among mycologists that the fungi are polyphyletic and reflect several distinct evolutionary lines. Evidence from studies on cell wall composition, cellular metabolism, genetics, and ultrastructure indicates that several groups of fungi evolved independently (Barr, 1992; Hawksworth,

Table 5.1. Classification of the Fungi

I. Kingdom Protista: Eukaryotic organisms with heterotrophic or autotrophic nutrition; includes protozoans, algae, and some fungal-like organisms.
  A. Division Myxomycota: true slime molds; plasmodial feeding stage; spores formed in a fruiting body.
  B. Division Oomycota: mycelial fungi with nonseptate hyphae; most form zoospores; some can form airborne propagules; sexual spores (oospores) are thick-walled resting spores.

II. Kingdom Fungi: true fungi (mycelial or yeast) with an absorptive, heterotrophic nutrition.
  A. Division Chytridiomycota: primitive fungi with flagellated asexual spores; no reports of airborne propagules.
  B. Division Zygomycota: mycelial fungi with nonseptate hyphae; sexual spores (zygospores) are thick-walled resting spores; asexual spores (sporangiospores) produced in a sporangium.
    1. Zygomycetes: common soil saprophytes; black bread molds.
  C. Division Ascomycota: mycelial fungi or yeasts; sexual spores (ascospores) formed in an ascus; asexual spores (conidia) abundant in some groups.
    1. Hemiascomycetes: yeasts or simple mycelial fungi; no fruiting body produced.
    2. Euascomycetes: mycelial fungi; asci produced in a fruiting body (ascocarp); powdery mildews, cup fungi, morels.
  D. Division Basidiomycota: mycelial fungi or yeasts; sexual spores (basidiospores) formed on a basidium; asexual spores abundant in two classes.
    1. Hymenomycetes: mushrooms and bracket fungi; basidia are exposed during development.
    2. Gasteromycetes: puffballs and allies; basidia enclosed during development.
    3. Teliomycetes: rust and smut fungi; plant pathogens; no fruiting body; abundant asexual urediospores (in the rusts) and teliospores (rusts and smuts).
  E. Asexual Fungi: mycelial fungi; no sexual stage; asexual spores are conidia; most are anamorphs of the Ascomycota.
    1. Hyphomycetes: mycelial fungi or yeast; conidia formed on the mycelium or conidiophores.
    2. Celomycetes: mycelial fungi; conidia formed on conidiophores on or in specialized structures.

Note: Classes listed are only those containing airborne fungi.

1991; Kendrick, 1992; Wainright et al., 1993). The classification scheme presented here (Table 5.1) reflects this evidence and places the Myxomycota (slime molds) and Oomycota (water molds) in the Kingdom Protista, while all the other divisions, or phyla, of true fungi (Chytridiomycota, Zygomycota, Ascomycota, and Basidiomycota) are included in the Kingdom Fungi (Kingdom Eumycota in some systems). Although "fungal-like" organisms are now classified in two kingdoms, the term "fungi" is still used in the traditional sense to refer to any and all organisms studied by mycologists (Hawksworth, 1991).

A. Kingdom Protista

Division Myxomycota. This division consists of slime molds, organisms that have affinities to both animals and fungi. The plasmodial (ameboid) feeding stage shows animal-like characteristics. It has no cell wall and it actively engulfs organic matter and bacteria in the environment. The reproductive stage, however, aligns the slime molds with fungi, since spores are produced within fruiting bodies. The spores are well adapted for wind dispersal and have been identified during air sampling (Gregory, 1973). Santilli et al. (1985) and Benaim-Pinto (1992) found that spore extracts of the slime mold *Fuligo septica* elicited positive skin test responses in a group of atopic patients.

Division Oomycota. The organisms included here are commonly known as water molds. Biflagellated zoospores are produced in sporangia and are normally dependent on water for spore dispersal. However, within this division, members of the order Peronosporales produce nonflagellated propagules (actually one-spored sporangia) that are dispersed by wind. Included in this group are some well-known plant pathogens such as the fungi that cause late blight of potato, downy mildew of grape, and the white rusts. Becker et al. (1992) recently reported on the sensitization of a greenhouse worker to *Plasmopara viticola* (downy mildew of grape).

B. Kingdom Fungi

Division Chytridiomycota. Organisms in this group are common in aquatic or moist habitats. The group includes both unicellular and mycelial species that exist as parasites of algae, vascular plants, invertebrates, and even other fungi. Saprobic chytridiomycetes are well known for their ability to degrade cellulose, chitin, and keratin (Powell, 1993). Asexual reproduction is by zoospores with a single posterior flagellum, while sexual reproduction is by gametic fusion. There are no reports of airborne dispersal in this group of primitive fungi.

Division Zygomycota. Members of this division are characterized by thick-walled sexual resting spores, called zygospores, which develop from the fusion of compatible gametangia on hyphal tips. Asexual spores, known as sporangiospores, develop within a sporangium and are dispersed by wind. The mycelium is nonseptate, with cross walls only present at the base of reproductive structures. Members of this division are usually saprobes that can occur in a variety of substrates, especially areas with high humidity. *Rhizopus* and *Mucor*, two common zygomycetes, are frequently identified from soil and house dust, well as both outdoor and indoor air samples.

Division Ascomycota. Members of the Ascomycota are characterized by the production of sexual spores (ascospores) within a saclike structure called an ascus. Within the ascus, karyogamy and meiosis occur followed by an additional mitotic division to produce eight haploid ascospores. Members of this division are commonly called "ascomycetes" although this term no longer has taxonomic rank. Classes within the division are determined by the morphology of the ascus and also the type of fruiting body. They range from simple yeasts, in which a naked ascus is produced by fusion of two compatible cells, to the morels, where asci line the convoluted depressions of a macroscopic fruiting body. Asexual reproduction is accomplished by the production of conidia. Both conidia and ascospores are adapted to airborne dispersal but normally occur in the air under different environmental conditions. The conidial stages of many ascomycetes (including species of *Aspergillus* and *Penicillium*) are common indoor contaminants. Most of the recognized fungal allergens are derived from conidial ascomycetes. Dutch elm disease, chestnut blight, and powdery mildews are among the devastating plant diseases caused by members of the Ascomycota.

Division Basidiomycota. This group includes mushrooms, puffballs, and bracket fungi, the largest and most conspicuous fungi in the environment. Also included are the rusts and smuts (two groups of plant pathogens that lack fruiting bodies) and some yeasts. The characteristic sexual spore in the basidiomycetes (the common name for members of this division) is the basidiospore, which forms externally on a basidium. Each basidium typically bears four spores on peglike appendages. Karyogamy and meiosis occur within the basidium, and the four haploid products of meiosis give rise to the four basidiospores. Basidia line the gills of mushrooms and the pores of bracket fungi, with the resulting basidiospores exposed to the atmosphere throughout development. In puffballs, the basidia develop enclosed within the fruiting body, and the basidiospores are only exposed to the atmosphere when fully mature. Basidiospores are dispersed by wind and are often significant contributors to the air spora. Asexual spores are also formed by basidiomycetes. The rusts and smuts (two classes of this division) produce asexual spores (urediospores and teliospores, respectively) that are the major airborne propagules.

The presence of basidiomycete spores in the indoor environment usually indicates the entrance of inadequately filtered fresh air, since few indoor substrates normally support growth of these fungi. Unusual indoor circumstances, such as a faucet leaking on a wooden cabinet or baseboard, could supply sufficient moisture for the growth and even fruiting body production by wood-rotting members of this division. In addition, some of the basidiomycetous yeasts (e.g., *Sporobolomyces*) can colonize indoor reservoirs.

Asexual fungi. The asexual fungi, sometimes called the deuteromycetes or the fungi imperfecti, constitute an artificial grouping classified on the basis of their asexual spores. These fungi are among the most abundant in both indoor and outdoor environments. At one time the sexual or teleomorphic stages of all these fungi were unknown; however, many fungi classified as deuteromycetes have been determined to be anamorphs or conidial stages of ascomycetes (or in a few cases, basidiomycetes). For some members of this group the sexual stage is therefore known, for others it has yet to be identified, while for others the ability to undergo sexual reproduction has apparently been lost (Alexopoulos and Mims, 1979).

When the sexual and asexual stages of fungi were identified separately, they were given separate names. For example, it is now recognized that some species of the common fungus *Aspergillus* are the conidial stages of an ascomycete genus known as *Eurotium*. However, different sexually defined fungi may have similar conidial stages, and other *Aspergillus* species are the asexual states of *Sartorya* or *Emericella*. To make matters even more complex, the reverse may also occur. Some species of the ascomycete genus *Pleospora* have asexual stages recognized as *Alternaria*, while other species of *Pleospora* have anamorphs identified as *Phoma* or *Stemphyllium*. The International Code of Botanical Nomenclature permits the use of both names, but rules that the name of the teleomorphic (sexual) state should be used when referring to the whole fungus (Hawksworth et al., 1983). Aerobiologists follow this practice, and use the sexually defined name for the ascospores and the anamorphic or asexual name for the conidia.

Various systems are in existence for the identification and classification of the asexual fungi. While some of these systems rely entirely on spore descriptions, others are based on conidial development and provide a more useful tool for determining evolutionary relationships (Burge, 1992).

V. AEROBIOLOGY OF FUNGI

A. Aerodynamics

Spores function as both reproductive and dispersal units of fungi, with the majority of spore types adapted for airborne dispersal. In still air, spores would fall to the ground in response to gravity at a rate (based on Stokes' law) that is proportional to the square of the spore radius for a spherical particle (Gregory, 1973). Aerodynamic behavior is also influenced by the shape and surface characteristics of spores. Spore ornamentation and nonspherical shape (especially irregular shape) increase surface drag. This effectively decreases aerodynamic size and delays deposition (Lacey, 1991b). Aerodynamic behavior is also altered by spore aggregation. Spores of many taxa, including *Cladosporium* and *Penicillium,* frequently occur as aggregates, either irregular masses or chains, in the atmosphere. Aggregation increases particle size, which tends to increase the rate of fall but may also increase surface drag (Lacey, 1991b). Factors influencing aerodynamic behavior are often considered (both theoretically and experimentally) in relation to still air; however, perfectly still air seldom occurs in the natural environment, and prevailing air currents delay spore deposition due to gravity while at the same time increasing the chance of impaction on surfaces.

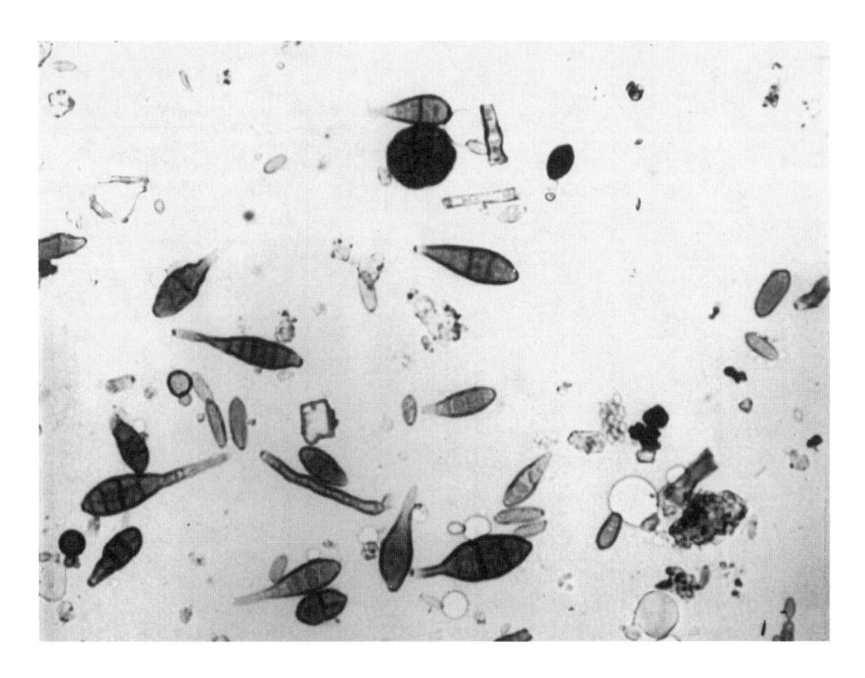

Figure 5.1. Typical components of the dry-air spora.

B. Release and Discharge Mechanisms

Spores are released into the air either passively or by active discharge mechanisms (Gregory, 1973; Aylor, 1990). Spores can be passively dispersed by air movement, or the mechanical action of rain drops or animals. Spores dispersed by air movement are usually hydrophobic, and the airborne concentration depends on the ease with which spores are detached from the mycelium (or fruiting structure) and by conditions such as wind speed and turbulence. This type of dispersal is enhanced by dry weather, and spores of fungi such as *Cladosporium, Alternaria, Epicoccum, Drechslera,* smut spores (teliospores), and rust spores (urediospores), referred to as the "dry air spora" (Figure 5.1), usually peak during the afternoon hours when humidity is low and wind speeds are highest (Hirst, 1953).

Spores that are passively dispersed by rain are usually hydrophilic and include conidia from species of *Fusarium, Gliocladium, Verticillium,* etc. These form part of the wet weather air spora. Puffballs are unique among the basidiomycetes because of their passively dispersed spores. These basidiospores are discharged in clouds or puffs as rain drops strike the mature fruiting body. Strong gusts of wind or small animals impacting the puffball surface can accomplish the same puffing action.

Active discharge mechanisms that propel spores into the turbulent layer are common in the fungi. Many ascospores and basidiospores are actively discharged by mechanisms that require moisture or high humidity (Figure 5.2).

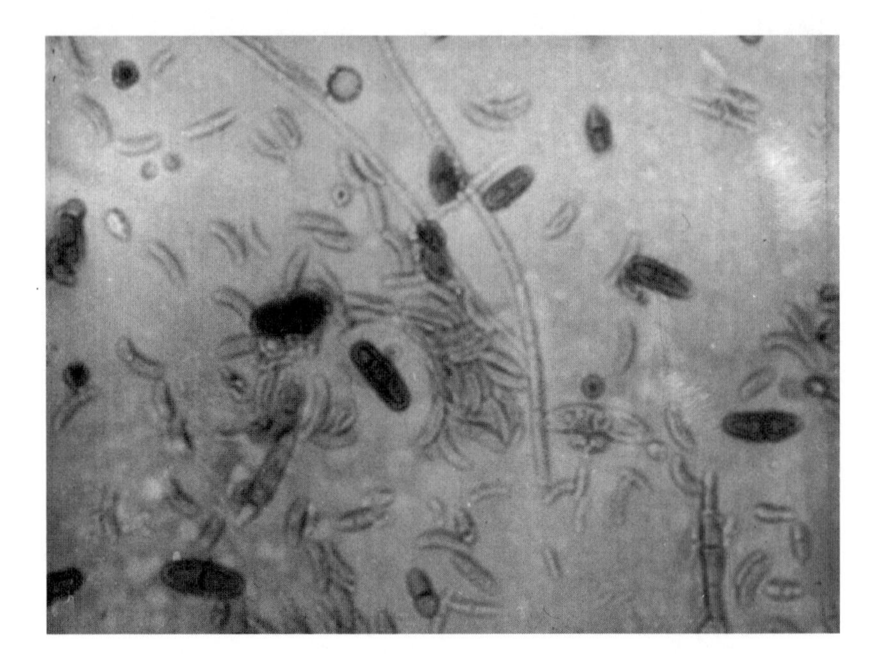

Figure 5.2. Typical components of the wet-weather air spora.

In the ascomycetes, high osmotic pressure develops within the ascus either by the direct absorption of rain water or by the swelling of mucilage within the ascus (Moore-Landecker, 1990). The resulting pressure causes the ascus tip to rupture, forcing the spores out explosively through the laminar layer and into the turbulent layer. Ascospores are often abundant in the atmosphere during and after rainfall (Hirst, 1953; Burge, 1986; Allitt, 1986).

Discharge of basidiospores from mushrooms and bracket fungi also requires atmospheric moisture, although the mechanism is not completely understood. Basidiospores are frequently seen in the predawn hours when the humidity is high. However, Haard and Kramer (1970) investigated spore discharge in 19 species of mushrooms or bracket fungi and found three basic patterns. In many species, a nighttime maximum was evident, with spore release increasing in the late afternoon or early evening, peaking during the middle of the night, and then decreasing in the morning. Other species, especially some bracket fungi, showed a double peak of spore release in the early morning and evening hours, while many small mushrooms showed continuous spore discharge with a peak 24 to 48 h after the onset of spore discharge and then a gradual decrease as the basidiocarp aged.

C. Airborne Transport

At the surface of the ground and the surface of objects an extremely thin film of still air exists; the air in this film is held by molecular forces to the surface. Above this is the laminar boundary layer where air flow is parallel to the surface. The laminar layer is usually thin (approximately 1 mm) but varies with wind speed and topography of adjacent surfaces. Under still air the thickness of laminar layer may increase exponentially, while at high wind speeds it thins down to a fraction of a millimeter (Gregory, 1973). Above the laminar layer is the turbulent layer, the part of the atmosphere in which airborne transport normally occurs. Air movement in this layer is constantly shifting and is sometimes unpredictable. The turbulence depends on wind speed and direction, temperature, and local eddies caused by projections and roughness of the terrain. The dispersive power of this layer is related to the intensity of the turbulence (Gregory, 1973; Mason, 1979; Cox, 1987).

Fungal spores are a normal component of the turbulent layer, and atmospheric concentrations may exceed 200,000 spores/m^3 of air. Carried by wind, spores are transported both vertically and horizontally. Spores carried upward by thermals have been recovered from altitudes of over 5000 m. Horizontally, spores have been carried for thousands of kilometers. The transport of rust urediospores from the southern United States and Mexico to the wheat belt in the northern United States and Canada has been well documented, as have other instances of long-distance dispersal (Nagarajan and Singh, 1990). Temperatures below freezing will significantly decrease spore production on substrates, and snow or ice cover will prevent dispersal of any previously formed spores in that area (Burge, 1985).

D. Influence of the Airborne State

Although many fungal spores are adapted to airborne dispersal and much more resistant to environmental stress than the parent hyphae, they are vulnerable to certain types of environmental damage while airborne. Exposure to harmful radiation and extremes of temperature and humidity are among the factors that can influence the survival, culturability, and infectivity of fungal spores. Pathogenic species may lose infectivity under harsh conditions, but it should be emphasized that most allergenic spores probably retain their allergenic properties even when the spore is no longer viable. Changing wind speeds may produce rapid changes in relative humidity, which may effect survival (Cox, 1987). This may be more important for thin-walled colorless spores, which may easily plasmolyze, with the hazards of desiccation greatest in daytime. At night and at high altitudes, conditions are apparently less stressful, and spores have even been reported to germinate in the clouds (Gregory, 1973). Ultraviolet radiation in the upper atmosphere may also affect spore survival. Again, thin-walled colorless spores may be more vulnerable than dark, melanin-containing spores. Low temperatures in the upper atmosphere, however, may protect the spores from UV damage (Gregory, 1973). Despite these environmental hazards, many fungal spores are well known to survive long-range transport under harsh conditions (Kramer, 1979; Gregory, 1973).

E. Outdoor Prevalence

While airborne pollen has well-known and well-defined seasons, fungal spores can be present virtually year-round in many areas of the world. Atmospheric prevalence patterns of the dry air spora, especially those of *Cladosporium, Alternaria, Drechslera*, and *Epicoccum*, are best known. The large, distinctive conidia of these fungi are successfully collected by most air sampling instruments and are easily identified. The literature on the airborne concentrations of these spores is extensive with many fine reviews (Gregory, 1973; Kramer, 1979; Solomon, 1980; Salvaggio and Aukrust, 1981; Lehrer et al., 1983; Burge, 1985, 1990; Bush, 1989; Lacey, 1991a).

In most parts of the world *Cladosporium* is the most abundant genus identified from atmospheric sampling. Concentrations may exceed 200,000 spores/m^3 and may form over 90% of the total spore load (Lacey, 1991a). Other members of the dry air spora are less prevalent but still prominent and may overtake *Cladosporium* in some regions at certain times of the year. Although spores can be present year-round, the highest overall concentrations often occur in summer and early fall in temperate areas. *Cladosporium* and *Alternaria* concentrations are frequently at yearly peaks during this period (Ebner et al., 1989; Larsen and Gravesen, 1991; Cosentino et al., 1990). In a several-year study at two cities in Sweden, total spore levels and *Cladosporium* levels peaked from July to September. Conidia from *Cladosporium* were the most abundant spore type in the air, representing 37% of total air spora (Rubulis, 1984).

During a 5-year study of airborne spores in Kuwait, Halwagy (1989) found that highest spore concentrations occurred in the spring with a lower peak in the autumn. *Cladosporium* conidia were the most abundant airborne spores, followed by *Ustilago* (smut teliospores), *Alternaria* and *Drechslera* conidia, and *Chaetomium* (ascospores). By contrast, *Curvularia* and *Nigrospora* tend to be important members of the air spora in tropical areas, with *Aspergillus* species also characteristic of the humid tropics (Lacey, 1991a). In the tropical climate of Singapore, Tan et al. (1992) found that *Nigrospora sphaerica* was the most abundant single species identified during cultural sampling at three urban sites. *Cladosporium* and *Alternaria* conidia were less frequently isolated; however, several species of *Curvularia* were identified, making this genus well represented in the atmospheric spore load. Abdalla (1988) found that *Aspergillus* species constituted 68% of the colonies isolated during a 12-month study in Khartoum, Sudan.

Far less is known about the atmospheric concentrations of ascospores and basidiospores. Many aerobiological studies have been performed using Rotorod samplers, which are reasonably efficient for spores 10 μm and larger but greatly underestimate smaller spores such as many basidiospores and ascospores. Suction samplers such as the Burkard Volumetric Spore Trap (Burkard Manufacturing Ltd., Rickmansworth, U.K.) are needed to assess the aerobiological contributions of these spores. Ascospore concentrations in the atmosphere

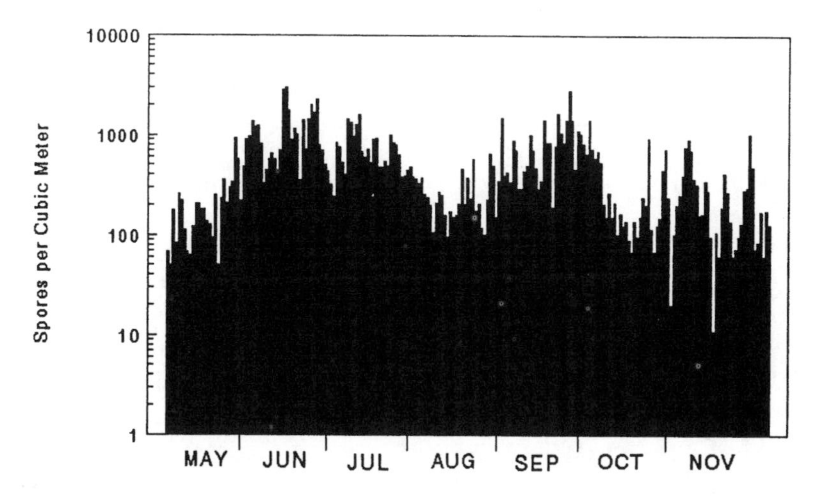

Figure 5.3. Average daily concentration of airborne basidiospores in Tulsa, OK (1987), showing late spring and fall peaks.

are generally associated with rainfall due to the discharge mechanism and reach high levels at these times. Many studies group all ascospores together or, at best, just identify a few easily recognizable genera. Spores of *Chaetomium, Leptosphaeria, Venturia*, and the family Xylariaceae fall into this category, and they have been identified in many aerobiological surveys (Gregory, 1973; Burge, 1985; Hamilton, 1959; Levetin, unpublished observations). Investigating other ascospores, Allitt (1986) found that hyaline-1-septate ascospores reached high levels following rainfall during July and August. Daily averages of these spores reached 9500/m^3 with hourly concentrations as high as 90,000 spores/m^3. The majority of these spores belonged to various species of *Didymella*.

The airborne basidiospores have also been ignored in many studies, yet high concentrations of basidiospores exist in the atmosphere at various locations around the world (Gregory and Hirst, 1952; Hamilton, 1959; Salvaggio et al., 1971; Gregory, 1973; Rubulis, 1984; Anderson, 1985; Hasnain et al., 1985; Misra, 1987; Levetin, 1990, 1991). In Sweden, basidiospores accounted for about 30% of the total air spora collected during a several year study (Rubulis, 1984). In Arctic and subarctic areas, the air spora tends to be dominated by basidiospores and ascospores during the short spore-producing season (Anderson, 1985; Lacey, 1991a). Fall peaks in the concentration of airborne basidiospores have been reported in some of these studies, while both late spring and fall peaks occur in Tulsa, Oklahoma (Figure 5.3). Previously, studies have categorized basidiospores as colored and colorless or identified one or two distinctive genera; recent work has identified airborne basidiospores from 18 genera (Levetin, 1991). Although *Coprinus* was the most abundant genus identified and at times represented over 60% of the basidiospore load, other genera were a consistent component of the basidiospore total. *Ganoderma* spores have also been found to be abundant in the atmosphere at various locations (Tarlo et al., 1979; Hasnain et al., 1984; Levetin, 1990). Colorless basidiospores that lack distinctive morphological features are, for the most part, still an unidentified but significant component of the air spora. Many common mushrooms and bracket fungi produce colorless basidiospores and undoubtedly contribute to this segment of the air spora.

Although the composition of the air spora is similar worldwide, at any one time the air spora may be dominated by nearby sources of spores (Lacey, 1991a). During crop harvesting or mowing, incredible levels of spores (up to 10^9 spores/m^3) may be dispersed into the atmosphere, with *Alternaria, Cladosporium*, or *Epicoccum* often the most abundant taxa. Near woodland areas or forests, basidiospores may dominate the spore load (Tarlo et al., 1979; Hasnain et al., 1984), or ascospore levels may be elevated or even dominate near standing or running water.

VI. INDOOR AEROBIOLOGY

The air in almost all indoor environments also contains fungal spores. Environmental factors that influence indoor fungus concentrations include outdoor concentrations, type and rate of ventilation, and indoor moisture levels (Burge, 1990). While not as unpredictable as in the atmosphere, indoor air flow tends to be turbulent and is subject to many variables including room size, design, heating, ventilation, furnishings, and activities of inhabitants. Ventilation, either natural from windows or from central HVAC (heating, ventilation, and air-conditioning) systems, generates turbulence in a room. Furnishings cause local eddies in the air flow, and heat sources produce vertical convection currents that may enhance the turbulence. In fact, convection currents alone are sufficient to disperse fungal spores throughout an enclosed space (Gregory, 1973; Chatigny and Dimmick, 1979; Cox, 1987).

All air in interior spaces originally comes from outdoors. In spaces primarily ventilated through open windows and doors, outdoor air with its entrained spores readily enters, and the indoor fungus air spora reflects that outdoors. Even when doors and windows are closed, at wind speeds greater than 1 mph, bioaerosols from the outside air can penetrate indoors through tiny cracks around windows, doors, and walls (Dingle, 1957; Gregory, 1973). In poorly ventilated interiors (i.e., when windows and doors remain closed), the air spora may gradually decrease in concentration due to settling and impaction, or may be dominated by spores from indoor sources. Mechanical ventilation usually includes filters, which remove many spores from the air. It is generally accepted that properly maintained central HVAC systems effectively reduce indoor spore concentrations and even reduce incidence of respiratory allergies or diseases (Vaughan and Cooley, 1933; Hirsch et al., 1978; Rose and Hirsch, 1979; Burge, 1990; Pan et al., 1992).

In warm or mild climates, the contribution from outdoor bioaerosols may not diminish in winter and, although subfreezing temperatures occasionally occur, outdoor fungus aerosols are present year-round (Lehrer et al., 1983; Levetin and Buck, 1987). In north temperate areas where long periods of subfreezing temperatures and snow cover are common, outdoor concentrations during the winter are greatly reduced. Many surveys of indoor airborne fungi have focused on sampling during this season to minimize the influence from outdoor taxa (Solomon, 1975, 1976; Pasanen et al., 1989; Wickman et al., 1992; Su et al., 1992). Studies during winter in these latitudes have generally found spore levels to be higher indoors than outdoors, indicating intramural sources of contamination.

Airborne concentrations of fungal spores in indoor environments also vary with the amount of mechanical and/or human activity. Large numbers of people or abundant activity stirs up dust (reentraining settled spores) and intensifies air currents, delaying deposition by gravity. In addition, fungal spores can be introduced when people enter the area, either on people themselves or on clothing (Pasenen et al., 1989; Burge, 1990).

A. Indoor Sources of Bioaerosols

Common substrates in the indoor environment serve as nutrient sources for saprobic and/or opportunistic pathogens and allow for growth and continued spore formation indoors. The most familiar indoor substrates include carpets (especially jute or other natural backings), components of upholstered furniture, soap films on shower walls, shower curtains, and other bathroom fixtures, wallpaper, water and scale in humidifiers, and soil and surfaces of containers for potted plants. HVAC systems can also serve as amplification and dissemination sites for fungal spores (Samson, 1985; Mahoney et al., 1979). Fungi have been found growing on air filters, cooling coils, and drip pans as well as in the ducts. Routine filter and drip-pan maintenance and control of relative humidity can usually prevent or minimize problems from this source.

Water availability appears to be the most critical factor controlling fungal colonization of indoor substrates. The extent of fungal amplification is closely related to indoor relative humidity (RH). Below 30% RH little interior mold growth usually occurs, while above 70% conditions may be optimal for fungal growth (Burge, 1985). High humidity causes moisture to condense on cool surfaces. This can be a problem in the winter when water condenses on cold windows and accumulates on moldings and sills to create a suitable habitat for fungal colonization. In addition, high humidity can allow hygroscopic materials such as skin scales in dust, leather, wool, etc. to absorb enough water to support fungal growth. Also, moisture seeping through walls, ceilings, basements, and concrete slabs can provide conditions suitable for fungi. In recent decades, increased use of household amenities (washing machines, dishwashers, and other moisture sources) coupled with the quest for greater energy efficiency (resulting in tightly sealed buildings) has added to this problem. Vaporizers and some humidifiers exacerbate conditions by actively spraying water droplets into the air. Often these droplets are already contaminated with microbial propagules (Solomon, 1976).

The soils and leaf surfaces of potted plants are also suitable substrates for saprobic fungi, including a number of opportunistic human pathogens (Summerbell et al., 1989). Evidence for the airborne dispersal of fungal propagules from plants is, however, contradictory. Burge et al. (1982) found that undisturbed plantings did not contribute significantly to the airborne spore load, while other studies showed that potted plants were sources of airborne fungi (Staib et al., 1978a; Smith et al., 1988; Malloy and Levetin, unpublished observations). It is possible that dispersal from plants may depend on airflow patterns existing within the room, on the condition of the plants, and on other unknown factors.

Air cleaners, either as part of a central system or as free-standing portable units, have been shown to be effective at removal of indoor airborne spores; however, not all cleaners are equally efficient (Scherr and Peck, 1977; Kooistra et al., 1978; Nelson et al., 1988; Resiman et al., 1990; Levetin et al., 1992). Air cleaners with HEPA (high-efficiency particulate air) filters or electrostatic precipitators are more efficient than other cleaning technologies (Levetin et al., 1992).

B. Indoor Prevalence

Much of our knowledge on concentrations of indoor fungal spores has resulted from case studies performed following occupant complaints or diagnoses of illness believed to originate from environmental exposure. As a result, very little baseline data exists on randomly selected homes, schools, or occupational environments in various parts of the world, and there is almost no systematically collected data on relationships between fungal prevalence and disease (Solomon, 1975, 1976; Dungby et al., 1986; Lacey and Crook, 1988; Pasenen et al., 1989; Tunon De Lara et al., 1990; Macher et al., 1991; Lacey, 1991a; Su et al., 1992; Ebner et al., 1992; Wickman et al., 1992).

In a study of airborne culturable fungi in 26 midwestern homes, indoor levels averaged 25% of the outdoor levels from spring through fall, with *Cladosporium* by far the most abundant genus (Solomon, 1975). Outdoors, other members of the dry air spora were also prominent; however, indoors *Penicillium* was the second most frequently isolated genus. During the winter *Penicillium, Aspergillus, Sporothrix, Oospora*, and yeast dominated the indoor samples at levels that often reached 1000 colony-forming units (cfu)/m^3. The corresponding outdoor concentrations never exceeded 230 cfu/m^3 during winter. In a larger study during winter in 150 homes, Solomon (1976) found indoor levels as high as 20,000 cfu/m^3. He also found a positive correlation between indoor spore concentrations and indoor relative humidity and suggested that humidifiers served as a reservoir for fungal growth dispersing spores into the air.

In the Netherlands, Waegemaekers et al. (1989) recovered approximately twice the mean airborne concentration of culturable fungi in damp than in dry houses, with *Penicillium, Aspergillus, Botrytis*, and *Cladosporium* the most prevalent fungi isolated. In 44 houses, 15 archives and libraries, 10 museums, and 4 offices and schools, van Reenen-Hoekstra et al. (1991) found that *Aspergillus versicolor* was the most common species, followed by *Cladosporium* spp. and *Penicillium chrysogenum.* In archives, libraries, and museums, xerophilic fungi were especially common in surface samples but were only occasionally isolated from the atmosphere.

Su et al. (1992) used factor analysis to establish correlations and associations with home factors for airborne fungi identified from 150 homes in Topeka, Kansas. Groups of associated fungal contaminants were established and related to housing conditions. *Cladosporium, Alternaria, Epicoccum*, and *Aureobasidium* were identified as a group of common decay fungi; these fungi were present in high concentrations in homes with a gas cooking stove. The authors suggest a possible relation between the increased humidity from combustion or the increased pentetration of outdoor air (due to venting requirements) and elevated levels of these taxa. *Penicillium* and *Aspergillus* were also separated as a group that was significantly associated with dirt-floor, crawl-space types of basements. High concentrations of *Fusarium* and other water-requiring fungi were associated with homes having water-collecting problems. The authors also found a direct relationship between the incidence of respiratory problems and the indoor concentration of *Epicoccum, Cladosporium, Aureobasidium*, and yeast. This study provided the first quantitative connections between biological characteristics of the fungi and both housing characteristics and health variables. Clearly, more research of this type is needed to establish large databases for reliable assessments of the possible health effects of contaminated indoor environments.

Dust analysis has been used as a surrogate for airborne exposure to fungi by Wickman et al. (1992), who investigated the culturable fungi isolated from house dust in the homes of 118 atopic children and 57 nonatopics in Stockholm. The study was conducted in midwinter during an extended period of subfreezing outdoor temperatures; thus the contribution from the outdoor air spora was minimal. *Penicillium, Alternaria*, and *Cladosporium* spp. were the most common taxa identified. Although no atmospheric sampling was conducted and no direct connection between results of dust analyses and respiratory symptoms could be found, high colony-forming units from house dust were associated with higher relative humidities.

Recent longitudinal studies have sought to evaluate changes in fungus prevalence over time in residential environments, for example, environments selected without regard to the presence or absence of complaints. Macher et al. (1991) examined the concentrations and types of culturable fungi present in the air and on interior surfaces of a new apartment at 3-month intervals over 2 years. *Cladosporium* was the most abundant genus identified from both indoor and outdoor samples, with peaks occurring in September outdoors and in December indoors. There was no significant difference in the concentrations collected during the first and second year of the study, but marked seasonal differences were present in both years. All indoor levels were less than those outdoors, but the differences were not significant during the June and December sampling periods, which were the times of the lowest and highest indoor concentrations, respectively.

Ebner et al. (1992) sampled biweekly for viable fungi at three indoor and two outdoor alpine locations for 1 year with an additional 10 months of sampling at two of the stations (one indoor and one outdoor). The low-altitude indoor sites had lower concentrations than the corresponding outdoor samples. At the higher elevations, the indoor levels were approximately twice those outdoors. At all sampling locations *Cladosporium* was the most abundant genus identified, followed by *Penicillium* in the indoor locations and *Epicoccum* at the outdoor sites. The seasonal distribution indoors roughly paralleled that outdoors with the exception of *Penicillium* and *Phoma,* which showed midwinter peaks indoors.

Crook and Lacey (1991) investigated the airborne microorganisms associated with respiratory diseases among workers involved in indoor mushroom cultivation. During composting, workers were exposed to high concentrations of thermophilic actinomycetes (a group of filamentous bacteria). During spawning and picking operations, however, concentrations of airborne fungi averaged 10^5 colony-forming units (cfu)/m^3. *Penicillium* spp. and *Aspergillus fumigatus* were the most common taxa with concentration 100 to 1000 times greater than other fungi (*Monilia sitophila, Trichoderma viride*, and *Peziza ostracoderma*). Other studies have also focused on the working environments where organic materials were processed and high airborne spore loads contributed to respiratory disease among the workers (Lacey and Crook, 1988; Lacey, 1991a).

Concentrations of fungal spores in farm buildings have also been studied in relation to respiratory disease (Lacey and Crook, 1988; Lacey, 1991a). Wherever crops or hay are stored, storage fungi are abundant and airborne spore concentrations up to 10^{10} spores/m^3 have been recorded. The taxa present depend upon actual storage conditions, but *Penicillium* and *Aspergillus* species frequently predominate, including thermophilic and thermotolerant *Aspergillus* species. Pasenen et al. (1989) compared the airborne spore load in urban and rural living environments during the winter in Finland. Cow barns were found to have total spore levels as high as 10^9 spores/m^3, with the highest concentrations associated with the handling of hay. Farmhouses had concentrations up to 10^5 spores/m^3, about 10 to 1000 times higher than concentrations measured in urban apartments. *Cladosporium, Penicillium*, and *Aspergillus* were the dominant genera in both the rural and urban indoor locations; however, other genera (*Alternaria, Acremonium, Botrytis*, and *Chrysosporium*) were not detected in the urban apartments. The authors suggest that many of the fungal spores were carried into the farmhouses on the person or clothing of occupants. As in many other winter prevalence studies, spore concentrations were higher at indoor sites than those outdoors.

VII. DISEASES CAUSED BY AIRBORNE FUNGI

The connection between exposure to airborne fungi and disease was first made by Blackley in the 1870s when he described chest tightness and "bronchial catarrh" after inhalation of *Penicillium* spores. It has since been clearly demonstrated that the exposure to airborne fungal spores, hyphal fragments, or metabolites can cause a variety of respiratory diseases. These range from allergic diseases including allergic rhinitis, asthma, and hypersensitivity pneumonitis to infectious diseases such as histoplasmosis, blastomycosis, and aspergillosis. In addition, acute toxicosis and cancer have been ascribed to respiratory exposure to mycotoxins.

A. Allergic Diseases

The fungi are well-known allergen sources. Fungus exposure is most commonly associated with hay fever and asthma, but has also been implicated in hypersensitivity pneumonitis (see Chapter 10). Many (and possibly all) fungal spores contain allergens that may cause one or more of the allergic diseases. Note that, unlike infectious agents, spores may not have to be viable to retain allergenic properties. For this reason, it is important to know the concentration and composition of the total air spora including nonculturable spores and allergen-containing fragments, which require immunochemical methods for detection (see Chapter 12).

Allergens have been reported from every major group in the Kingdom Fungi (Table 5.2), and from both outdoor and indoor exposures. Numerous studies have confirmed the importance of allergens of both sexual and asexual fungi as a cause of asthma, hay fever, and hypersensitivity pheumonitis (HP). Many of the most widely recognized sources include asexual taxa recognized as abundant in the atmosphere: *Cladosporium, Alternaria, Epicoccum,* and *Curvularia* (Gregory, 1973; Solomon, 1980; Salvaggio and Aukrust, 1981; Lehrer et al., 1983; Chapman and Williams, 1984; Burge, 1985; Lacey, 1991a). Gregory and Hirst (1952) suggested airborne basidiospores as possible allergen sources, and clinical evidence has been accumulating on the incidence of basidiospore-induced allergic disease (Herxheimer et al., 1969; Giannini et al., 1975; Tarlo et al., 1979; Hasnain et al., 1984, 1985; Lopez et al., 1985, Santilli et al., 1985; Burge, 1986; Lehrer et al., 1986; Butcher et al., 1987; Davis et al., 1988; O'Neil et al., 1988; Horner et al., 1989; De Zubiria et al., 1990). Far less is known about the allergenicity of ascospores (Gregory, 1973; Burge, 1986). Allitt (1986) reviewed the literature on the incidence of allergy related to high

Table 5.2. Allergenic Fungi

- Myxomycota
  - *Fuligo*
  - *Lycogala*
  - *Stemonitis*
- Oomycota
  - *Plasmopara*
- Zygomycota
  - *Absidia*
  - *Mucor*
  - *Rhizopus*
- Ascomycota
  - *Chaetomium*
  - *Claviceps*
  - *Daldinia*
  - *Erisyphe*
  - *Eurotium*
  - *Leptosphaeria*
  - *Microsphaeria*
  - *Saccharomyces*
  - *Xylaria*
- Basidiomycota
  - Hymenomycetes
    - *Agaricus*
    - *Amanita*
    - *Armillaria*
    - *Boletus*
    - *Boletinellus*
    - *Cantharellus*
    - *Chlorophyllum*
    - *Coprinus*
    - *Dacrymyces*
    - *Ganoderma*
    - *Hypholoma*
    - *Inonotus*
    - *Merulius*
    - *Naematoloma*
    - *Pleurotus*
    - *Psilocybe*
    - *Polyporus*
    - *Stereum*
  - Gasteromycetes
    - *Calvatia*
    - *Geastrum*
    - *Lycoperdon*
    - *Pisolithus*
    - *Podaxis*
    - *Scleroderma*
  - Teliomycetes
    - *Puccinia*
    - *Tilletia*
    - *Urocyctis*
    - *Ustilago*
- Asexual fungi
  - *Alternaria*
  - *Aspergillus*
  - *Aureobasidium*
  - *Botrytis*
  - *Candida*
  - *Cephalosporium*
  - *Cladosporium*
  - *Coniosporium*
  - *Curvularia*
  - *Dicoccum*
  - *Dreschlera*
  - *Epicoccum*
  - *Epidermophyton*
  - *Fusarium*
  - *Gliocladium*
  - *Helminthosporium*
  - *Monilia*
  - *Nigrospora*
  - *Paecilomyces*
  - *Penicillium*
  - *Phoma*
  - *Rhodotorula*
  - *Spondylocladium*
  - *Sporobolomyces*
  - *Sporotrichum*
  - *Stemphylium*
  - *Tilletiopsis*
  - *Torula*
  - *Trichoderma*
  - *Trichophyton*
  - *Trichothecium*

Note: Adapted from Burge, H. A., *Clin. Rev. Allergy,* 3, 319, 1985.

atmospheric concentrations of *Didymella* ascospores and identified the species of *Didymella* involved. Overall, the importance of both the ascospores and basidiospores as aeroallergens has probably been greatly underestimated (Lacey, 1991a).

Many working environments have been identified where the spores of specific fungi overwhelm the air spora and cause occupational asthma and HP (Lacey and Crook, 1988; Lacey, 1991a). Among the many occupational settings where fungus-related asthma have been reported are mushroom cultivation

(*Aspergillus funigatus, Lentinus elodes, Pleurotus ostreatus, Oidiodendon* sp.), cheese manufacturing (*Penicillium camembertii*), flour mills (*Alternaria* spp., *Aspergillus* spp.), and food processing (*Aspergillus* spp., *Penicillium* spp., *Alternaria,* spp., *yeasts*). In addition, HP has been reported in conjunction with exposure to *Alternaria, Aspergillus, Aureobasidium, Cladosporium, Eurotium, Penicillium, Pleurotus, Pholiota, Phoma, Rhizopus,* and *Trichoderma.*

It is widely accepted that there is a cause-and-effect relationship between aeroallergen exposure and allergic disease; however, threshold levels either for sensitization or symptom development have not been clearly defined for any fungal allergen. Reported atmospheric concentrations normally represent an average daily concentration for a large geographic area. However, during a given 24-h period, short concentration peaks may trigger symptoms in sensitive individuals. In addition, atmospheric concentrations close to a source will be significantly higher than the daily average for a given area, and sensitive individuals near this source may experience symptoms.

Although these variables influence individual exposure levels, attempts have been made, nevertheless, to correlate some atmospheric concentrations with symptoms. It has been suggested that grass and ragweed pollen concentrations of 20 grains/m^3 and 100 grains/m^3 (24-h average), respectively, could elicit symptoms in sensitive individuals (Solomon, 1980). Similarly, there has been speculation upon the threshold concentrations for fungal spores. Published reports suggest that concentrations of 100 *Alternaria* conidia/m^3 and 3000 *Cladosporium* conidia/m^3 are levels that may induce symptoms (Dhillon, 1991).

No estimates have been suggested for basidiospore or ascospore concentrations; however, Salvaggio et al. (1971) have shown a positive correlation between increased hospital admissions due to asthma and high atmospheric levels of basidiospores. Likewise, several studies have associated late summer asthma with high concentrations of ascospores identified as *Leptosphaeria* species (Gregory, 1973), and "barley asthma" has been associated with high levels of *Didymella* (Allitt, 1986). While there may be threshold concentrations below which no symptoms are experienced (Dhillon, 1991), this is probably not an absolute value but a gradient based on individual sensitivities.

B. Infectious Diseases

Although the majority of fungal spores are probably capable of causing allergic responses, only a small group of fungi are commonly considered human pathogens (Al-Doory, 1980; Drouhet, 1988). Possibly 85 to 90% of all mycoses are recurring infections of the skin or mucous membranes (i.e., dermatophytic diseases such as athlete's foot and *Candida* infections such as thrush) (Male, 1991). Only a few fungi are considered primary, systemic human pathogens that can infect any nonimmune person (e.g., *Histoplasma, Blastomyces*, and *Cryptococcus*). However, many common saprobes that are ubiquitous air contaminants can cause serious human infections in immunocompromised or high-risk patients (Weitzman, 1986). Invasive fungal infections have been reported in 25% of these patients (Drouhet, 1988). These diseases are of special concern because they do not respond to conventional antibiotic treatment, and mortality rates are high. Because fungal spores are ubiquitous, strict isolation of high-risk patients is often necessary to prevent exposure.

Infection rates following hip and knee replacement surgery can be as high as 10% (Lidwell et al., 1982), and infections can surface months or even years after the operation (Charnley, 1973). The increasing use of bone marrow transplants to treat patients with acute leukemia and other forms of cancer, severe aplastic anemia, and congenital and acquired immunodeficiencies, as well as other disorders, is associated with an increase of infections due to a variety of pathogens (Arlet et al., 1989). Prior to and following transplants, patients are given immunosuppressive drugs to prevent rejection. In addition, in the case of leukemia, the patient may also be given total body irradiation to destroy malignant cells. All during this time, but especially just following transplant, the patient is at high risk for many opportunistic infections (Streifel et al., 1989; Young et al., 1970). Epidemics of nosocomial opportunistic fungal infection have occurred that provide proof of airborne spread. The fungi involved are ubiquitous saprobes in the environment and are usually considered nonpathogenic. The normal method of dissemination of these fungi is through airborne spores. *Aspergillus* is the most frequently reported opportunistic pathogen, while *Fusarium* infections have only only recently been identified. Because of the widespread occurrence of these genera, the remainder of this discussion focuses on these opportunistic infections.

The genus *Aspergillus* is widely distributed and is common in the soil and on decaying vegetation, dust, and other organic debris. Unlike most saprobes, many species of *Aspergillus* are tolerant of temperatures at or above 37°C. The spores are passively aerosolized when spore clusters are disturbed. The small spores are extremely buoyant and remain airborne for long periods of time (Streifel et al., 1989). Most opportunistic infections are caused by *Aspergillus fumigatus*. However, *A. flavus, A. glaucus, A. nidulans, A. niger,* and *A. terreus* have also been reported as causes of infections. Invasive aspergillosis is common in patients with neoplastic disease (especially leukemia) and in patients who have undergone transplants (Weitzman, 1986; Leedom and Loosli, 1979). The lung is the most frequent site of infection, followed by the gastrointestinal tract, brain, liver, and kidneys (Young et al., 1970).

The literature on aspergillosis is extensive, and in many cases the environmental source of the infection was located. Aisner et al. (1976) reported a cluster of eight cases of pulmonary aspergillosis in leukemia patients who acquired the disease shortly after a move into a new hospital facility. *Aspergillus fumigatus, A. flavus*, and *A. niger* were found growing on the fireproofing material that had been sprayed on pipes and ceiling panels in seven rooms.

During a 2-month period in 1977, five children with lymphoblastic leukemia were admitted to a Texas hospital where they subsequently developed invasive *Aspergillus* infections (Mahoney et al., 1979). All died within a short time, with the *Aspergillus* infection being the direct cause of death in three of the five children. The patients were housed in four contiguous rooms either simultaneously or sequentially. A subsequent inspection of the ventilation system revealed several sites with heavy mold growth on moist coils. Although the potential for airborne contamination by the air-conditioning system was demonstrated, no cultures were obtained to confirm the environmental source. Lentino et al. (1982) described a clustering of 10 cases of aspergillosis confirmed at autopsy. Contaminated window air conditioners were the source of the fungus for at least 7 of the 10 patients.

By contrast, a comparison of a naturally ventilated old hospital with a mechanically ventilated new hospital revealed a difference in the cases of aspergillosis over consecutive 5-year periods (Rose, 1972). Eleven cases of pulmonary aspergillosis occurred in the old hospital but no cases were present in the new one. Also, colonies of *Aspergillus fumigatus* were isolated from the old hospital using settle plates, but none were isolated from the new facility. These results led to the suggestion that the mechanical ventilation removed spores from the air. This hypothesis was supported by subsequent volumetric air sampling (Rose and Hirsch, 1979). This contrast stresses the importance of mechanical ventilation in reducing exposure and also emphasizes the need to maintain clean mechanical ventilation systems.

Members of the genus *Fusarium* are also common soil saprobes, but are best known as important plant pathogens. They have a fairly wide host range and are known to cause vascular wilt in tomatoes and other crops as well as other plant diseases. Several species of *Fusarium* are known to produce mycotoxins that led to outbreaks of alimentary toxic aleukia following ingestion of contaminated grain (Richardson et al., 1988). *Fusarium* infections in humans are rare and normally cause only localized keratitis or skin infections. Recently, however, a number of cases of invasive as well as cutaneous *Fusarium* infections in immune-compromised hosts has been reported (Anaissie et al., 1986; Blazar et al., 1984; Cho et al., 1973; Kiehn et al., 1985; McNeely et al., 1981; Richardson et al., 1988; Weitzman, 1986). In those instances when the infection became disseminated rather than localized, the patients died despite antifungal therapy. No *Fusarium* epidemics have been reported and there have been no environmental studies that have allowed location of the sources of the infections in individual cases. Although airborne exposure resulting in infection with *Fusarium* has not been shown, it should be expected, based on the ubiquitous nature of the genus and previous studies that have isolated *Fusarium* in other indoor settings (Solomon, 1975, 1976; Hirsch and Sosman, 1976; Levetin, unpublished observations).

The increased incidence of nosocomial opportunistic fungal infections has stimulated an increased interest in the environmental sources of the fungal propagules. Attention has been focused on ornamental potted plants present in hospital corridors, lounges, and even patient rooms. Both leaf surfaces and soils can be reservoirs of potentially pathogenic organisms. Staib et al. (1978a) consistently isolated *Aspergillus fumigatus* and *A. niger* from the soil of two potted plants during a 4-month study. *A. fumigatus* was found in the soil of a cactus that was kept at 38°C during the study, and *A. niger* was isolated from a clivia plant that was growing at ambient temperatures. Air samples showed airborne dissemination only when the soil was dry and air movement was present. In a later study, Staib et al. (1978b) found positive correlation between the antigens in *Aspergillus fumigatus* isolates from the soil of ornamental plants with the antigens present in strains of the fungus from patients with aspergillosis. However, it should be noted that all *Aspergillus fumigatus* strains may share common antigens.

Summerbell et al. (1989) investigated the soils of 5 potted plants from a Toronto hospital and isolated 16 species of potentially pathogenic fungi including *Aspergillus fumigatus*, 3 other *Aspergillus* species, and 2 *Fusarium* species. Smith et al. (1988) demonstrated that potted plants infested with *Aspergillus* species gave rise to a significant number of airborne propagules. In a limited study in a Tulsa, Oklahoma hospital, the airborne concentrations of culturable spores in patient rooms with potted plants were significantly higher than comparable rooms without plants (Malloy and Levetin, unpublished data). The results of the study support the idea that indoor plants should be considered potentially hazardous to immunosuppressed patients.

VIII. FUTURE RESEARCH NEEDS

Although fungal spores are well recognized as bioaerosols, much basic research is still needed on fungi to understand their ability to survive and proliferate in indoor environments, to elucidate factors controlling airborne dispersal, human exposure, and human disease, and to document effective methods for exposure control.

Aerobiologists need a broad database of indoor fungal spore levels for homes, occupational settings, schools, and vehicles in different parts of the country and at different seasons. This information needs to be correlated with outdoor spore concentrations, with indoor environmental conditions, with the types of ventilation systems (natural or mechanical), and with instances of respiratory disease. The sampling needs to be done with volumetric sampling instruments using standardized methods for both total and viable spores. Only then will there be sufficient evidence to fully document the health effects of respiratory exposure and establish nationwide standards for those organisms that pose serious health hazards.

Efficient sampling and analytical methods are needed for airborne mycotoxins, antigens, and other fungal metabolites, and more research is required on the health effects of respiratory exposure to these compounds. Prevalence studies of mycotoxins are also needed, as well as basic research on the chronic effects of inhalation of low levels of mycotoxins combined with effects of low levels due to ingestion. Since many mycotoxins are potent carcinogens, it is possible that inhalation of even low levels could have significant long-range effects. Mycotoxins are also known to have immunosuppressive action, which could be another important factor in chronic health effects (Autrup and Autrup, 1992).

Research is needed to determine the control measures that will most effectively minimize health risks to building occupants. Approaches that should be studied include (1) adjusting environmental conditions to inhibit growth or reduce viability, (2) reducing available substrates, (3) using environmentally safe biocides that are non-toxic to humans, (4) increasing ventilation rates to dilute aerosols, and (5) reducing exposure with personal filters.

Often, moisture control can be an effective means to prevent amplification of fungal growth. It is generally accepted that indoor relative humidity (RH) levels over 70% are optimal for active growth and that levels below 30% permit little or no growth. However, many indoor sites fall in between these levels. Studies should focus on the exact water availability or a_w requirements for fungi, especially for xerophilic and xerotolerant species. Hyphae of a number of leaf surface fungi (*Alternaria, Cladosporium, Epicoccum*, and *Aureobasidium*) are remarkably resistant to desiccation. After extended periods of drying, desiccated hyphal tips were found to resume growth within 1 h after return to moist conditions (Deacon, 1984). There are many unanswered questions on the tolerance mechanisms of these fungi and the implications of their presence as contaminants in the indoor environment.

Increased ventilation rates will dilute airborne contaminants including fungi; however, at air flow rates from conventional mechanical ventilation systems, contaminant removal is usually slow (Whitfield, 1988). Increasing air flow can cause turbulence, which results in eddies as well as pockets of static air. Also, isolated increases in levels of airborne contaminants may occur when the turbulence recycles air near sources (Hughes, 1974; Whitfield, 1988). Laminar flow systems, which are used in some areas within hospitals and microelectronic and pharmaceutical cleanrooms, provide up to 600 air changes per hour with very little turbulence, and dramatically increases contaminant removal (Whitfield, 1988). Laminar flow applications have expanded significantly in the past 20 years, and additional research is needed to determine if this technology can be adapted for control of airborne fungi in other settings.

Most laminar flow systems are coupled with HEPA filters to provide clean air to contaminated areas; however, more research is needed on the effectiveness of these filters alone, as well as other types of air cleaners in contaminated areas. While cleaners with HEPA filters and electrostatic precipitators are efficient for particle removal in chamber studies, more data is needed on the long-term effectiveness of portable air cleaners in contaminated buildings.

IX. CONCLUSION

Fungal spores are ever-present bioaerosols in the natural environment. Most spores are adapted for airborne dispersal, and those present in the atmosphere will be introduced indoors along with fresh air. Once indoors, many saprobic fungi are able to colonize countless substrates. Any time that moisture is available, spores can germinate, grow, and proliferate using organic matter in these sites. Asthma, hypersensitivity pneumonitis, and airborne epidemics of aspergillosis have all resulted from exposure to airborne spores. However, their importance in the indoor environment has generally been underestimated. Much research is still needed on many aspects of fungal biology to fully understand their ability to survive indoors and cause respiratory diseases. Threshold levels also need to be established to adequately comprehend the health risks to occupants.

REFERENCES

Abdalla, M. H., Prevalence of airborne *Aspergillus flavus* in Khartoum (Sudan) airspora with reference to dusty weather and inoculum survival in simulated summer conditions, *Mycopathologia,* 104, 137, 1988.

Aisner, J., Schimpff, S. C., Bennett, J. E., Young, V. M., and Wiernik, P. H., *Aspergillus* infections in cancer patients — association with fireproofing materials in a new hospital, *JAMA,* 235, 411, 1976.

Al-Doory, Y., *Laboratory Medical Mycology*, Lea & Febiger, Philadelphia, 1980.

Alexopoulos, C. J. and Mims, C. W., *Introductory Mycology*, 3rd ed., John Wiley & Sons, New York, 1979.

Allitt, U., Identity of airborne hyaline, one-septate ascospores and their relation to inhalant allergy, *Trans. Br. Mycol. Soc.*, 87, 147, 1986.

Anaissie, E., Kantarjian, H., Jones, P., Barlogie, B., Luna, M., Lopez-Berenstein, G., and Bodey, G. P., *Fusarium*: a newly recognized fungal pathogen in immunosuppressed patients, *Cancer,* 51, 2141, 1986.

Anderson, J. H., Allergenic airborne pollen and spores in Anchorage, Alaska, *Ann. Allergy,* 54, 390, 1985.

Arlet, G., Gluckman, E., Gerber, F., Perol, Y., and Hirsch, A., Measurement of bacterial and fungal air counts in two bone marrow transplant units, *J. Hosp. Infect.*, 13, 63, 1989.

Autrup, H. and Autrup, J. L., Human exposure to aflatoxins-biological monitoring, in *Handbook of Applied Mycology,* Vol. 5, *Mycotoxins in Ecological Systems*, Bhatnagar, D., Lillihoj, E. B., and Arora, D. K., Eds., Marcel Dekker, New York, 1992.

Aylor, D. E., The role of intermittent wind in the dispersal of fungal pathogens, *Annu. Rev. Phytopathol.*, 28, 73, 1990.

Barr, D. J. S., Evolution and kingdoms of organisms from the perspective of a mycologist, *Mycologia,* 84, 1, 1992.

Becker, W.-M., Mazur, G., Schaubschlager, W. W., and Borstel, M. D., Occupational sensitization to *Plasmopara viticola:* molecular characterization of IgE reactive antigens, *J. Allergy Clin. Immunol.*, 89, 203, 1992.

Benaim-Pinto, C., Sensitization to basidiomycetes and to *Fuligo septica* (Myxomycetae) in Venezuelan atopic patients suffering from respiratory allergy, *J. Allergy Clin. Immunol.*, 89, 282, 1992.

Blazar, B. R., Hurd, D. D., Snover, D. C., Alexander, J. W., and McGlave, P. B., Invasive *Fusarium* infections in bone marrow transplant recipients, *Am. J. Med.*, 77, 645, 1984.

Burge, H. A., Fungus allergens, *Clin. Rev. Allergy,* 3, 319, 1985.

Burge, H. A., Some comments on the aerobiology of fungus spores, *Grana*, 25, 143, 1986.

Burge, H. A., Bioaerosols: prevalence and health effects in the indoor environment, *J. Allergy Clin. Immunol.*, 86, 687, 1990.

Burge, H. A., Classification of the fungi, *Clin. Rev. Allergy,* 10, 153, 1992.

Burge, H. A., Solomon, W. R., and Muilenburg, M. S., Evaluation of indoor plantings as allergen exposure source, *J. Allergy Clin. Immunol.*, 70, 101, 1982.

Burge, H. A., Hoyer, M. E., Solomon, W. R., Simmons, E. G., and Gallup, J., Quality control factors for *Alternaria* allergens, *Mycotaxon*, 34, 55, 1989.

Bush, R. K., Aerobiology of pollen and fungal allergens, *J. Allergy Clin. Immunol.*, 84, 1120, 1989.

Butcher, B. T., O'Neil, C. E., Reed, M. A., Altman, L. C., Lopez, M., and Lehrer, S. B., Basidiomycete allergy: measurement of spore-specific IgE antibodies, *J. Allergy Clin. Immunol.,* 80, 803, 1987.

Chapman, J. A. and Williams, S., Aeroallergens of the southeast Missouri area: a report of skin test frequencies and air sampling data, *Ann. Allergy,* 52, 411, 1984.

Charnley, J., Clean air in the operating room, *Cleveland Clin. Q.,* 40, 99, 1973.

Chatigny, M. A. and Dimmick, R. L., Transport of aerosols in the intramural environment, in *Aerobiology: The Ecological Systems Approach*, Edmonds, R. L., Ed., Dowden, Hutchinson, and Ross, Stroudsburg, PA, 1979.

Cho, C. T., Vats, T. S., Lowman, J. T., Brandsberg, J. W., and Tosh, F. E., *Fusarium solani* infection during treatment for acute leukemia, *J. Pediatr*, 6, 1028, 1973.

Cosentino, S., Pisano, P. L., Fadda, M. E., and Palmas, F., Pollen and mold allergy: aerobiologic survey in the atmosphere of Cagliari, Italy (1986-1988), *Ann. Allergy,* 65, 393, 1990.

Cox, C. S., *The Aerobiological Pathway of Microorganisms*, John Wiley & Sons, New York, 1987.

Crenshaw, R., Esch, R., Portnoy, J., Upadrasha, B., Pacheco, F., and Barnes, C., Specific allergen content, enzymatic profile, and morphology of eleven strains of *Alternaria, J. Allergy Clin. Immunol.*, 89, 242, 1992.

Croft, W. A., Jarvis, B. B., and Yatawara, C. S., Airborne outbreak of trichothecene toxicosis, *Atmos. Environ.*, 20, 549, 1986.

Crook, B. and Lacey, J., Airborne allergenic microorganisms associated with mushroom cultivation, *Grana*, 30, 445, 1991.

Davis, W. E., Horner, W. E., Salvaggio, J. E., and Lehrer, S. B., Basidiospore allergens: analysis of *Coprinus quadrifidus* spore, cap, and stalk extracts, *Clin. Allergy*, 18, 261, 1988.

Deacon, J. W., *Introduction to Modern Mycology*, 2nd ed., Blackwell Scientific Publications, Oxford, 1984.

De Zubiria, A., Horner, W. E., and Lehrer, S. B., Evidence for cross-reactive allergens among basidiomycetes: immunoprint-inhibition studies, *J. Allergy Clin. Immunol.*, 86, 26, 1990.

Dhillon, M., Current status of mold immunotherapy, *Ann. Allergy,* 66, 385, 1991.

Dingle, A., Meteorological considerations in ragweed hay fever research, *Fed. Proc. Am. Soc. Exp. Biol.*, 16, 615, 1957.

Drouhet E., Overview of fungal antigens, in *Fungal Antigens: Isolation, Purification, and Detection,* Drouhet, E., Cole, G. T., de Repentigny, L., Latge, J.-P., and Dupont, B., Eds., Plenum Press, New York, 1988.

Dungby, C. I., Kozak, P. P., Gallup, J., and Galant, S. P., Aeroallergen exposure in the elementary school setting, *Ann. Allergy,* 56, 218, 1986.

Ebner, M. R., Haselwandter, K., and Frank, A., Seasonal fluctuations of airborne fungal allergens, *Mycol. Res.*, 92, 170, 1989.

Ebner, M. R., Haselwandter, K., and Frank, A., Indoor and outdoor incidence of airborne fungal allergens at low- and high-altitude alpine environments, *Mycol. Res.*, 96, 117, 1992.

Fentress, C. L., A study of variation in morphology and extracted proteins of *Epicoccum nigrum.* M.S. Thesis, University of Tulsa, OK, 1991.

Giannini, E. H., Northey, W. T., and Leathers, C. R., The allergenic significance of certain fungi rarely reported as allergens, *Ann. Allergy,* 35, 372, 1975.

Gregory, P. H., *The Microbiology of the Atmosphere*, 2nd ed., Halstead Press, New York, 1973.

Gregory, P. H. and Hirst, J. M., Possible role of basidiospores as air-borne allergens, *Nature*, 170, 414, 1952.

Griffin, D. H., *Fungal Physiology*, John Wiley & Sons, New York, 1981.

Haard, R. T. and Kramer, C. L., Periodicity of spore discharge in the hymenomycetes, *Mycologia*, 62, 1145, 1970.

Halwagy, M., Seasonal airspora at three sites in Kuwait 1977-1982, *Mycol. Res.,* 93, 208, 1989.

Hamilton, E. D., Studies on the air spora, *Acta Allergol.,* 13, 143, 1959.

Hasnain, S. M., Newhook, F. J., Wilson, J. D., and Corbin, J. B., First report of *Ganoderma* allergenicity in New Zealand, *N. Z. J. Sci.*, 27, 261, 1984.

Hasnain, S. M., Wilson, J. D., Newhook, F. J., and Segedin, B. P., Allergy to basidiospores: immunologic studies, *N. Z. Med. J.*, 98, 393, 1985.

Hawksworth, D. L., Sutton, B. C., and Ainsworth, G. C., *Ainsworth and Bisby's Dictionary of the Fungi*, 7th ed., Commonwealth Mycological Institute, Kew, England, 1983.

Hawksworth, D. L., The fungal dimension of biodiversity: magnitude, significance, and conservation, *Mycol. Res.*, 95, 641, 1991.

Herxheimer, H., Hyde, H. A., and Williams, D. A., Allergic asthma caused by basidiospores, *Lancet,* 2, 131, 1969.

Hirsch, D. J., Hirsch, S. R., and Kalbfleisch, J. H., Effect of central air conditioning and meteorologic factors on indoor spore counts, *J. Allergy Clin. Immunol.*, 62, 22, 1978.

Hirsch, S. R. and Sosman, J., A one-year survey of mold growth inside twelve homes, *Ann. Allergy,* 36, 29, 1976.

Hirst, J. M., Changes in atmospheric spore content: diurnal periodicity and the effects of weather, *Trans. Br. Mycol. Soc.,* 36, 375, 1953.

Horner, W. E., Ibanez, M. D., and Lehrer, S. B., Immunoprint analysis of *Calvatia cyathiformis* allergens. I. Reactivity with individual sera, *J. Allergy Clin. Immunol.*, 83, 784, 1989.

Hughes, D., Clean rooms — keeping the fresh air flowing, *Chem. Br.*, 10, 84, 1974.

Kendrick, B., *The Fifth Kingdom*, 2nd ed., Mycologue Publications, Newburyport, MA, 1992.

Kiehn, T. E., Nelson, P. E., Bernard, E. M., Edwards, F. F., Koziner, B., and Armstrong, D., Catheter-associated fungemia caused by *Fusarium chlamydosporum* in a patient with lymphocytic lymphoma, *Clin. Microbiol.*, 21, 501, 1985.

Kooistra, J. B., Pasch, R., and Reed, C. E., The effects of air cleaners on hay fever symptoms in air conditioned homes, *J. Allergy Clin. Immunol.*, 61, 315, 1978.

Kramer, C. L., Fungus, moss and fern spores, in *Aerobiology, The Ecological Systems Approach*, Edmonds, R. L., Ed., Dowden, Hutchinson, and Ross, Stroudsburg, PA, 1979.

Lacey, J., Aerobiology and health: the role of airborne fungal spores in respiratory disease, in *Frontiers in Mycology*, Hawksworth, D. L., Ed., C.A.B. International, Wallingford, Oxon, United Kingdom, 1991a.

Lacey, J., Aggregation of spores and its effect on aerodynamic behavior, *Grana,* 30, 437, 1991b.

Lacey, J. and Crook, B., Fungal and actinomycete spores as pollutants of the workplace and occupational allergens, *Ann. Occup. Hyg.*, 32, 515, 1988.

Larsen, L. and Gravesen, S., Seasonal variation of outdoor viable microfungi in Copenhagen, Denmark, *Grana,* 30, 467, 1991.

Latge, J.-P. and Paris, S., Allergens of *Alternaria* and *Cladosporium* in *Fungal Antigens: Isolation, Purification, and Detection,* Drouhet, E., Cole, G. T., de Repentigny, L., Latge, J.-P., and Dupont, B., Eds., Plenum Press, New York, 1988.

Leedom, J. M. and Loosli, C. G., Airborne pathogens in the indoor environment with special reference to nosocomial (hospital) infections, in *Aerobiology: The Ecological Systems Approach,* Edmonds, R. L., Ed., Dowden, Hutchinson, and Ross, Stroudsburg, PA, 1979.

Lehrer, S. B., Aukrust, L., and Salvaggio, J. E., Respiratory allergy induced by fungi, *Clin. Chest Med.,* 4, 23, 1983.

Lehrer, S. B., Lopez, M., Butcher, B. T., Olson, J., Reed, M., and Salvaggio, J. E., Basidiomycete mycelia and spore-allergen extracts: skin test reactivity in adults with symptoms of respiratory allergy, *J. Allergy Clin. Immunol.*, 78, 478, 1986.

Lentino, J. R., Rosenkranz, M. A., Michaels, J. A., Kurup, V. P., Rose, H. D., and Rytel, M. W., Nosocomial aspergillosis, *Am. J. Epidemiol.*, 116, 430, 1982.

Levetin, E., Studies on airborne basidiospores, *Aerobiologia,* 6, 177, 1990.

Levetin E., Identification and concentration of airborne basidiospores, *Grana,* 30, 123, 1991.

Levetin, E. and Buck, P., A re-evaluation of aeroallergens of northeast Oklahoma, *Proceedings of the 8th Conference of Biometeorology and Aerobiology*, Preprint Vol. 283, 1987.

Levetin, E., Shaughnessy, R., and Sublette, K., Portable indoor air cleaners: a comparison of cleaning technologies, *J. Allergy Clin. Immunol.*, 89, 257, 1992.

Lidwell, O. M., Lowbury, E. J. L., Whyte, W., Blowers, R., Stanley, S. J., and Lowe, D., Effect of ultraclean air in operating rooms on deep sepsis in the joint after total hip or knee replacement: a randomised study, *Br. Med. J.*, 285, 10, 1982.

Lopez, M., Butcher, B. T., Salvaggio, J. E., Olson, J. A., Reed, M. A., McCants, M. W., and Lehrer, S. B., Basidiomycete allergy: what is the best source of antigen?, *Int. Arch. Allergy Appl. Immunol.*, 77, 169, 1985.

Lysek, G., Physiology and ecology of rhythmic growth and sporulation in fungi, in *The Ecology and Physiology of the Fungal Mycelium*, Jennings, D. H. and Rayner, A. D. M., Eds., Cambridge University Press, Cambridge, U.K., 1984.

Macher, J. M., Huang, F.-Y., and Flores, M., A two-year study of microbiological indoor air quality in a new apartment, *Arch. Environ. Health,* 46, 25, 1991.

Mahoney, D. H., Steuber, C. P., Starling, K. A., Barrett, F. F., Goldberg, J., and Fernbach, D. J., An outbreak of aspergillosis in children with acute leukemia, *J. Pediatr.*, 95, 70, 1979.

Male, O., The significance of mycology in medicine, in *Frontiers in Mycology*, Hawksworth, D. L., Ed., C.A.B. International, Wallingford, Oxon, United Kingdom, 1991.

Mason, C. J., Atmospheric transport, in *Aerobiology: The Ecological Systems Approach,* Edmonds, R. L., Ed., Dowden, Hutchinson and Ross, Stroudsburg, PA, 1979.

McNeely, D. J., Vas, S. I., Dombros, N., and Orepoulos, D. G., *Fusarium* peritonitis: an uncommon complication of continuous ambulatory peritoneal dialysis, *Peritoneal Dial. Bull.*, 1, 94, 1981.

Misra, R. P., Studies on seasonal and diurnal variation in the occurrence of airborne spores of basidiomycetes, *Perspect. Mycol. Res.,* 1, 243, 1987.

Moore-Landecker, E., *Fundamentals of the Fungi*, 3rd ed., Prentice-Hall, Englewood Cliffs, NJ, 1990.

Moss, M. O., The mycelial habit and secondary metabolite production, in *The Ecology and Physiology of the Fungal Mycelium*, Jennings, D. H. and Rayner, A. D. M., Eds., Cambridge University Press, Cambridge, U.K., 1984.

Nagarajan, S., and Singh, D. V., Long-distance dispersion of rust pathogens, *Annu. Rev. Phytopathol.,* 28, 139, 1990.

Nelson, H. S., Hirsch, S. R., Ohman, J. L., Platts-Mills, T. A. E., Reed, C. E., and Solomon, W. R., Recommendations for the use of residential air-cleaning devices in the treatment of allergic respiratory diseases, *J. Allergy Clin. Immunol.,* 82, 661, 1988.

O'Neil, C. E., Hughes, J. M., Butcher, B. T., Salvaggio, J. E., and Lehrer, S. B., Basidiospore extracts: evidence for common antigenic/allergenic determinants, *Int. Arch. Allergy Appl. Immunol.*, 85, 161, 1988.

Pan, P. M., Burge, H. A., Su, H. J., and Spengler, J. D., Central vs. room air conditioning for reducing exposure to airborne fungus spores, *J. Allergy Clin. Immunol.*, 89, 258, 1992.

Pasenen, A.-L., Kalliokoski, P., Pasenen, P., Salmi, T., and Tossavainen, A., Fungi carried from farmers' work into farm homes, *J. Am. Ind. Hyg. Assoc.*, 50, 631, 1989.

Powell, M. J., Poisonous and medicinal fungi, in *Poisonous and Medicinal Plants*, Blackwell, W. H., Ed., Prentice-Hall, Englewood Cliffs, NJ, 1990.

Powell, M. J., Looking at mycology with a Janus face: A glimpse at chytridiomycetes active in the environment, *Mycologia,* 85, 1, 1993.

Pritchard, H. and Bradt, P., *Biology of Nonvascular Plants*, C. V. Mosby, St. Louis, MO, 1984.

Reijula, K. E., Kurup, V. P., and Fink, J. N., Ultrastructural demonstration of specific IgG and IgE antibodies binding to *Aspergillus fumigatus* from patients with aspergillosis, *J. Allergy Clin. Immunol.,* 87, 683, 1991.

Resiman, R. E., Mauriello, P. M., Davis, G. B., Georgitis, J. W., and DeMasi, J. M., A double-blind study of the effectiveness of a high-efficiency particulate air (HEPA) filter in the treatment of patients with perennial allergic rhinitis and asthma, *J. Allergy Clin. Immunol.*, 85, 1050, 1990.

Richardson, S. E., Bannatyne, R. M., Summerbell, R. C., Milliken, J., Gold, R., and Weitzman, S. S., Disseminated *Fusarium* infection in the immunocompromised host, *Rev. Infect. Dis.,* 10, 1171, 1988.

Rose, H. D., Mechanical control of hospital ventilation and *Aspergillus* infections, *Am. Rev. Respir. Dis.,* 105, 306, 1972.

Rose, H. D. and Hirsch, S. R., Filtering hospital air decreases *Aspergillus* spore counts, *Am. Rev. Respir. Dis.,* 119, 511, 1979.

Rubulis, J., Airborne fungal spores in Stockholm and Eskilstuna, central Sweden, *Nordic Aerobiol.*, 85, 1984.

Ruiz-Herrara, J., *Fungal Cell Wall: Structure, Synthesis, and Assembly*, CRC Press, Boca Raton, FL, 1991.

Salvaggio, J., Seabury, J., and Schoenhardt, E. A., New Orleans asthma. V. Relationship between Charity Hospital asthma admissions rates, semiquantitative pollen and fungal spore counts, and total particulate aerometric sampling data, *J. Allergy Clin. Immunol.,* 48, 96, 1971.

Salvaggio, J. and Aukrust, L., Mold-induced asthma, *J. Allergy Clin. Immunol.*, 68, 327, 1981.

Samson, R. A., Occurrence of moulds in modern living and working environments, *Eur. J. Epidemiol.,* 1, 54, 1985.

Santilli, J., Rockwell, W. J., and Collins, R. P., The significance of the spores of the basidiomycetes (mushrooms and their allies) in bronchial asthma and allergic rhinitis, *Ann. Allergy*, 55, 469, 1985.

Scherr, M. S. and Peck, L. W., The effects of high efficiency air filtration system on nighttime asthma attacks, *W. Va. Med. J.,* 73, 144, 1977.

Sharma, R. P. and Salunke, D. K., Introduction to mycotoxins, in *Mycotoxins and Phytoalexins*, Sharma, R. P. and Salunkhe, D. K., Eds., CRC Press, Boca Raton, FL, 1991.

Smith, V., Streifel, A., Rhame, F. S., and Juni, B., Potted plant fungal spore shedding, *Abstr. Annu. Meet. Am. Soc. Microbiol.,* 414, 1988.

Solomon, W. R., Assessing fungus prevalence in domestic interiors, *J. Allergy Clin. Immunol.*, 56, 235, 1975.

Solomon, W. R., A volumetric study of winter fungus prevalence in the air of midwestern homes, *J. Allergy Clin. Immunol.*, 57, 46, 1976.

Solomon, W. R., Common pollen and fungus allergens, in *Allergic Diseases of Infancy, Childhood and Adolescence*, Bierman, C. W. and Pearlman, D. S., Eds., W. B. Saunders, Philadelphia, 1980.

Staib, F., Tompak, B., Thiel, D., and Blisse, A., *Aspergillus fumigatus* and *Aspergillus niger* in two potted ornamental plants, cactus (*Epiphyllum truncatum*) and *clivia (Clivia miniata*). Biological and epidemiological aspects, *Mycopathologia,* 66, 27, 1978a.

Staib, F., Folkens, U., Tompak, B., Abel, Th., and Thiel, D., A comparative study of antigens of *Aspergillus fumigatus* isolates from patients and soil of ornamental plants in the immunodiffusion test, *Zentralbl. Bakteriol. Parasitenk. Infektionskr. Hyg, Abt.* I *Orig. Reihe A*, 242, 93, 1978b.

Streifel, A. J., Vesley, D., Rhame, F. S., and Murray, B., Control of airborne fungal spores in a university hospital, *Environ. Int.,* 15, 221, 1989.

Su, H. J., Rotnitzky, A., Burge, H. A., and Spengler, J., Examination of fungi in domestic interiors by using factor analysis: correlations and associations with home factors, *Appl. Environ. Microbiol.*, 58, 181, 1992.

Summerbell, R. C., Krajden, S., and Kane, J., Potted plants in hospitals as reservoirs of pathogenic fungi, *Mycopathologia,* 106, 13, 1989.

Tan, T. K., Teo, T. S., Tan, H., Lee, B. W., and Chong, A., Variations in tropical airspora in Singapore, *Mycol. Res.*, 96, 221, 1992.

Tarlo, S. M., Bell, B., Srinivasan, J., Dolovich, J., and Hargreave, F. E., Human sensitization to *Ganoderma* antigen, *J. Allergy Clin. Immunol.*, 64, 43, 1979.

Tunon de Lara, J.-M., Tessier, J.-F., Lafond-Grellety, J., Domblides, P., Mary, J., Faugere, J.-G., and Taytard, A., Indoor moulds in asthmatic patients homes, *Aerobiologia,* 6, 98, 1990.

van Reenen-Hoekstra, E. S., Samson, R. A., Verhoeff, A. P., van Wijnen. J. H., and Brunekreef, B., Detection and identification of moulds in dutch houses and non-industrial working environments, *Grana,* 30, 418, 1991.

Vaughan, W. T. and Cooley, L. E., Air conditioning as a means of removing pollen and other particulate matter and of relieving pollinosis, *J. Allergy*, 5, 37, 1933.

Waegemaekers, M., van Wageningen, N., Brunekreef, B., and Boleij, J. S. M., Respiratory symptoms in damp houses, *Allergy,* 44, 192, 1989.

Wainright, P. O., Hinkle, G., Sogin, M. L., and Stickel, S. K., Monophyletic origins of the metazoa: An evolutionary link with fungi, *Science,* 260, 340, 1993.

Weitzman, I., Saprophytic molds as agents of cutaneous and subcutaneous infection in the immunocompromised host, *Arch. Dermatol.*, 122, 1161, 1986.

Whitfield, W., The clean room: technical aspects of its evolution, *9th ICCCS Proceedings,* 301, 1988.

Wickman, M., Gravesen, S., Nordvall, S. L., Pershagen, G., and Sundell, J., Indoor viable dust-bound microfungi in relation to residential characteristics, living habits, and symptoms in atopic and control children, *J. Allergy Clin. Immunol.*, 89, 752, 1992.

Wogan, G. N., Introduction, in *Mycotoxins in Food Stuffs*, Wogan, G. N., Ed., MIT Press, Cambridge, MA, 1965.

Young, R. C., Bennett, J. E., Vogel, L. C., Carbone, P. P., and DeVita, V. T., Aspergillosis: the spectrum of the disease in 98 patients, *Medicine*, 49, 147, 1970.

6 THE POTENTIAL IMPACT ON HUMAN HEALTH FROM FREE-LIVING AMOEBAE IN THE INDOOR ENVIRONMENT

R. L. Tyndall and A. A. Vass

CONTENTS

I. INTRODUCTION

Pathogenic protozoa have always been one of the most serious microbial threats to human health. Most protozoan diseases occur primarily in tropical and Third World countries. Examples of parasitic protozoa include *Entamoeba, Giardia, Trypanosoma, Plasmodia,* and *Leishmania*. While most populations are asymptomatic, estimates of human infection by *Entamoeba histolytica* range as high as 600 million worldwide (Elsdon-Dew, 1968). Children are the most likely victims of the severe diarrhea resulting from *E. histolytica* infections. Similarly, effects of *Giardia* infection, such as diarrhea, abdominal pain, and weight loss, are also more likely expressed in children. *Giardia* is widely

0-87371-724-4/95/$0.00+$.50

distributed and appears in at least 10% of the U.S. population. *Trypanosomas*, which are carried by the tsetse fly and cause African sleeping sickness, are responsible for millions of infections. Additionally, millions of humans are affected by species of *Plasmodium,* which are spore-forming protozoa with complex life cycles and are the causative agents of malaria. Symptoms similar to those of malaria also occur during infection with *Leishmania*. Like malaria, leishmaniasis affects millions of people, particularly in Third World nations.

While protozoa globally infect millions and are a major public health problem, their impact on health in the indoor environment is generally restricted to effects of the free-living amoebae. These amoebae can act directly as pathogens and allergens, or they can interact with bacteria and amplify bacterial pathogens. This is particularly true with free-living amoebae such as *Acanthamoeba, Hartmannella*, and *Naegleria*.

II. THE NATURE AND CLASSIFICATION OF PROTOZOA

Free-living amoebae and other protozoa are eukaryotes and, as such, have true nuclei and divide by binary fission, which is preceded by DNA synthesis, replication, and mitosis. Free-living amoebae also have mitochondria, ribosomes, Golgi structures, endoplasmic reticulum, and microtubules. These organelles are not found in cells of prokaryotic organisms such as bacteria.

Classification of free-living amoebae based solely on morphology is difficult because of their amorphous nature. Nevertheless, extensive studies to classify these amoebae by their appearance have been undertaken. Page (1976, 1987) divided the order Amoebida into a variety of families, with the genus *Naegleria* assigned to the family Vahlkampfiidae; *Hartmannella* to Hartmannellidae; and *Acanthamoeba* to Acanthamoebidae. At present, this classification is the most widely accepted.

A. *Acanthamoeba*

Acanthamoeba trophozoites (i.e., the ameboid form) are characterized by, and named for, the spiked projections (acanthapodiae) that characterize their cell surface. The trophozoites of *Acanthamoeba* have a sterol-containing cell membrane (Pfiffner, 1991) and a granular endoplasm containing typical eukaryotic organelles (Rondanelli, 1987b). *Acanthamoeba* DNA is replicated by mitotic division similar to that in mammalian cells. Initially, the centrosphere divides and the spindle is formed between the centrospheres. The chromosomes cluster at the center of the spindle and replicate. The two nuclei then separate and the cell divides.

Acanthamoebae can also form cysts during times of environmental stress. The cyst forms by rounding of the trophozoite, shrinking of the nucleus, and formation of a double-layered cyst wall. When environmental conditions improve, the trophozoite reemerges from the cyst through a pore in the cyst wall.

B. *Hartmannella*

Unlike *Acanthamoeba, Hartmannella* is a member of the limax, or "slug-like", group of amoebae. Movement is by projection of a single pseudopod. Acanthapodia are absent. *Hartmannella* contains the same organelles as *Acanthamoeba,* and cell division is also preceded by true mitosis. Cyst formation also occurs. *Hartmannella* trophozoites and cysts are smaller than their *Acanthamoeba* counterparts.

C. *Naegleria*

Like *Hartmannella, Naegleria* is also a limax amoeba and moves by extension of a single large pseudopodium. *Naegleria*'s various cytoplasmic organelles resemble those seen in other free-living amoebae, as does the large "bull's-eye" nucleolus. An unusual feature of pathogenic *Naegleria* is the suckerlike structure, the amoebaestome (John et al., 1985), which is involved in attachment to and phagocytosis of host cells. Mitosis in *Naegleria* is not well defined. In the early stages of mitosis the nucleolus separates into polar masses, with concomitant formation of a mitotic spindle. Subsequently, the chromatin divides and congregates at the spindle, followed by the appearance of an interzonal body between the polar masses. Nuclear division and cell division then occur. An unusual feature of *Naegleria* is its ability to form two to four flagella under conditions of hypotonicity. The flagellar form is transitory, and it can either revert to the trophozoite or, like *Acanthamoeba* and *Hartmannella*, can encyst. The double-walled cysts, as with other free-living amoebae, have pores through which the trophozoite can reemerge on excystment.

III. DISEASES CAUSED BY FREE-LIVING AMOEBAE

Some species of free-living amoebae cause infections. *Acanthamoeba* is the causative agent of two distinct syndromes. The introduction of *Acanthamoeba* into wounds or the bloodstream, particularly in immunocompromised individuals, may result in the hematogenous spread of these microbes from the primary site of infection to the brain, resulting in fatal chronic granulomatous meningoencephalitis. Epidermal, mucosal, or pulmonary granulomas may also result from introduction of *Acanthamoeba* into traumatized sites.

Severe eye infections (i.e., keratitis) may also occur when Acanthamoebae are introduced into traumatized eyes. Several species of *Acanthamoeba* have been implicated in such eye infections, including *A. castellani* and *A. polyphaga* (Pearl et al., 1981; Visvesvara et al., 1975). The relative ineffectiveness of antibiotic therapy in *Acanthamoeba* keratitis coupled with the organism's ability to encyst in soft tissues may result in a continuing viable presence, which may necessitate surgical removal of the infected tissues. *Acanthamoebae* are ubiquitous in soil and water, which increases the chances of their introduction into eye and other tissues from injuries and trauma occurring during outdoor activities.

In the last 15 years the cumulative number of eye infections from the contamination of contact lenses with *Acanthamoeba* has increased from 2 to about 200 cases (Wilhelmus, 1991). Studies of the home environment demonstrated the presence of acanthamoebae in some of the faucets that were used to wash contact lenses. Since therapies for controlling amoebic keratitis are generally ineffective, contact lens users should clean and store their lenses as hygienically as possible. Analogous to the contamination of contact lenses in the home environment is the recent observation that eye-wash stations in the industrial environment may also harbor acanthamoebae (Tyndall et al., 1987; Paszko-Kolva et al., 1991). Since such devices would only be used to flush traumatized eyes, the presence of acanthamoebae, whose proclivity for infection often depends on trauma, is contraindicated.

Pathogenic *Naegleria*, that is, *N. fowleri*, is the etiologic agent of fatal primary amoebic meningoencephalitis (PAME). Over 50 cases of PAME have occurred in the United States during the last 20 years. In such cases *N. fowleri* gains entry into the host via the nasal mucosa. A host's participation in swimming or other sports in water laden with the pathogen can result in such an infection. In a susceptible individual, *Naegleria* traverses the nasal mucosa, follows the olfactory nerve through the cribriform plate, and then enters the cerebrum. The resulting infection proceeds rapidly, causing death within 2 weeks after exposure.

Hartmannella is not obviously infectious for humans, but some species may be infectious for invertebrates (Noble and Noble, 1976).

IV. ECOLOGY OF FREE-LIVING AMOEBAE

As previously indicated, free-living amoebae are abundant in soil and water and are widely distributed in the environment. In addition, they have been isolated from a variety of wild and domestic animals (Rondanelli, 1987a). The most well-documented environmental factor that enhances the emergence of pathogenic amoebae is thermal addition. The ability of pathogenic acanthamoebae and *Naegleria* to grow at higher temperatures than many nonpathogenic species has been demonstrated in laboratory studies (Griffin, 1972). This has proven to be a useful tool when isolating pathogenic free-living amoebae from environmental samples. Various studies have shown a clear association between the presence of pathogenic *Naegleria* and warm water (Tyndall et al., 1984). In one study, the distribution of thermophilic and pathogenic *Naegleria* spp. in thermal discharges of electric power plants in the United States relative to ambient source waters was documented. Analysis of over 300 samples from 17 cooling systems showed amplification of thermophilic amoebae, the presence of thermophilic *Naegleria* spp., and, particularly, of *N. fowleri* in heated versus unheated systems. These differences were all significant at the $p < 0.05$ level by chi-square analysis (Tyndall and Ironside, 1990). Similar data were reported for two northern impoundments sampled in 1983 (Tyndall, 1984). One site had received thermal additions from an electric power plant since 1970, and water temperatures at the sampling locations ranged from 34.8 to 41.5°C. The second site was an unheated control lake in which the water never exceeded 33°C. Statistical comparisons between the heated and unheated sites showed significant differences in the presence of thermophilic amoebae, thermophilic *Naegleria* spp., and *N. fowleri* at the $p < 0.01$ level.

A dramatic impact of thermal additions on thermophilic amoebae, including *Naegleria,* was also seen in a newly created cooling lake at the Savannah River Plant (SRP) in Aiken, South Carolina, where intermittent thermal additions resulted in corresponding blooms of thermophilic amoebae and *Naegleria*. Densities of thermophilic amoebae and *Naegleria* in water were significantly correlated with temperature at a statistical level of $p < 0.00001$. Densities of thermophilic amoebae, *Naegleria,* and pathogenic *Naegleria* in the sediments were also significantly correlated with temperature at $p < 0.00001$, $p < 0.00001$, and $p < 0.0001$, respectively (Tyndall et al., 1989).

Aerosol samples taken proximal to the water of the SRP cooling lake with high concentrations of *Naegleria* were generally negative for culturable amoebae. High-volume Litton, impinger, and Andersen samplers were used to collect samples with volumes of air sampled ranging from 720 to 24,000 l. Only 1 of 36 samples was positive. The one positive sample produced an organism that appeared to be *Hartmannella,* based on its trophozoite and cyst morphology (Tyndall et al., 1989). The low aerosolization potential of culturable free-living amoebae was also illustrated in recent studies of home humidifiers deliberately seeded with acanthamoebae and in studies of humidifier discharge from homes in which humidifier reservoirs contained various free-living amoebae (Tyndall et al., 1985). In spite of the prevalence of amoebae in the humidifier reservoirs, few culturable amoebae were recovered from effluents (aerosols) from either cool mist or ultrasonic humidifiers. Conversely, the same humidifiers readily aerosolized culturable bacteria (Tyndall et al., 1985). Likewise, in a study of bioaerosols at a sewage plant, neither *Naegleria* nor acanthamoebae was recovered in air samples taken proximal to the agitated sludge that contained these amoebae (Seyfried, 1990).

While aerosolization of living amoebae per se apparently poses a minimal threat, aerosolization of amoebic antigens may be harmful. In 1976, an outbreak of fever in workers in an industrial setting prompted an investigation into the possible involvement of *Naegleria*. Air sampling was conducted using the Andersen impaction sampler fitted with non-nutrient agar plates spread with a lawn of *Escherichia coli*. After many days of incubation, amoebic outgrowth was observed on some plates. Morphologic analysis indicated that the amoebae were *Naegleria* (Edwards, 1980). The amoebae were then isolated and extracts were prepared. Serologic tests were used to show that the isolates were indeed *Naegleria* and that they may have contributed to worker illness. The source of the amoebae was then traced to the humidification system, where successful remedial action was undertaken.

V. BACTERIAL/AMEOBIC INTERACTIONS

Even though some free-living amoebae can be pathogenic and aerosolization of amoebic antigens may be harmful and part of the "humidifier fever" syndrome, the interaction of free-living amoebae and bacteria may create a greater potential impact on human health.

There are many examples of biological symbiosis (e.g., bacterial consortia facilitating digestion of cellulose in the gut of termites). In protozoa, photosynthetic symbiosis results from the interaction between *Chlorella* and *Paramecium*. *Paramecium* can also interact symbiotically with Gram negative bacteria.

A. Overview of Ameobic/Microbial Interactions

A variety of symbiotic relationships between free-living amoebae and microorganisms has also been described. In a series of studies, Schuster and Dunnebacke showed the presence of both unidentified viruslike and bacterialike particles in *Naegleria* (Schuster and Dunnebacke, 1974a, 1974b). The viruslike particles could be transmitted to "uninfected" amoebae, after which both the newly infected amoebae and the originally infected *Naegleria* could not encyst or flagellate. Hall and Voelz (1985) reported the presence of endosymbionts in Acanthamoebae by demonstrating the presence of intercytoplasmic bacteria-shaped rods. These rods were detected by both histologic and microscopic analyses. The bacterial cell walls were typical of Gram negative bacteria. In addition, the bacteria were retained during encystment and excystment, and their presence was not inhibited by exposure of the amoebae to a wide spectrum of antibiotics. Despite extensive efforts, the endosymbionts could not be grown outside of the amoebae. A most interesting and unusual example of ameobic/bacterial interaction has been described by Jeon and co-workers (Jeon, 1987) in which extensive cocultivation of the microbes resulted in the establishment of an endosymbiotic relationship in which the amoebae could no longer survive without their bacterial symbionts. Nuclei of infected amoebae transplanted into the enucleated cytoplasm of uninfected amoebae produced nonviable amoebae whose viability was restored after infection with the endosymbionts.

Extremely complex ameobic/bacterial consortia have been isolated from a contaminated aquifer. Water samples were obtained from three wells that were used to monitor the aquifer at a waste disposal site (Tyndall et al., 1991). Water from all three test wells yielded free-living amoebae on mineral salts agar plates spread with *E. coli* as a food source. That the amoebae harbored methanotrophic bacteria was demonstrated when bacterial growth occurred in a 20% methane-in-air atmosphere along the area of amoebic migration. Numerous heterotrophic bacteria were also isolated. Bacterial and amoebic growth subsequently occurred when the amoebic populations were transferred to fresh mineral salts agar plates without *E. coli* in a methane-in-air atmosphere. The consortia have been maintained in this fashion for over 5 years. Heterotrophic and methanotrophic bacteria and amoebae continually coexist in the consortia. Microscopic and enzymatic analyses showed that the resultant populations were a mixture of *Hartmannella* and at least 23 different bacterial isolates, including, but not limited to, *Pseudomonas, Alcaligenes, Bacillus, Cytophaga,* and *Hyphomicrobium* (Tyndall et al., 1991).

The nature of these stable ameobic/bacterial associations is not completely understood, but they would appear to be in part symbiotic. This is indicated by the continuing presence of both amoebae and heterotrophs on a minimal salt medium in a methane-in-air atmosphere. Neither the heterotrophic bacteria nor the free-living amoebae alone can grow and persist under these conditions. The most likely explanation for the stability of the amoebic/bacterial consortia on mineral salts agar in a methane atmosphere is the growth of the methanotrophic bacterium (*Hyphomicrobium),* which obtains energy by converting CH_4 to CO_2. The *Hyphomicrobium* then provides the growth factors that are required by the other bacteria in the consortia (Tyndall et al., 1991). The possible pathogenicity of the bacteria and amoebae in the consortium has not yet been evaluated.

B. Pathogens in Protozoan/Bacterial Consortia

The foregoing examples illustrate the complexity of protozoan/bacterial consortia. The role and potential impact on human health in the indoor environment, however, is best illustrated by the involvement of known bacterial pathogens in such consortia. Thus far, these pathogens include *Mycobacterium, Listeria,* and *Legionella* species.

Jadin observed that mycobacteria can be sequestered within free-living amoebae. Moreover, viable mycobacteria could survive within the amoebic cyst and be released after extended periods upon excystation of the amoebae. Free-living amoebae isolated from the nasal mucosa of leprosy patients harbored *Mycobacterium leprae* (Rondanelli, 1987a). More recently, Panikov et al. (1992) have described the growth of *Listeria monocytogenes* within *Tetrahymena.* The recent emergence of antibiotic-resistant strains of *Mycobacterium tuberculosis* has stimulated interest in possible sequestering of this organism by free-living amoebae. Undoubtedly, however, the most extensive studies of the interaction between bacterial pathogens and amoebae are those dealing with *Legionella* (see Chapter 3).

The interaction of *Legionella* and free-living amoebae was first described by Rowbotham (1980). He described the deleterious effects on amoebae that sequester *Legionella,* suggesting growth of *Legionella* within the amoebae. Tyndall and Dominque (1982) subsequently found that the interaction of *Legionella* with *Naegleria* and *Acanthamoeba* resulted in a relationship in which the amoebae did indeed support the growth of *Legionella* under conditions where the *Legionella* could not replicate in the absence of the amoebae. These studies showed that *Legionella* concentrations as high as 10^{10}/ml could be supported by the amoebae. That amoebae could greatly amplify concentrations of *Legionella* was confirmed and expanded by a variety of subsequent studies (see Chapter 3). Fields et al. (1984) showed that protozoa other than free-living amoebae (e.g., *Tetrahymena)* could also support the replication of amoebae.

Recent studies involving the interaction of *Legionella* and amoebae have tried to relate laboratory studies to the clinical or home environment. The best example of such a relationship is nosocomial Legionnaires' Disease and its possible dependence on *Hartmannella*-amplified *Legionella* populations. These studies were initiated when faucets and shower heads fed by hot water were identified as sources of nosocomial legionellosis (or *Legionella* per se) in the home environment. In a study by Alary and Joly (1991) a variety of heating and plumbing characteristics were positively correlated with the presence of culturable *Legionella* in home hot water tanks. While none of the gas-fueled water heaters yielded *Legionella*, *Legionella* was isolated from 39% of the electric water heaters. This difference between gas and electric hot water heaters was attributed to the stratification of water temperatures in the electric heaters. Other factors related to the presence of *Legionella* included houses located in older districts of the test cities and the age of the hot water heater.

Recognizing that free-living amoebae can be isolated from potable water and that amoebae containing *Legionella* can be found in environmental waters, Centers for Disease Control investigators have tried to evaluate the importance of amoebae in maintaining *Legionella* populations in hot water heaters. Fields et al. (1989) studied the ability of *Legionella* to multiply in potable water samples obtained from hospitals with a history of nosocomial legionellosis. *Legionella* multiplied in 3 of 14 water samples on incubation at 35 or 45°C. All three samples were from hot water tanks where both *L. pneumophila* and free-living amoebae were isolated. However, if the amoebae population was removed from the water by filtration, *Legionella* would not grow when it was reinoculated into the amoebae-free water. Subsequent studies showed that amoebae populations isolated from six hot water tanks could support the growth of *Legionella*. These studies implicated free-living amoebae and, in particular, *Hartmannella* in the growth or continuing presence of *Legionella* in hot water tanks from which *Legionella* could be subsequently aerosolized by faucets or shower heads (Fields et al., 1989).

Breiman et al. (1990) investigated the associations between protozoa and *Legionella* in hot water systems of hospitals with a ongoing nosocomial legionellosis compared to hospitals without such infections. *Hartmannella* was detected in 71% of water samples from hospitals with ongoing legionellosis, and only 15% of water samples from hospitals without ongoing legionellosis. Of those samples yielding *L. pneumophila* serogroup I, the correlation between the presence of *L. pneumophila* and *Hartmannella* was highly significant at

$p < 0.001$. In the same study, Breiman et al. (1990) reconfirmed an association between showering with *Legionella*-infected water and contracting legionellosis and demonstrated the difficulty of ridding hot water systems of *Hartmannella*. They concluded that control of *Legionella* populations and their aerosolization may depend on controlling the amoeba population in the potable water system.

The presence of both *Legionella* and free-living amoebae has been documented in home humidifiers (Tyndall et al., 1985). As previously mentioned, amoebae in the humidifier reservoirs were not readily aerosolized, at least in a culturable state. However, bacteria and bacterial endotoxins were readily aerosolized from both cool mist and ultrasonic humidifiers. Amoebae were detected in approximately 50% of the humidifier reservoirs. *Legionella* were isolated from 2 of the 25 units tested and were presumably aerosolized, since the isolates were from cool mist humidifiers, which readily produce culturable bacterial aerosols.

C. Rescue and Recombination

While analyzing the amoebae populations from the humidifiers in order to characterize the associated bacterial populations, a peculiar "blebbing" of some of the amoebae-derived bacterial colonies was observed. Continued passage of isolated, single blebbing colonies rarely resulted in several distinct colonies. The blebbing of these amoebae-derived bacterial colonies was different from the traditional satelliting sometimes seen in mixed bacterial cultures where one bacterial colony is dependent on products from a different bacterial colony. The blebbing cultures were inoculated into axenized *Acanthamoeba* and *Hartmannella* cultures in the hopes that some of the colony types in the mixture might be eliminated by the amoebae, thereby purifying the mixture. Surprisingly, more, not fewer, colony types were often isolated from the inoculated amoebae. Consequently, environmentally and clinically derived *Legionella* cultures, purified by single-colony transfers, were inoculated into axenic *Acanthamoeba* and *Hartmannella* cultures. After 5 days the cocultures were streaked onto a variety of antibiotic-free media (Tyndall et al., 1993).

As expected, the inoculated strain of *Legionella* was the predominating bacterium recovered from all infected amoebae. However, in each of eight experiments, a small subpopulation of colonies not indicative of *Legionella* also appeared on the streak line of the test agar plates. These bacteria could grow on media that does not usually support *Legionella*. While initially reactive with *Legionella* antibody, the bacteria lost their reactivity on subsequent subculture. These isolates ranged in pigmentation from white to pink and morphologically from Gram negative rods to Gram positive cocci. This result occurred when either clinically or environmentally derived *Legionella* was used. In 13 separate experiments involving five different species of *Acanthamoeba*, bacteria other than *Legionella* were isolated after *Legionella* infection. Likewise, with *Pseudomonas* as the inoculum, *Acanthamoeba* yielded bacteria other than *Pseudomonas* in 11 of 15 tests. Conversely, in two tests where *Hartmannella vermiformis* was inoculated with *Legionella* the test mix yielded only *Legionella*. Similarly, in four experiments when *Hartmannella* was inoculated with *Pseudomonas*, only *Pseudomonas* was reisolated from the amoebae. Subsequent experiments were then conducted with Costar™* plates in which the amoebae and bacterial populations could be combined or separated by a 0.1-μm membrane. These experiments showed that an intimate association between the amoebae and the bacteria was necessary to produce the newly derived bacterial cultures (Tyndall et al., 1993).

A number of scenarios may explain these different observations. Since nonculturable bacterialike and viruslike particles have been seen in *Acanthamoeba* by electron microscopy, infection with *Acanthamoeba* by one bacterial species may rescue these nonviable bacterial endosymbionts from the amoebae. Another possibility is that genetic recombination between the inoculum and endosymbionts, endosymbiontic fragments, or viral fragments may result in bacteria that were not originally present. Alternatively, a combination of amoebic and bacterial DNA may result in the formation of new bacteria. A nonculturable bacterial contaminant could exist in conjunction with the bacterial inoculum, and this contaminant could be subsequently rescued by the amoebae. However, this is unlikely because the bacterial inocula were exhaustively tested on a variety of media as well as microscopically analyzed for the presence of contaminants (Tyndall et al., 1993). Whatever the mechanisms, the results to date indicate that some protozoan populations, such as *Acanthamoeba*, may provide a mechanism for rapid genetic rescue or recombination resulting in newly derived microbes (Tyndall et al., 1993).

The instability of amoeba-derived bacterial populations was originally observed from home humidifiers and indicated that rescue or recombination of genetic information in bacterial-infected amoebae was not just a laboratory curiosity. Recent studies by Rowbotham (1993) reported detection of *Legionella*-like bacteria in *Acanthamoeba* and *Hartmannella* that could not be cultured on any laboratory medium. These organisms were recovered from water tanks, whirlpool baths, showers, and industrial cooling equipment, all reservoirs from which bacteria are readily aerosolized, and human exposure is likely. Therefore, it is imperative to delineate the pathogenicity or toxicity of these newly derived bacteria and their products. Since free-living amoebae can harbor and or amplify known bacterial pathogens such as *Legionella, Mycobacterium* and *Listeria*, it is imperative that other as yet unidentified bacteria that are harbored, amplified, and or derived from such amoebae be isolated and their possible role as human pathogens be determined. Only after protozoan interaction with potential human bacterial and viral pathogens is fully understood can the impact of protozoa on human health in the indoor environment be fully assessed.

* Registered trademark of Costar Scientific, One Alewife Center, Cambridge, MA 02140.

REFERENCES

Alary, M., Joly, J. R., Risk factors for contamination of domestic hot water systems by *Legionellae*, *Appl. Environ. Microbiol.* 57, 2360, 1991.

Breiman, R. F., Fields, B. S., Sanden, G. N., Volmer, L., Meier, A., Spika, J. S., Association of shower use with Legionnaire's disease — possible role of amoebae, *JAMA* 263, 2924, 1990.

Edwards, J. H., Microbial and immunological investigations and remedial action after an outbreak of humidifier fever, *Br. J. Ind. Med.* 37, 55, 1980.

Elsdon-Dew, R., The epidemiology of amoebiasis, *Adv. Parasitol.* 6, 1, 1968.

Fields, B. S., Shotts, E. B., Jr., Feeley, J. C., Gorman, G. W., Martin, W. T., Proliferation of *Legionella pneumophila* as an intracellular parasite of the ciliated protozoan *Tetrahymena pyriformis*, pp. 327-328. *Legionella, Proceedings of the 2nd International Symposium*, Editors, Thornsberry, C., Balows, A., Feeley, J. C., Jakubowski, W., American Society for Microbiology, Washington, D.C., 1984.

Fields, B. S., Sanden, G. N., Barbaree, J. M., Morrill, W. E., Wadowsky, R. M., White, E. H., Feeley, J. C., Intracellular multiplication of *Legionella pneumophila* in amoebae isolated from hospital hot water tanks, *Curr. Microbiol.* 18, 131, 1989.

Griffin, J. L., Temperature tolerance of pathogenic and nonpathogenic free-living amoebas, *Science* 178, 869, 1972.

Hall, J., Voelz, H., Bacterial endosymbionts of *Acanthamoeba* sp., *J. Parasitol.* 71(1), 89, 1985.

Jeon, K. W., Change of cellular "pathogens" into required cell components, *Ann. N.Y. Acad. Sci.* 503, 359, 1987.

John, D. T., Cole, T. B., Bruner, R. A., Amoebostomes of *Naegleria fowleri*, *J. Protozool.* 32, 12, 1985.

Noble, E. R., Noble, G. A., *Parasitology: The Parasitology of Animal Parasites*, Lea & Febiger, Philadelphia, p. 56, 1976.

Page, F. C., *An Illustrated Key to Freshwater and Soil Amoebae.* Freshwater Biological Association Scientific Publication 34, 1, 1976.

Page, F. C., The classification of "naked" amoebae (phylum Rhizopoda), *Arch. Protistenkd.* 133, 199, 1987.

Panikov, N. S., Merkurov, A. E., Tartakovskii, I. S., Kinetic studies of *Legionella* interactions with protozoa, pp. 153–156. *Legionella: Current Status and Emerging Perspectives*, Editors, Barbaree, J. M., Breiman, R. F., Dufour, A. P., American Society for Microbiology, Washington, DC, 1993.

Paszko-Kolva, C., Hiroyuki, Y., Shahamat, M., Sawyer, T. K., Morris, G., Colwell, R. R., Isolation of amoeba and *Pseudomonas* and *Legionella* spp. from eyewash stations, *Appl. Environ. Microbiol.* 57, 163, 1991.

Pearl, M., Willaert, E., Juechter, K. B., Stevens, A. R., A case of keratitis due to *Acanthamoeba* in New York, New York, and features of 10 cases, *J. Infect. Dis.* 143(5), 662, 1981.

Pfiffner, S. M., Lipid characterization and ecological studies on free-living amoebae and a *Legionella*-amoebic association, Florida State University College of Arts and Sciences, Ph.D. thesis, Tallahassee, FL, 1991.

Rondanelli, E. G., Ed., *Infections Diseases Color Atlas Monographs Amphizoic Amoebae Human Pathology*, Piccin Nuova Libraria, Italy, Jadin, J. B., Chapter 1 — History, 1987a.

Rondanelli, E. G., Ed., *Infectious Diseases Color Atlas Monographs Amphizoic Amoebae Human Pathology,* Piccin Nuova Libraria, Italy, Rondanelli, E. G., Carosi, G., Lanzarini, P., Filice, G., Chapter 4 — Ultrastructure of *Acanthamoebae-Naegleria* free-living amoebae, 1987b.

Rowbotham, T. J., Preliminary report on the pathogenicity of *Legionella pneumophila* for freshwater and soil amoebae, *J. Clin. Pathol.* 33, 1179, 1980.

Rowbotham, T. J., Leeds, P. H. L., *Legionella*-like amoebal pathogens (LLAPs), pp. 137–140. *Legionella: Current Status and Emerging Perspectives,* Editors, Barbaree, J. M., Breiman, R. F., Dufour, A. P., American Society for Microbiology, Washington, DC, 1993.

Schuster, F. L., Dunnebacke, T. H., Growth at 37°C of the EGs strain of the amoeboflagellate *Naegleria guberi* containing virus-like particles. I. Nuclear changes, *J. Invertebr. Pathol.* 23, 172, 1974a.

Schuster, F. L., Dunnebacke, T. H., Growth at 37°C of the EGs strain of the amoeboflagellate *Naegleria guberi* containing virus-like particles. II. Cytoplasmic changes, *J. Invertebr. Pathol.* 23, 182, 1974b.

Seyfried, P. L., Microorganismes, parasites et endotoxines en suspension dans la section de deshydratation d'une usine de traitement des eaux usées, *Sci. Tech. de L'Eau aout,* 275, 1990.

Tyndall, R. L., Domingue, E. L., Cocultivation of *Legionella pneumophila* and free-living amoebae, *Appl. Environ. Microbiol.* 44, 954, 1982.

Tyndall, R. L., Environmental isolation of pathogenic *Naegleria, Crit. Rev. Environ. Control* 13, 3, 195, 1984.

Tyndall, R. L., Dudney, C. S., Katz, D. S., Ironside, K. S., Jernigan, R. A., Identification of bioaerosols and certain particulate emissions into enclosed environments from home humidifiers, vaporizer, and other appliances found in homes, CPSC Phase I Final Report, 1985.

Tyndall, R. L., Lyle, M. M., Ironside, K. S., The presence of free-living amoebae in portable and stationary eye wash stations, *Am. Ind. Hyg. Assoc. J.* 48, 11, 933, 1987.

Tyndall, R. L., Ironside, K. S., Metler, P. L., Tan, E. L., Hazen, T. C., Fliermans, C. B., Effect of thermal additions on the density and distribution of thermophilic amoebae and pathogenic *Naegleria fowleri* in a newly created cooling lake, *Appl. Environ. Microbiol.* 55, 3, 722, 1989.

Tyndall, R. L., Ironside, K. S., Free-living amoebae: health concerns in the indoor environment, in *Biological Contaminants in Indoor Environments,* STP 1071, ASTM, Philadelphia, Morey, P. R., Feeley, J. C., Sr., Otten, J. A., Eds., p. 163, 1990.

Tyndall, R. L., Ironside, K. S., Little, C. D., Katz, D. S., Kennedy, J. R., Free-living amoebae used to isolate consortia capable of degrading trichloroethylene, *Appl. Biochem. Biotech.* 28/29, 917, 1991.

Tyndall, R. L., Vass, A. A., Fliermans, C. B., Mixed bacterial populations derived from *Legionella*-infected free-living amoebae, pp. 142–145. *Legionella: Current Status and Emerging Perspectives,* Editors, Barbaree, J. M., Breiman, R. F., Dufour, A. P., American Society for Microbiology, Washington, DC, 1993.

Visvesvara, F. S., Jones, D. B., Robinson, N. M., Isolation, identification, and biological characterization of *Acanthamoebae polyphaga* from a human eye, *Am. J. Trop. Med. Hyg.* 24 (5), 784, 1975.

Wilhelmus, K. R., The increasing importance of *Acanthamoeba, Rev. Infect. Dis.* 13, 5, S367, 1991.

7 ALLERGENS OF ARTHROPODS AND BIRDS

Susan Pollart Squillace

CONTENTS

I. THE ARTHROPOD CLASSES

The phylum Arthropoda includes several species of organisms that are important in causing diseases via the indoor air. Of the nine classes of extant forms in the Arthropod phylum, the Hexapoda (or Insecta) and the Arachnida contain species that are known to produce proteins potentially allergenic to humans. Several species of the Arachnida class (the house dust mites) are primarily indoor organisms that require a controlled climate for growth and development and depend on mammalian by-products for food. In most homes these organisms coexist with human inhabitants and are unnoticed. Members of the Hexapoda are generally outdoor organisms, some of which can be household pests.

0-87371-724-4/95/$0.00+$.50

II. THE ARACHNIDA (MITES, SPIDERS)

A. Mites

The organisms and their ecology. The Acarina order of the Arachnida class includes a range of genera that inhabit a wide variety of indoor and outdoor microenvironments. The majority of species feed on detritus of animal or vegetable origin and therefore live in close association with their hosts.

Mites of the genera *Lepidoglyphus* and *Tyrophagus* rely on decaying vegetation as a food source and are often labeled "storage mites" (van Hage-Hamsten et al., 1987). Although not a significant problem in homes, these mites are prominent in agricultural environments and have been reported to be a cause of allergic occupational rhinitis among Finnish dairy farmers (Terho et al., 1985, 1987).

Acarina of the family Pyroglyphidae rely on a human food source and are ubiquitous in homes in temperate and tropical climates. These "house dust mites" require a warm temperature (70° to 80°F) and a moist environment (absolute humidity >7 g water/g dry air) for survival (Korsgaard, 1983; Voorhorst et al., 1967, 1969). House dust mites are between 250 and 500 μm in size, are not easily visible without magnification, and are usually ignored by a homes' human inhabitants. The specific genera and species of mites in homes are apparently related to climate and geography. Mites of the genus *Dermatophagoides* are most common in North America and Europe (Platts-Mills and de Weck, 1989), whereas *Euroglyphus* and *Blomia* are found more commonly in Central and South America (Arruda et al., 1991; Fernandez-Caldas et al., 1992; Leby et al., 1992).

Although house dust mites have been found in schools, offices, and geriatric nursing homes, their density is apparently highest in living rooms and bedrooms of traditional homes (Dybendal et al., 1989; Platts-Mills et al., 1986a; Tovey et al., 1981; Vyszenski-Moher et al., 1986). Carpets, upholstered surfaces, and bedding are repositories of human skin scales and retain moisture, providing ideal nesting sites for mite growth and development. Studies of homes in the central and southern United States have found a seasonal rise in mite bodies during the warm, humid summer months and a decrease during dry winter months (Platts-Mills et al., 1986a). Tropical areas that do not experience significant temperature or humidity fluctuations are more likely to sustain a mite population indoors on a year-round basis.

Mites and allergic disease. Of all the arthropods, the pyroglyphid mites have been the most extensively studied in terms of allergens and allergic disease. Immunochemical techniques, including the production of monoclonal antibodies specific for mite allergens, have allowed the identification, purification, and quantitation of several major mite allergen groups produced by *Dermatophagoides farinae*, *Dermatophagoides pteronyssimus*, *Dermatophagoides microceras,* and

Euroglyphus maynei. The Group I mite allergens from these species (*Der f* I, *Der p* I, *Der m* I, and *Eur m* I) show an amino acid sequence homology to cystein protease, and the identified Group III allergens (*Der f* III and *Der p* III) are likely to be trypsins. *Der f* II and *Der p* II do not show homology to any proteins in sequencing data banks but are also likely to be mite proteases (Platts-Mills et al., 1992).

Epidemiologic studies have demonstrated a strong correlation between exposure to mites and mite allergen, immediate hypersensitivity to dust mite allergens, and atopic disease (Gelber et al., 1993; Pollart et al., 1989; Sporik et al., 1990). Asthma, allergic rhinitis, and atopic dermatitis have all been associated with elevated levels of immunoglobulin E (IgE) antibody and/or positive skin tests to pyroglyphid mites (Mitchell et al., 1982, 1984, 1986; Platts-Mills and de Weck, 1989; Platts-Mills et al., 1992). Additionally, studies of avoidance of mite allergens through environmental control have demonstrated remission of symptoms and, in the case of asthma, reversal of the underlying bronchial hyperreactivity (Murray and Ferguson, 1983; Platts-Mills et al., 1982).

Environmental assessment. The availability of monoclonal antibodies to specific dust mite allergens has allowed the measurement of mite allergen levels in the indoor environment (Lind et al., 1987; Luczynska et al., 1989, Chapter 12). Most studies have focused on the concentration of the major mite allergens in dust samples collected from known mite reservoirs within homes (e.g., bedding, carpeting, upholstered furniture). Dust is typically collected using a modified hand-held vacuum cleaner to obtain a 2-min sampling from 1 m^2 of surface. Once collected, dust samples are sieved to remove large particles and a standard quantity of fine dust is extracted in a buffered saline solution. The concentration of allergen in the resulting extract is expressed as micrograms of allergen per gram of collected dust. Using this method and correlating allergen levels with patient symptoms, "safe" and "unsafe" levels of dust mite allergen have been defined. Currently, levels of Group I less than 2 μg/g of dust are considered safe, levels between 2 and 10 μg/g of dust are considered likely to sensitize genetically predisposed individuals, and levels greater than 10 μg/g of dust are expected to provoke exacerbations of allergic disease in previously sensitized individuals (Arruda et al., 1991; Charpin et al., 1991; Lau et al., 1989; Peat et al., 1987; Wood et al., 1989). Although the usefulness of reservoir measurements has been recently questioned (Price et al., 1990; Sporik et al., 1990; Warner and Warner, 1991), it is the concensus of two international workshops that these methods are relevant and remain the most practical measurements currently available (Platts-Mills et al., 1989, 1992).

Airborne mite allergens have been difficult to measure. Several techniques have been used to sample air in rooms during and after disturbances. Volumetric sampling with membrane filters has repeatedly demonstrated that mite allergen levels are very low or undetectable under normal circumstances (deBlay et al., 1991; Platts-Mills et al., 1986b; Sakaguchi et al., 1989; Swanson et al., 1985; Tovey et al., 1981). One study designed to investigate the distribution of mite and cat allergen on wall surfaces found that dust mite allergen was rarely detectable on adjacent walls even in homes with high levels of dust mite allergen in settled dust. In contrast, cat allergen (which is carried on very small particles) was readily recovered from both floor dust and wall surfaces (Wood et al., 1992). The Group I dust mite allergens are carried on mite fecal particles, which are 10 to 20 μm in size. The relatively large size of these particles causes them to become airborne only after disturbances and to settle rapidly (within 30 min) after disturbances (Platts-Mills et al., 1982; Swanson et al., 1989; Yoshizawa et al., 1991). Group II allergens are probably associated with mite bodies and also appear to fall rapidly when they do become airborne (deBlay et al., 1991). Not surprisingly, most previous studies attempting to correlate measurements of airborne mite allergen with sensitization or symptoms of allergic disease have been unsuccessful. It is possible that future studies using personal air samplers may be more successful in defining patterns of exposure to airborne mite allergen.

Environmental control. Environmental control targets reservoirs of the pyroglyphid mites and their allergens. Any surface that can collect human skin scales and that maintains a moderate temperature and humidity will provide an environment suitable for mite growth and allergen accumulation. Bedding (including mattresses), carpeting, and upholstered furniture are therefore addressed most aggressively in dust mite avoidance regimens. The most beneficial interventions have focused on modifying bedrooms and include covering mattresses and pillows with a zippered, vinyl cover; washing bedding (including sheets and blankets) weekly in hot (>130°F) water; and removing carpeting. Replacing upholstered furniture with vinyl or leather and removing carpeting throughout the home is also recommended (Murray and Ferguson, 1983; Platts-Mills et al., 1982). Carpet that lies directly on top of a ground-level concrete slab is particularly problematic (Rose et al., 1992). In this situation, sufficient moisture remains in the slab to keep the carpet moist almost constantly and, if the indoor temperature is adequate, may create a microenvironment that is suitable for mite growth year-round.

Covering mattresses and washing bedding are fairly easy and inexpensive procedures. However, carpet removal and installation of vinyl or wood flooring are more expensive and less acceptable to patients and their families. For this reason, investigators have studied the use of acaracides and chemicals that denature mite allergen on carpeted and upholstered surfaces (Bischoff et al., 1990; Colloff, 1990). Benzyl benzoate is currently available as a moist powder and a foam for treating carpets and upholstered furniture, respectively. The product, marketed by Fisons as Acarosan, kills mites in vitro and has been shown to reduce mite numbers and mite allergen levels in vivo (Charpin et al., 1991). However, a significant number of carpets studied show no change in mite numbers or mite allergen levels. This may be related to the type of carpet fiber, its length, or the amount of dirt and dust that has accumulated. Additionally, it is unclear how often carpets need to be retreated with benzyl benzoate. It is likely that some carpets in particularly humid environments may need to be treated monthly.

Tannic acid denatures the Group I and Group II dust mite allergens. Because tannic acid does not eliminate dust mites, when used alone it must be reapplied every several weeks to control allergen levels (Green et al., 1989; Miller et al., 1989). A 3% solution is most effective, but may stain light-colored carpeting and fabrics. The use of tannic acid on upholstered furniture is much less well studied, and no recommendations about proper and effective use can be made at this time.

Research needs. Intensive research on allergy to pyroglyphid mites continues worldwide. The Second International Workshop recently published its report on dust mite allergens and asthma and made the following recommendations regarding research needs:

1. Recent studies have confirmed that, in the absence of disturbance, the level of airborne mite allergen is very low, and because of this finding, it is very difficult to standardize measurements of airborne exposure. At the present time, measurements of airborne mite allergen cannot be recommended as a routine method for determining exposure. However, further studies in this area should be encouraged.
2. Rapid progress has been made in cloning and sequencing mite allergens. This progress has allowed the production of fragments that can now be tested for reactivity with IgE antibodies, monoclonal antibodies (mAbs), and T cells. It may well be possible to develop peptides or modified molecules specifically reactive with T cells, which could be used for immunotherapy.
3. Avoidance of mite allergens should be further investigated as a primary treatment for mite-sensitive patients. Concern has arisen about the appropriateness of measurements of "gram of allergen per gram of dust" because of the potential dilutional effect when acaracides are added to a upholstered or carpeted surface. Therefore, methodology for the determination of exposure level warrants further standardization.

B. Spiders

Spiders and other members of the arachnida are not commonly reported as sources for aeroallergens, and have not been studied.

III. INSECTS (THE HEXAPODA)

All insects belong to the class Hexapoda. This class consists of approximately 750,000 species and contains the largest number of pest species. The Hexapoda class also has the greatest medical and veterinary significance. Members of at least eight different orders of Hexopoda have been reported to cause disease via the indoor and/or the outdoor air in the form of respiratory allergy. These insects are encountered in homes and in workplaces, in schools and museums, and in hospitals and laboratories. Their potential for causing allergic disease is primarily limited by the presence of a sensitized person.

A. The Cockroach

The organism and its ecology. Of the insects, cockroaches, members of the order Cursoria, have been studied most extensively in relation to respiratory allergy. Of the approximately 55 species of cockroaches that inhabit the continental United States, only 5 to 10 are important indoor insects. The two most common cockroaches inside homes in the temperate areas of the United States are *Periplaneta americana* (the American cockroach) and *Blatella germanica* (the German cockroach). These two species are the most extensively studied in terms of allergic human disease.

Entomologic studies suggest that the cockroach is likely to be a year-round inhabitant of homes as long as sufficient moisture is available (Horsfall, 1962). Cockroaches are able to survive low ambient humidity much better than mites and, unlike mites, are able to search actively for water. The requirement for moisture implies that cockroaches will be found in the highest concentrations in kitchens (e.g., around the sink and under the refrigerator) and in bathrooms. Patient observation supports this assumption, and recently developed monoclonal antibody-based assays for cockroach allergen have found the lowest levels of cockroach allergen in bedrooms and 10- to 20-fold higher levels in the kitchens and bathrooms of cockroach infested homes (Pollart et al., 1991a; Schou et al., 1991). Although cockroach infestation is usually obvious, in one recent study up to 15% of homes with no visible evidence of cockroaches had measurable levels of cockroach allergen (Gelber et al., 1993).

Cockroach allergy has been reported consistently in studies of patients with asthma in lower socioeconomic conditions in urban areas (Bernton and Brown, 1964; Bernton et al., 1972; Kang, 1976; Kivity et al., 1989; Morris et al., 1986). Cockroach sensitivity (with or without symptomatic allergic disease) is widespread even in nonurban areas, possibly because of the ubiquitous nature of the insect and the multiple opportunities for exposure (Frayh, 1990; Lan et al., 1988; Mendoza and Snyder, 1970; Pollart et al., 1989). Several groups of investigators have defined two major allergens for the German cockroach (*Bla g* I and *Bla g* II) and one for the American cockroach (*Per a* I) (Pollart et al., 1991b; Schou et al., 1990). The source of these allergens is not completely clear. Investigators have pointed to cast skins, whole bodies, egg shells, fecal particles, and saliva as potent sources (Lehrer et al., 1991). The recent demonstration that cockroach washings and washes of "clean" (i.e., no visible fecal material or debris) cockroach environments contain significant amounts of *Bla g* I and *Bla g* II suggests that cockroaches may secrete their allergen onto their bodies or onto surfaces in their environment (Vailes et al., 1990).

Environmental assessment. A monoclonal antibody-based enzyme-linked immunosorbent assay (ELISA) has been developed for the German cockroach allergens, and a monospecific rabbit antibody ELISA has been developed that identifies determinants shared by *Bla g* I and *Per a* I (Pollart et al., 1991b; Schou et al., 1990). These assays have been used to measure cockroach allergen levels in homes, schools, and day care centers. Data are accumulating that will allow the determination of safe and hazardous levels of cockroach allergen in settled dust (Pollart et al., 1991b; Schou et al., 1991). These assays have not been extensively used to determine airborne cockroach allergen levels or the conditions under which cockroach allergen becomes airborne. Good data are available to show that cockroach allergen can become airborne in homes. However, particle size and source have not been determined.

Cockroach control. Pesticides are widely available for the control of cockroaches; however, reinfestation even after professional extermination is common, and dead cockroaches and/or excreta may remain in the house for long periods, depending on the degree of infestation and the hygienic status of the house. A major feature of cockroach infestation is the rapid movement of cockroaches from one part of a building to another, allowing them to avoid the effects of spraying in one site.

Insecticides do kill cockroaches. However, their effectiveness in reducing cockroach allergen levels is unknown. Currently, available assays for the major allergens of the German and American cockroaches provide the opportunity for such studies. Thus, it is difficult to define a simple approach to reducing cockroach exposure (Swanson et al., 1989).

In most urban areas where cockroach infestation and cockroach-associated allergy are problems, many families would have considerable difficulty effectively eradicating roaches. However, if a level of cockroach allergen could be defined as a risk factor for asthma, then it might be possible to recommend levels of cockroach infestation that should be regarded as a health hazard. This in turn could lead to a systematic approach to reducing cockroach allergen levels.

B. Other Insects

Asthma and/or rhinitis have been reported in association with many other insects including waterfleas, bed bugs, mayflies, aphids, houseflies, honey bees, bean weevils, several species of moths, butterflies, silkworms, caddis flies, and chronomid midges (Horsfall, 1962). Unfortunately, epidemiologic data to support causation for the majority of these suspected allergens is minimal.

The opportunity for exposure to insects other than cockroaches tends to be seasonal and most often occurs outdoors. An investigation of the concentration of insect allergens in outdoor air in Ohio found that ant, aphid, cricket, house fly, fruit fly, mosquito, and moth allergens were all present in outdoor air in higher concentrations than *Alternaria* or ragweed pollen allergens (Lierld, 1992). Body fragments and insect debris may readily move into homes on the inhabitants' clothing or with the influx of outdoor air. Additionally, workplaces may sustain a large insect population that would not be tolerated in most homes. A recent study of workers at a caddis fly-infested hydroelectric plant found 61% of those skin tested were positive to caddis fly. The majority of the skin test responders complained of work-related rhinitis, conjunctivitis, or asthma (Silviu-Dan et al., 1992).

A group of investigators in New Delhi evaluating the allergenic significance of 13 common insects found that 36% of patients skin tested showed markedly positive reactions to at least one insect whole-body extract. Two thirds of the allergic asthmatic patients undergoing bronchial provocation testing with insect whole-body extracts (moths, mosquito, cockroach, or pulse beetle) had a positive response (Gupta et al., 1990).

In Japan, skin test reactivity against extracts of silkworm wing, caddis fly wing, and chironomid whole body among asthmatic patients ranges from 54 to 70%. Antigenic cross-reactivity has been demonstrated between members of the Lepidoptera order including butterfly, moth, and silkworm (Kino and Oshima, 1989). Wing components have been implicated as the antigen source from silkworms and caddis flies, but it is likely that a body component of the chironomid midge is the main sensitizer of allergic asthmatic patients.

The epidemiology of diseases due to exposure to insect allergens via the indoor air needs to be explored in considerably more detail before recommendations can be made regarding environmental control. It is likely that the diseases will be region specific and reflect seasonal fluctuations in insect populations outdoors. With the definition of clinically significant insect allergens, assays can be developed for their detection inside homes and methods for avoidance can be tested.

IV. BIRDS

Like mammals, birds probably evolved from reptiles, although there is no fossil evidence of the stages through which the evolutionary change took place. The first bird, *Archaeopteryx lithographica*, demonstrates a skeleton that would pass for that of a reptile but clearly demonstrates the presence of feathers (Swinton, 1960). From that one common ancestor has arisen the avian class with its multiple superorders, orders, and suborders.

Birds are the only class of vertebrates that consist exclusively of oviparous forms. Although there are other examples of egg-laying species, every other class contains at least some species in which the mother retains the offspring for a gestational period and subsequently bears them free from the egg covering at an advanced stage of development (Storer, 1960).

Birds of the Psittacidae family are the most common household pets. However, humans come into contact with chickens, ducks, geese, turkeys, pigeons, and doves in domestic and industrial indoor environments. Birds contribute to human disease via the indoor air in several different ways. A number of diseases are shared by birds and humans, and many of these are true zoonosis. Additionally, some humans exposed to bird proteins may develop an immunoglobulin G (IgG) mediated hypersensitivity pneumonitis. Finally, humans may develop allergy to arthropods that inhabit and feed on bird feathers.

A. Hypersensitivity Pneumonitis

The inhalation of organic dust from numerous birds is known to induce a delayed hypersenstivity reaction in some exposed people. Pigeon breeder's disease, resulting from inhalation of aerosolized pigeon droppings, is the most well studied of these bird-related reactions. In addition, duck fever, turkey handler's disease, budgerigar fancier's disease, and (chicken) feather plucker's disease are hypersensitivity diseases associated with exposure to avian antigens. In addition, contact with goose and duck feathers, doves, and lovebirds has caused IgG-mediated disease (Liu et al., 1989; Majima et al., 1990; Salvaggio, 1987).

The pigeon serum protein immunoglobulin A (IgA) is the most important antigenic material (Goudswaard et al., 1977) and is the predominant antigen found in pigeon droppings (Fredricks, 1978). Exposure to this antigen may be associated with accumulation of droppings in poorly cleaned lofts, especially during cleaning activities. Pigeon bloom, a waxy substance that coats the feathers, is another potent source of pigeon-related IgE (Banham et al., 1982). The bloom, derived from the feather, is composed of keratin granules of respirable size (average size 1 μm), which are coated with the antigenic material. Bloom is found on all birds, is readily airborne, and is found in highest quantities in flying birds. Bloom is therefore likely to be the antigen source for the hypersensitivity pneumonitis associated with other flying birds such as budgerigars.

Pigeon breeders come into contact with the causative antigen in their own lofts and in other situations such as pigeon shows where exposure to birds is increased (Boyd et al., 1982). Clinical disease has also been reported in relation to indirect exposure and may be caused by deposition of antigen onto the hands and clothing of those carrying directly for the pigeons (Boyd, 1978). One study of particle size in a pigeon loft using an Andersen sampler found the highest levels of soluble antigen on particles between 0.5 and 5 μm in size (Edwards et al., 1991). A comparison between lofts that used litter materials to dehydrate voided pigeon droppings and lofts cleaned regularly with no litter agent found that the use of litter materials significantly increased the number of airborne particles and the antigen levels.

The pulmonary disease resulting from hypersensitivity to pigeon antigen may be acute and progressive or acute, intermittent, and nonprogessive (Boyd et al., 1982; Schmidt et al., 1988). Pigeon breeders often attempt to reduce exposure to pigeon antigen by reducing contact or wearing masks during exposure. If partial avoidance does not result in improvement in symptoms, complete avoidance of pigeons and pigeon lofts is advised.

While pigeon breeder's disease is fairly well defined, information on hypersensitivity pneumonitis from other birds, particularly to birds that are more typically found inside homes, is not as complete. The form in which antigen from these domestic birds becomes airborne needs to be defined and methods for reducing exposure inside of homes delineated.

B. IgE-Mediated Allergies

Skin testing with feather extract has long been used in the evaluation of perennial rhinitis and asthma. The fact that allergic symptoms are most often associated with aged feathers suggests that feather-filled pillows and furniture acquire antigenicity through breakdown of organic material or accumulation of new antigenic sources. Investigators studying skin test reactivity to feather extract and house dust extract found them to be identical and have concluded that infestation of feather sources by Pyroglyphid mites is the cause of "feather" allergy (Voorhorst et al., 1969). It has been demonstrated in vitro that feathers are a suitable media for the growth of the house dust mite, *Dermatophagoides pteronyssimus* (Spieksma, 1976). Other investigators have identified active allergens from feathers that are different from those identified in house dust (Berrens, 1970).

Live caged birds can also be infested with mites. Mites of the species *Pterolichus* have been noted in high density on the primary feathers of budgerigars and seem to rely on the soft keratin at the base of the feather shaft as their primary source of nutrition (Colloff et al., 1992). Mites made up between 1 and 10% of the weight of the feathers studied and were likely to be a major source of soluble protein from the feathers. Immediate hypersensitivity to *Pterolichus* mites in humans has yet to be demonstrated, however. In vitro assays have demonstrated no cross-reactivity of *Pterolichus* species with the Group I allergens of *Dermatophagoides* species.

Contact with feather allergen or allergens occurs through exposure to feather-stuffed pillows, bedding, or furniture or through live birds. Prevention of symptoms upon exposure to feather-stuffed materials can be accomplished through routine dust-mite avoidance procedures including covering the material with a zippered plastic covering. Because washing with water is not recommended for feather-stuffed materials, avoidance is the only other currently available treatment.

In England, approximately 20% of budgerigar and canary owners suffering from rhinitis or asthma have IgE antibodies to the feathers of their pets (Van Toorenenbergen et al., 1985). If feather allergy in bird owners is found to be due to sensitivity to mites infesting the birds, acaracides are a potential method of eliminating the antigenic source. Fowl mites that infect poultry are sensitive to a number of acaracides including permetrin, tetrachlorvinphos, carbaryl, and coumaphos (Arthur and Axtell, 1983). The efficacy of these acaracides on pet birds and the toxicity of the chemical to the bird need to be established before recommending them as treatments.

C. Infectious Disease

Parrots, members of the Psittacidae family, and other pet birds can act as vectors of infectious diseases via the indoor air. Zoonotic disease has taken on increasing importance as urbanization and decreased living space have led to an increase in the practice of keeping birds as pets. Since immunocompromised hosts are particularly susceptible to zoonosis, the rising prevalence of HIV infection and AIDS (acquired immune deficiency syndrome) have been associated with a drastic increase in zoonotic diseases (Harris, 1991).

Psittacosis, tuberculosis, toxoplasmosis, and colibacillosis can all be transmitted from birds to humans through inhalation of the offending agent. Psittacosis, the most common of the bird zoonoses, is caused by *Chlamydia psittaci* (Hayashi et al., 1990; Macfarlane and Macrae, 1983). This organism infects a wide variety of avian species but is most common in Psittacidae. Humans become infected via the respiratory route after direct bird contact or inhalation of *Chlamydia*-containing aerosolized material from bird feces or nasal secretion. In the adult host with a normal immune system, the disease is easily treated with antibiotics. However, in the immunocompromised population, mortality can be significant. The diagnosis of psitticosis in birds can be difficult, and presumptive treatment of birds before placement with an immunocompromised pet owner may be wise to prevent transmission.

Mycobacterium avium intracellulare (MAI) and *Mycobacterium tuberculosis* both infect humans and birds. *Mycobacterium tuberculosis* in birds is acquired from infected humans via contaminated food, water, and soil. *Mycobacterium avium* is common in grey cheek parakeets, mature Amazon parrots, and domestic poultry. The organism is shed in large numbers in feces and infects humans through aerosolization and inhalation. MAI in birds and humans is relatively resistant to pharmacotherapy and is a serious problem in immunocompromised humans. Acid-fast smears of feces can diagnose MAI in birds, and some veterinarians feel a positive stain should preclude placement of an infected bird as a pet (Harris, 1991).

Toxoplasmosis, caused by an infection with the protozoan *Toxoplasma gondii,* is most often associated with contact with feces from infected cats. However, *Toxoplasma* can also infect domestic birds including canaries, pigeons, and domestic fowl, and there is one case of human disease in the caretaker of an infected mynah bird. The protozoa produce necrotic foci in the liver, spleen, and lungs of infected birds.

Pathogenic strains of *Escherichia coli* may infect domestic fowl and lead to gastrointestinal or respiratory disease. Transmission to humans occurs via the fecal-oral route or from respiration of *E. coli*-containing droplets. The disease is easily treated in birds and human with appropriate antibiotics.

REFERENCES

Arruda, L. K., Rizzo, M. C., Chapman, M. D., Fernandez-Caldas, E., Baggio, D., Platts-Mills, T. A. E., and Naspitz, C. K., Exposure and sensitization to dust mite allergens among asthmatic children in São Paulo, Brazil, *Clin. Exp. Allergy,* 21, 433, 1991.

Arthur, F. H. and Axtell, R. C., Susceptibility of northern fowl mites in North Carolina to five acaricides, *Poult. Sci.,* 62, 428, 1983.

Banham, S. W., McKenzie, H., McSharry, C., Lynch, P. P., and Boyd, G., Antibody against a pigeon bloom extract: a further antigen in pigeon fancier's lung, *Clin. Allergy,* 12, 173, 1982.

Bernton, H. S. and Brown, H., Insect allergy — preliminary studies of the cockroach, *J. Allergy*, 35, 506, 1964.

Bernton, H. S., McMahon, T. F., and Brown, H., Cockroach asthma, *Br. J. Dis. Chest,* 66, 61, 1972.

Berrens, L., Structural studies of house dust allergens, *Clin. Exp. Immunol.*, 6, 71, 1970.

Bischoff, E., Fischer, A., and Lienbenberg, B., Assessment and control of house dust mite infestation, *Clin. Ther.*, 12, 216, 1990.

Boyd, G., Clinical and immunological studies in pulmonary extrinsic allergic alveolotis, *Scott. Med. J.*, 23, 267, 1978.

Boyd, G., McSharry, C. P., Banham, S. W., and Lynch, P. P., A current view of pigeon fancier's lung, *Clin. Allergy*, 12, 53, 1982.

Charpin, D., Birnbaum, J., Haddi, E., N'Guyen, A., Fondarai, J., and Vervloet, D., Evaluation of the acaricide, ACARDUST in the treatment of allergy to acarids, *Rev. Fr. Allergol.*, 30, 149, 1990.

Colloff, M. J., House dust mite. II. Chemical control, *Pestic. Outlook,* 1, 3, 1990.

Colloff, M., Rees, J., Merrett, J., and Merrett, T., Airborne allergens — budgerigar feather mites, *J. Allergy Clin. Immunol.*, 89, 313, 1992.

deBlay, F., Heymann, P. W., Chapman, M. D., and Platts-Mills, T. A. E., Airborne dust mite allergens: comparison of group II allergens with group I allergen and cat-allergen *Fel d* I, *J. Allergy Clin. Immunol.*, 88, 919, 1991.

Dybendal, T., Vik, H., and Elsayed, S., Dust from carpeted and smooth floors, *Allergy,* 44, 401, 1989.

Edwards, J. H., Trotman, D. M., Mason, O. F., Davies, B. H., Jones, K. P., and Alzubaidy, T. S., Pigeon breeders' lung — the effect of loft litter materials on airborne particles and antigens, *Clin. Exp. Allergy*, 21, 49, 1991.

Fernandez-Caldas, E., Mercado, R., Peurta, L., Lockey, R. F., and Caraballo, L. R., House dust mite sensitivity and mite fauna in the tropics, *J. Allergy Clin. Immunol.,* 89, 257, 1992.

Frayh, A. A., The pattern of skin test reactivity to aeroallergens in asthmatic children in Riyadh, *J. Asthma*, 27, 315, 1990.

Fredricks, W., Antigens in pigeon droppings extracts, *J. Allergy Clin. Immunol.*, 61, 221, 1978.

Gelber, L. E., Seltzer, L. H., Bouzoukis, J. K., Pollart, S. M., Chapman, M. D., and Platts-Mills, T. A. E., Sensitization and exposure to indoor allergens as risk factors for asthma among patients presenting to hospital, *Am. Rev. Resp. Dis.,* 147, 573, 1993.

Goudswaard, J., Vaerman, I. P., and Heremans, I. F., Three immunoglobin classes in the pigeon (*Columbia livia*), *Int. Arch. Allergy Appl. Immunol.,* 53, 409, 1977.

Green, W. F., Nicholas, N. R., Salome, C. M., and Woolcock, A. J., Reduction of house dust mites and mite allergens: effects of spraying carpets and blankets with Allersearch DMS, an acaricide combined with an allergen-reducing agent, *Clin. Exp. Allergy*, 19, 203, 1989.

Gupta, S., Jain, S., Chaudhry, S., and Agarwal, M. K., Role of insects as inhalant allergens in bronchial asthma with special reference to the clinical characteristics of patients, *Clin. Exp. Allergy*, 20, 519, 1990.

Harris, J. M., Zoonotic diseases of birds, *Vet. Clin. North Am.*, 21, 1289, 1991.

Hayashi, Y., Kato, M., Ito, G., Yamamoto, K., Kuroki, H., Matsuura, T., Yamada, Y., Goto, A., and Takeuchi, T., A case report of psittacosis and chlamydial isolation from a dead pet bird, *Jpn. J. Thorac. Med.*, 28, 535, 1990.

Horsfall, W. R., *Medical Entomology. Arthropods and Human Disease*, Ronald Press, New York, 1962.

Kang, B., Study on cockroach antigen as a probably causative agent in bronchial asthma, *J. Allergy Clin. Immunol.*, 58, 357, 1976.

Kino, T. and Oshima, S., Environmental factors affecting pathogenesis of bronchial asthma, *J. Med. Jpn.*, 28, 544, 1989.

Kivity, S., Struher, D., Greif, J., Schwart, Y., and Topilsky, M., Cockroach allergen: an important cause of perennial rhinitis, *Allergy*, 11, 291, 1989.

Korsgaard, J., Mite asthma and residency. A case-control study on the impact of exposure to house-dust mites in dwellings, *Am. Rev. Respir. Dis.,* 128, 231, 1983.

Lan, J. L., Lee, D. T., Wu, C. H., Chang, C. P., and Yen, C. L., Cockroach hypersensitivity: preliminary study of allergic cockroach asthma in Taiwan, *J. Allergy Clin. Immunol.*, 82, 736, 1988.

Lau, S., Falkenhorst, G., Weber, A., et al., High mite-allergen exposure increases the risk of sensitization in atopic children and young adults, *J. Allergy Clin. Immunol.*, 84, 718, 1989.

Lehrer, S. B., Horner, W. E., Menon. P., and Stankus, R. P., Comparison of cockroach allergenic activity in whole body and fecal extracts, *J. Allergy Clin. Immunol.*, 87, 571, 1991.

Leby, D. A., Lemao, J., Fains, A., Leynadier, F., and Guerin, B., Mites and allergy on the island of Mauritius, *J. Allergy Clin. Immunol.*, 89, 256, 1992.

Lierld, M. B., Concentration of insect allergens in outside air, *J. Allergy Clin. Immunol.*, 89, 314, 1992.

Lind, P., Norman, P. S., Newton, M., Lowenstein, H., and Schwartz, B., The prevalence of indoor allergens in the Baltimore area: house-dust mite and animal-dander antigens measured by immunochemical techniques, *J. Allergy Clin. Immunol.*, 80, 541, 1987.

Liu, Y. N., Chen, L. A., Zhang, Z. Y., and Li, Q. S., Parrot breeder's lung: first case report in China, *Chin. Med. J.*, 102, 947, 1989.

Luczynska, C. M., Arruda, L. K., Platts-Mills, T. A. E., Miller, J. D., Lopez, M., and Chapman, M. D., A two-site monoclonal antibody ELISA for the quantification of the major *Dermatophagoides* spp. allergens, *Der p* I and *Der f* I, *J. Immunol. Methods*, 118, 227, 1989.

Macfarlane, J. T. and Macrae, A. D., Psittacosis, *Br. Med. Bull.*, 39, 163, 1983.

Majima, T., Kohara, F., Akiyama, Y., Ooshima, N., Katoh, H., Aihara, H., Hosokawa, Y., Yamaguchi, M., Hayashi, H., Horie, T., et al., Two cases of budgerigar breeder's lung, *Jpn. J. Thorac. Dis.*, 28, 756, 1990.

Mendoza, J. and Snyder, F. D., Cockroach sensitivity in children with bronchial asthma, *Ann. Allergy*, 28, 159, 1970.

Miller, J. D., Miller, A., Luczynska, C., Rose, G., and Platts-Mills, T. A. E., Effect of tannic acid spray on dust mite allergen levels in carpets (abstract), *J. Allergy Clin. Immunol.*, 83, 262, 1989.

Mitchell, E. B., Crow, J., Chapman, M. D., Jouhal, S. S., Pope, F. M., and Platts-Mills, T. A. E., Basophils in allergen-induced patch test sites in atopic dermatitis, *Lancet*, 1, 127, 1982.

Mitchell, E. B., Crow, J., Rowntree, S., Webster, A. D. B., and Platts-Mills, T. A. E., Cutaneous basophil hypersensitivity to inhalant allergens: local transfer of basophil accumulation with immune serum but not IgE antibody, *J. Invest. Dermatol.*, 83, 290, 1984.

Mitchell, E. B., Crow, J., Williams, G., and Platts-Mills, T. A. E., Increase in skin mast cells following chronic house dust-mite exposure, *Br. J. Dermatol.*, 114, 65, 1986.

Morris, E. C., Smith, T. F., and Kelly, L. B., Cockroach is a significant allergen for inner city children (abstract), *J. Allergy Clin. Immunol.*, 77, 206, 1986.

Murray, A. B. and Ferguson, A. C., Dust-free bedrooms in the treatment of asthmatic children with house dust mite allergy: a controlled trial, *Pediatrics*, 71, 418, 1983.

Peat, J. K., Britton, W. J., Salome, C. M., and Woolcock, A. J., Bronchial hyperresponsiveness in two populations of Australian school-children. III. Effect of exposure to environmental allergens, *Clin. Allergy*, 17, 297, 1987.

Platts-Mills, T. A. E. and de Weck, A. L., Dust mite allergens and asthma — A world wide problem, *J. Allergy Clin. Immunol.*, 83, 416, 1989.

Platts-Mills, T. A. E., Chapman, M. D., Heymann, P. W., and Luczynska, C. M., Measurement of airborne allergen using immunoassays, *Immunol. Allergy Clin. North Am.*, 9, 269, 1989.

Platts-Mills, T. A. E., Hayden, M. L., Chapman, M. D., and Wilkins, S. R., Seasonal variation in the dust mite and grass pollen allergens in dust from the houses of patients with asthma, *J. Allergy Clin. Immunol.*, 79, 781, 1986a.

Platts-Mills, T. A. E., Heymann, P. W., Longbottom, J. L., and Wilkins, S. R., Airborne allergens associated with asthma: particle size carrying dust mite and rat allergens measured with a cascade impactor, *J. Allergy Clin. Immunol.*, 77, 850, 1986b.

Platts-Mills, T. A. E., Thomas, W. R., Aalberse, R. C., Vervloet, D., and Chapman, M. D., Dust mite allergens and asthma: report of a second international workshop, *J. Allergy Clin. Immunol.*, 89, 1046, 1992.

Platts-Mills, T. A. E., Tovey, E. R., Mitchell, E. B., Moszoro, H., Nock, P., and Wilkins, S. R., Reduction of bronchial hyperreactivity during prolonged allergen avoidance, *Lancet*, 2, 275, 1982.

Pollart, S. M., Fiocco, G., Hayden, M. L., Chapman, M. D., and Platts-Mills, T. A. E., Inhalant allergens as a risk factor in emergency room asthma, *J. Allergy Clin. Immunol.*, 83, 875, 1989.

Pollart, S. M., Smith, T. F., Gelver, L. E., Platts-Mills, T. A. E., and Chapman, M. D., Environmental exposure to cockroach allergens: analysis with monoclonal antibody-based enzyme immunoassays, *J. Allergy Clin. Immunol.*, 87, 505, 1991a.

Pollart, S. M., Mullins, D. E., Vailes, L. D., Hayden, M. L., Platts-Mills, T. A. E., Sutherland, W. M., and Chapman, M. D., Identification, quantification and purification of cockroach allergens using monoclonal antibodies, *J. Allergy Clin. Immunol.*, 87, 511, 1991b.

Price, J. A., Pollock, I., Little, S. A., Longbottom, J. L., and Warner, J. O., Measurement of airborne mite antigen in homes of asthmatic children, *Lancet*, 336, 895, 1990.

Rose, G., Woodfolk, J. A., Hayden, M. L., and Platts-Mills, T. A. E., Testing of methods to control mite allergen in carpets fitted to concrete slabs, *J. Allergy Clin. Immunol.*, 89, 315, 1992.

Sakaguchi, M., Inouye, S., Yasueda, H., Irie, T., Yoshizawa, S., and Shida, T., Measurement of allergens associated with dust mite allergy. II. Concentrations of airborne mite allergens (*Der* I and *Der* II) in the house, *Int. Arch. Allergy Appl. Immunol.*, 90, 190, 1989.

Salvaggio, J. E., Hypersensitivity pneumonitis, *J. Allergy Clin. Immunol.*, 79, 558, 1987.

Schmidt, C. D., Jensen, R. L., Christensen, L. T., Crapo, R. Q., and Davis, J. J., Longitudinal pulmonary function changes in pigeon breeders, *Chest*, 93, 359, 1988.

Schou, C., Fernandez-Caldas, E., Lockey, R. F., and Lowenstein, H., Environmental assay for cockroach allergens, *J. Allergy Clin. Immunol.*, 87, 828, 1991.

Schou, C., Lind, P., Fernandez-Caldas, E., Lockey, R. F., and Lowenstein, H., Identification and purification of an important cross-reactive allergen from American (*Periplaneta americana*) and German (*Blatella germanica*) cockroach, *J. Allergy Clin. Immunol.*, 86, 935, 1990.

Silviu-Dan, F., Sloan, J., Kraut, A., and Warrington, R. J., Inhalant allergy caused by caddis fly, *J. Allergy Clin. Immunol.*, 89, 204, 1992.

Spieksma, F. T., Cultures of house-dust mites on animal skin scales, *Allergol. Immunopathol.*, 4, 419, 1976.

Sporik, R., Holgate, S. T., Platts-Mills, T. A. E., and Cogswell, J., Exposure to house dust mite allergen (*Der p* I) and the development of asthma in childhood: a prospective study, *N. Engl. J. Med.*, 323, 502, 1990.

Sporik, R., Chapman, M., and Platts-Mills, T., Airborne mite antigen (letter), *Lancet*, 336, 1507, 1990.

Storer, R. W., The classification of birds in *Biology and Comparative Physiology of Birds,* Marshall, A. J., Ed., Academic Press, New York, 1960, pp. 70, 72, 77.

Swanson, M. C., Agarwal, M. K., and Reed, C. E., An immunochemical approach to indoor aeroallergen quantitation with a new volumetric air sampler: studies with mite, roach cat, mouse, and guinea pig antigens, *J. Allergy Clin. Immunol.*, 76, 721, 1985.

Swanson, M. C., Campbell, A. R., Klauck, M. J., and Reed, C. E., Correlations between levels of mite and cat allergens in settled and airborne dust, *J. Allergy Clin. Immunol.,* 83, 776, 1989.

Swinton, W. E., The origin of birds, in *Biology and Comparative Physiology of Birds,* Marshall, A. J., Ed., Academic Press, New York, 1960, pp. 1-3.

Terho, E. O., Husman, K., Vohlonen, I., Rautalahti, M., and Tukiainen, H., Allergy to storage mites or cow dander as a cause of rhinitis among Finnish dairy farmers, *Allergy*, 40, 23, 1985.

Terho, E. O., Bohlonen, I., Husman, K., Rautalahti, M., Tukiainen, H., and Viander, M., Sensitization to storage mites and other work-related and common allergens among Finnish dairy farmers, *Eur. J. Respir. Dis.* (Suppl.), 152, 165, 1987.

Tovey, E. R., Chapman, M. D., Wells, C. W., and Platts-Mills, T. A. E., The distribution of dust mite allergen in the houses of patients with asthma, *Am. Rev. Respir. Dis.*, 124, 630, 1981.

Vailes, L., Glime, T., Pollart, S., and Chapman, M. D., Cockroach washes — a potent source of asthma associated allergens, *J. Allergy Clin. Immunol.* (abstract), 85, 171, 1990.

van Hage-Hamsten, M., Johansson, S. G., Johansson, E., and Wiren, A., Lack of allergenic cross-reactivity between storage mites and *Dermatophagoides pteronyssimus, Clin. Allergy,* 17, 23, 1987.

Van Toorenenbergen, A. W., Gerth van Wijk, A., van Dooremalen, G., and Dieges, P. H., Immunoglobulin E antibodies against budgerigar and canary feathers, *Int. Arch. Allergy Appl. Immunol.*, 77, 433, 1985.

Voorhorst, R., Spieksma, F. Th. M., Varekamp, H., Leupen, M. J., and Lyklema, A. W., The house dust mite (*Dermatophagoides pteronyssimus*) and the allergens it produces: identity with the house dust allergen, *J. Allergy,* 39, 325, 1967.

Voorhorst, R., Spieksma, F. Th. M., and Varekamp N., House dust atopy and the house dust mite *Dermatophagoides pteronyssimus*, Stafleu's Scientific Publishing Co., Leiden, 1969.

Vyszenski-Moher, D. L., Arlian, L. G., Bernstein, I. L., and Gallagher, J. S., Prevalence of house dust mites in nursing homes in southwest Ohio, *J. Allergy Clin. Immunol.*, 77, 715, 1986.

Warner, J. O. and Warner, J. A., Airborne mite allergen (letter), *Lancet*, 337, 1038, 1991.

Wood, R. A., Eggleston, P. A., Mudd, K. E., and Adkinson, N. F., Indoor allergen levels as a risk factor for allergic sensitization, *J. Allergy Clin. Immunol.*, 83, 199, 1989.

Wood, R. A., Mudd, K. E., and Eggleston, P. A., The distribution of cat and dust mite allergens on wall surfaces, *J. Allergy Clin. Immunol.*, 89, 126, 1992.

Yoshizawa, S., Sugawra, F., Yasueda, H., Shida, T., Irie, T., Sakaguchi, M., and Inouye, S., Kinetics of the falling of airborne mite allergens (*Der* I and *Der* II), *Jpn. J. Allergol.,* 10, 135, 1991.

8 MAMMALIAN AEROALLERGENS

Christina M. Luczynska

CONTENTS

I. INTRODUCTION

A. The Mammalian Class

The animal kingdom is divided into 21 phyla with over a million species identified and described. Mammals are one of the classes of the phylum Chordata, which includes fish, amphibia, reptiles, and birds. Animals with a backbone are commonly referred to as vertebrates, and have been investigated more completely than any other animals because of their direct structural and functional relationship with humans.

Mammals are warm-blooded animals that feed their young with secretions (milk) from the mother's mammary glands. There are around 4500 species of mammals, ranging from the tiny Kitti's hog-nosed bat to the 12-ton African

0-87371-724-4/95/$0.00+$.50

elephant. Mammals have been successful in a wide variety of habitats and include aquatic whales and dolphins, hoofed ungulates, flesh-eating carnivores, rodents and insectivores, the aerial bats, and the primates, which include humans.

B. An Overview of the Problem

A wide range of materials derived from mammals contains potentially allergenic material. These materials include hair, dander, serum, saliva, and urine, all of which humans can be exposed to during contact with animals. This review focuses on airborne allergens. However, it should be noted that mammalian proteins can also cause severe reactions in allergic individuals on skin contact (e.g., saliva) and ingestion (e.g., milk proteins).

The mammalian source allergens that have been described are acidic proteins that fall within the 16 to 50 kDa range (Dreborg et al., 1986; Lowenstein et al., 1987). The nature of mammalian allergens has recently been reviewed by Schou (1993). Sensitization to mammalian allergens can occur in susceptible individuals following contact with animals either in the home environment or through occupational exposure. Dogs and cats are extremely popular companion animals that are commonly kept as pets. Various rodents (mice, gerbils, hamsters, guinea pigs) are also kept indoors as pets in both homes and schools. Rodents can also occur as pests in indoor environments.

In addition to exposure from animals that occur in the home, 2.7% of the population in the United States own an average of three horses, so dander from horses and farm animals is brought into the home on clothing and contributes greatly to the antigenic composition of house dust (Mansfield and Nelson, 1982). Therefore, in homes, animal-derived allergen can either be shed indoors from pets or pests or carried in on clothing and body surfaces from pets or farm animals kept outdoors.

Occupational exposure to animals is also common, involving farmers, zookeepers, pet-shop workers, and technicians exposed, in particular, to small mammals in research laboratories. Studies have shown that 11 to 30% of exposed individuals develop laboratory animal allergy (Cockcroft et al., 1981). In one study, Beeson et al. (1983) showed that 22% of workers in a large pharmaceutical company were allergic to laboratory animals, and of these 67% were atopic (see Chapter 10).

C. Approaches to the Measurement of Mammalian Allergens

Over the last 10 years, several kinds of immunoassay have been used to detect allergen in air samples and dust extracts (see Chapter 11). Histamine release assays provide an indication of the presence of allergens in general, but are not specific for any one source or any one specific allergen. These assays can be used to determine whether or not people associated with exposure to some unknown potential allergen source have developed sensitivity. The RAST (radioallergosorbent test) can be used in a similar way, except only those substances (usually proteins) that can bind to the solid phase are detected. RAST inhibition is a crude measure of allergen mixtures in environmental samples, and depends on a supply of sensitized human serum. It is especially useful for situations where the offending allergen is not available in purified form. Inhibition RIA (radioimmunoassay) and two-site RIA and ELISAs (enzyme-linked immunosorbent assays) measure specific proteins, usually specific major allergens, and are especially useful for surveys of the prevalence of these particular allergens. The RIA and ELISA assays are sensitive for the specific allergen, but it must be assumed that the presence of these single proteins reflects the distribution of all proteins from the source. All of these methods have been used for the measurement of mammalian allergens in the environment or for evaluating sensitization of exposed people. Many animal allergens have been well defined, although two-site ELISAs are available for only a few (Table 8.1).

II. THE MAMMALS ASSOCIATED WITH ALLERGIC DISEASE

A. Pets

Cats. Cats are the second most popular companion animals after dogs, being kept as pets in an estimated 28% of American homes (Beck and Meyers, 1987). Approximately 2% of the U.S. population have been found to be allergic to cats (Gergen et al., 1987). In the United States and northern Europe many cat owners tend to own more than one cat and the trend of keeping cats housebound is increasing, particularly in urban areas. Exposure to cat allergen is almost unavoidable, and some individuals have positive skin tests to cat dander extracts even though they have not lived in a house with a cat.

The domestic cat produces a number of different allergens, but the most abundant in terms of quantity of allergen produced is *Fel d* I (*Felis domesticus* I), formerly known as Cat 1. *Fel d* I is a 36-kDa acidic protein dimer that was first purified and characterized by Ohman et al. (1977). Immunochemical studies have shown that *Fel d* I is present in parotid tissue and may be transferred through saliva during grooming (Ohman et al., 1974). Bartholome et al. (1985) suggested that *Fel d* I might originate from sebaceous glands, and Charpin et al. (1991) went on to show that antigen is produced in the intradermal hair follicles and accumulates in the hair roots before spreading to the tip of the hair shaft. It has also been shown that the antigen is produced by sebaceous glands and to a lesser extent by basal epithelial cells and that high concentrations are found on the fur and skin (Dabrowski et al., 1990). Morgenstern et al. (1991) suggested that the form of *Fel d* I produced by the salivary glands has a different molecular weight than that derived from the skin.

Table 8.1. Animal Allergens

	Exposure situation	Defined allergens	Sources	Airborne particle size
Cats	House pets	*Fel d* I	Saliva Sebaceous glands	<2.5 μm
Dogs	House pets	*Can f* I	Dander	—
Mice	Pests Laboratories	*Mus n* I	Urine	—
Rats	Pests Laboratories	*Rat n* IA *Rat n* IB	Urine	<10 μm
Guinea pigs	House pets Laboratories	*Cav p* I *Cav p* II	Saliva Urine	—
Rabbits	House pets Laboratories	*Ory c* I	Saliva	<2 μm
Cows	Farm	*Bos d* II	Urine	—
Horse	Farm Occupational	*Equ c* I *Equ c* II *Equ c* III	Dander Dander Dander	—
Pigs	Farm		Dander Urine	—
Bats	Occupational		Guano	—
Reindeer	Occupational		Epidermis	—

Wentz et al. (1990) found that cat allergen shedding varied not only between cats but also in the same cat during the course of a day and between days. Male cats were found to shed more allergen than female cats. Many patients do in fact report more severe allergic symptoms on exposure to some cats than to others. Monoclonal antibodies to *Fel d* I and an assay to detect the allergen in environmental samples were described by Chapman et al. (1988) and Luczynska et al. (1990). Detectable levels of *Fel d* I were present in virtually all houses tested by Wood et al. (1988). In another study, houses with cats had in excess of 10 μg *Fel d* I/g of dust, while most houses without cats had levels below 1 μg/g (Chapman et al., 1988; Ohman and Lorusso, 1987). Studies have shown that houses with cats contained between 2 and 20 ng/m^3 airborne *Fel d* I. Overall, a provisional value of 8 μg *Fel d* I/g of dust has been proposed as being indicative of significant exposure to cat allergen (Gelber et al., 1993). However, very high levels of more than 1 mg/g have often been measured, and individuals entering such a house will become sufficiently contaminated to carry significant allergen back to their own homes.

Cat allergen is carried on extremely small particles of less than 2.5 μm aerodynamic diameter, which easily become airborne, so that a patient with asthma entering a house with a cat will often be immediately aware of the presence of the cat (Luczynska et al., 1990).

Dogs. Dogs are the most popular companion animals and are kept as household pets worldwide, although working dogs in farming areas may be restricted to the outdoor environment. Dogs are found in an estimated 43% of American homes (Beck and Meyers, 1987), and dog antigen has been detected in 63% of homes in one area of the United States (Lind et al., 1987). It is also found to be abundantly distributed in other indoor environments where dogs are not normally kept, such as schools (Dybendal et al., 1989). De Groot et al. (1991) reported that dog dander allergy occurred in 17% of the population in one region studied.

The allergen content of extracts prepared from different dogs varies, probably due to differences in the relative concentrations of shared allergens (Knysak, 1989). Ford et al. (1989) identified 28 different allergens (Ag) in dog hair and dander extracts, 11 of which were also present in dog serum. Binding of immunoglobulin E (IgE) antibody was strongest to Ag23 (dog IgG) and Ag3 (dog serum albumin), and Ag8 and Ag1 were shown to be of slightly lesser importance. Schou et al. (1991b) purified and characterized an important dog hair- and dander-specific allergen equivalent to Ag8 as reported by Ford et al. (1989) and designated it *Can f* I. *Can f* I is a 25-kDa protein to which most dog allergic patients tested by Schou et al. (1991b) reacted. De Groot et al. (1991) reported that saliva is also an important source of *Can f* I.

Recently, de Groot et al. (1991) produced monoclonal antibodies to *Can f* I and developed assays to measure *Can f* I in allergen extracts and dust samples. Schou et al. (1991a), using an ELISA specific for *Can f* I, found that houses with dogs had 120 μg *Can f* I/g of dust. In general, most homes without a dog had less than 10 μg/g. It is not yet known on what particle size the allergen is carried, or whether it is airborne in significant quantities in houses with dogs.

B. Laboratory Animals

Although mice, rats, guinea pigs, and rabbits are kept as pets and mice and rats can be pests, exposure to these is most intense in laboratory animal vivaria.

Mice. Mice are found in houses, although more often as pests than as pets. Mice are kept as pets in only 2% of homes in the United States (Beck and Meyers, 1987). Their prevalence as pests is unknown, as is the prevalence of sensitization related to mouse allergen exposure. The allergenic activity of proteins excreted in the urine of mice was recognized by Newman-Taylor et al. (1977). Two major allergens (Ag1 and Ag3) were detected in all mouse dust extracts studied by Price and Longbottom (1987). Ag1 was shown to be antigenically identical to urinary pre-albumin, which constitutes the major urinary protein (MUP) complex (Finlayson et al., 1965) and is now designated *Mus m* I. It has a molecular mass of 18 kDa (Lorusso et al., 1986). Price and Longbottom (1990) reported that Ag3 was detectable in hair follicles and hence, in fur and dander extracts, and therefore was also important. High levels of airborne antigen in laboratory mouse rooms were shown by Twiggs et al. (1982), up to 825 ng/m^3 of mouse pelt extract protein (Ag3) and 59 ng/m^3 urine protein (*Mus m* I).

Rats. Allergy to rats was first described by Frankland in 1974. Since then, a number of investigators have reported cases of allergy to rats used as laboratory animals (Newman-Taylor et al., 1977; Cockcroft et al., 1981), although the prevalence of the disease remains unknown. Two major allergens from rat urine have been characterized, a-2 euglobulin (Ag13, now known as *Rat n* IB) and pre-albumin (Ag4, now known as *Rat n* IA) (Longbottom, 1980). The *Rat n* IB is a 17-kDa protein that is a major component of male rat urine and is barely detectable in female rat urine (Eggleston et al., 1989). Platts-Mills et al. (1986) sampled air in rooms housing laboratory rats and found that, although a very large quantity of rat allergen was produced (up to 4 g per cage), only a small quantity became airborne. The amount airborne depends on the type of litter and bedding used. Most of the airborne allergen was present on particles less than 10 μm. Eggleston et al. (1989) found that airborne *Rat n* I was significantly higher during cage-cleaning activities (19 to 310 ng/m^3) than during quiet activity (less than 1.5 to 9.7 ng/m^3).

Guinea pigs. The guinea pig is popular as a pet, but it is also widely used as a laboratory animal. Guinea pigs were first thought to provoke asthmatic reactions in sensitized subjects by Salter in 1968, who thought that the reaction was due to the smell of the animals. Studies have shown that guinea pig pelt, fur, serum, and urine can all provoke allergic reactions (Cockcroft et al., 1981; Beeson et al., 1983; Ohman et al., 1975). Walls et al. (1984) compared the allergen sources and found the allergic potency to be greatest in pelt. Two major allergens were found, Ag2 (*Cav p* I) and Ag3 (*Cav p* II). A subsequent study showed that salivary proteins shed in dander could represent the most important source of inhalant allergens (Walls et al., 1985). On the other hand, Swanson et al. (1984) showed that airborne allergen levels in animal rooms ranged from 0.5 to 15 μg/m^3 guinea pig allergen, with urinary proteins being the most important allergens.

Rabbits. Rabbits are relatively common domestic pets but are also important laboratory animals, although prevalence of rabbit-related allergy is unknown. The major rabbit antigen Ag R1 (*Ory c* I) is an important component of rabbit dust, fur, and saliva, but not of urine or dander (Price and Longbottom, 1988a). Warner and Longbottom (1991) reported many other possible allergens in urine, and identified serum albumin as the major allergen in dander. In one study, *Ory c* I appeared to be carried on small particles and was likely to remain airborne for long periods of time (Price and Longbottom, 1988b).

C. Farm Animals

Farmers and farm workers are exposed to a number of different aeroallergens, which include pollens, grain dust, and crop contaminants, such as fungi and storage mites, as well as animal-derived allergens. Although storage mites are thought to be the most important occupational sensitizers in farmers (Iversen et al., 1990), mammalian allergens have also been shown to cause respiratory disease in exposed individuals.

Cows. Cow epithelium has been shown to be an important inducer of occupational asthma in some localized farming communities. Terho et al. (1985) found that 20% of Finnish dairy farmers with allergic rhinitis showed positive reactions on nasal challenge with cow dander extract. A 20-kDa protein has been identified in both cow dander and urine and is designated *Bos d* II (Ylonen et al., 1990). Farmers with bovine induced allergic rhinitis were shown to have both IgG and IgE antibodies to this allergen (Ylonen et al., 1992). Although a number of other allergenic components have also been described (Prahl and Lowenstein, 1978; Prahl, 1981), *Bos d* II appears to be the most important. Virtanen et al. (1986) used an inhibition ELISA to measure airborne bovine epidermal antigen in cowsheds and found that levels of allergen varied up to 19.8 $\mu g/m^3$.

Horses. Exposure to horse allergens can occur in the occupational environment or through horse riding as a sport. The horse is known as a source of a very potent allergen to which severe allergic reactions have been noted. Allergenic activity has been found in horse serum and epithelium (Makussen et al., 1976), and epithelium has been used as a source of material for allergen extract production (Franke et al., 1990). Horse hair and dander have been found to contain three major allergens, Ag6 (*Equ c* I), Ag9 (*Equ c* II), and Ag11 (*Equ c* III) (Lowenstein et al., 1976). In addition, horse albumin is considered to be an allergen of minor importance (Wahn et al., 1982).

Pigs. Sensitization to pigs has been reported by Katila et al. (1981). Donham et al. (1986) found antibodies against swine dander, epithelium, and urine in the sera of swinery workers, but these antibodies were also present in controls. Virtanen (1990) used an ELISA technique to measure airborne swine epithelial antigens, where up to 300 μg of antigen/m^3 could be detected in workers' breathing zones. However, pig antigens are still poorly characterized.

D. Unusual Mammal Exposures

A more unusual type of occupational sensitization to mammals was described by El-Ansary et al. (1987), who described asthma associated with exposure to bat guano in workers exposed to bats in indoor working environments in the Sudan. Allergy to yellow hairy and black bat occurred when bat droppings accumulating from bats roosting in the roof spaces were inhaled.

Allergic reactions to reindeer epithelial extracts have recently been demonstrated among reindeer herders in northern Finland (Larmi et al., 1988). In addition, Reijula et al. (1992) developed an inhibition ELISA to reindeer epithelial antigen and detected concentrations of 0.1 to 3.9 μg/m^3 reindeer antigens in the air of a workshop used to process reindeer leather.

III. CONTROL OF EXPOSURE TO ANIMAL ALLERGENS

A. The Home Environment

Ideally, to eliminate exposure to allergens released from pets, the pet animal should be removed from the house. Removal of the animal may not result in immediate relief even when followed by vigorous cleaning, since allergen has been shown to remain in the home for many months (Wood et al., 1989). However, it is possible that rigorous vacuuming (using effective vacuum cleaners that do not increase airborne allergen levels) may be beneficial (Woodfolk et al., 1993). In addition, there is some suggestion that low concentrations of *Fel d* I in carpets, such as may be found after a cat has been removed, may be reduced by treatment with tannic acid (Miller et al., 1990).

However, allergy to pets constitutes a particular problem in that the allergic patient is often unwilling to part with the pet and consequently risks chronic respiratory disease. In surveys carried out in the United States, over half the allergists questioned believed that patients had emotional attachments to the pet. More than half of the physicians reported observing guilt feelings, emotional reactions, or psychological trauma following elimination of pets (Baker, 1979; Baker and McCulloch, 1983). In a survey of 22 families, which included allergic members, 73% said they would not give up their pets even if advised to do so by their physicians (Herring et al., 1981). If the pet cannot be removed from the house, it should be kept out of the allergic person's bedroom.

Because of the very high concentration of cat allergen in houses with cats, measures to reduce allergen levels are of limited value unless the cat is removed from the house. Soft furnishings are an important reservoir of cat allergen, so treating or reducing the number of soft furnishings may be beneficial in reducing the allergen load. Van der Brempt et al. (1991) showed that as mattresses were an important source of cat allergen in the home, mattress covers may be of some benefit. Air filtration is only effective if the carpet is either cleaned or removed first (Luczynska et al., 1990). Recently, de Blay et al. (1991) reported that if the carpet is removed and the cat is washed regularly, it may be possible for the cat to remain in the house of a cat-sensitive patient.

B. The Occupation Environment

It is usually impossible to deal with occupational exposure to mammalian allergens by removing the source. Hence, protection is only possible by the use of masks and protective clothing, or the use of air filtration and forced ventilation, or local exhaust ventilation. The concentration of any substance in the air is determined by the ratio of the rate of production to the rate of removal.

Thus, stock density has been shown to be an important factor in determining the amount of airborne rat urinary allergen in rat rooms (Gordon et al., 1991). Swanson et al. (1990) showed that male rats shed about 20 ng of allergen/min when the ventilation rate was about 15 air changes per hour, and that in a room of 300 rats the ventilation rate would need to be 172 air changes per hour to achieve reasonable control.

Solomon (1987) suggested the following points for investigating the animal-associated environment. The species, strain, and sex of the animals, stock density, and how the animals are bedded and fed should be assessed. The humidity, temperature, and air exchange rates should all be measured, as these can influence the particle size distribution and therefore the ease with which allergens can become airborne.

REFERENCES

Baker, E., A veterinarian looks at the allergy animal problem, *Ann. Allergy*, 43, 214, 1979.

Baker, E. and McCulloch, M. J., Allergy to pets: problems for the allergist and the pet owner, in *Perspectives on Our Lives with Companion Animals*, Katcher, A. H. and Beck, A. M., Eds., University of Pennsylvania Press, Philadelphia, 1983, 341.

Bartholome, K., Kissler, W., Baer, H., Kopietz-Schulte, E., and Wahn, U., Where does cat allergen 1 come from?, *J. Allergy Clin. Immunol.*, 76, 503, 1985.

Beck, A. M. and Meyers, N. M., The pet owner experience, *N. Engl. Regul. Allergy Proc.*, 8, 185, 1987.

Beeson, M. F., Dewdney, J. M., Edwards, R. G., Lee, D., and Orr, R. G., Prevalence and diagnosis of laboratory animal allergy, *Clin. Allergy*, 13, 433, 1983.

Chapman, M. D., Aalberse, R. C., Brown, M. J., and Platts-Mills, T. A. E., Monoclonal antibodies to the major feline allergen *Fel d* I. II. Single step affinity purification of *Fel d* I, N-terminal sequence analysis and development of a sensitive two-site immunoassay to assess *Fel d* I exposure, *J. Immunol.*, 140, 812, 1988.

Charpin, C., Mata, P., Charpin, D., Lavant, M. N., Allasia, C., and Vervloet, D., *Fel d* I allergen distribution in cat fur and skin, *J. Allergy Clin. Immunol.*, 88, 77, 1991.

Cockcroft, A., Edwards, J., McCarthy, P., and Andersson, N., Allergy in laboratory animal workers, *Lancet,* i, 827, 1981.

Dabrowski, A. J., Van der Brempt, X., Soler, M., Seguret, N., Lucciani, P., Charpin, D., and Vervloet, D., Cat skin as an important source of *Fel d* I allergen, *J. Allergy Clin. Immunol.*, 80, 462, 1990.

de Blay, F., Chapman, M. D., and Platts-Mills, T. A. E., Airborne cat allergen (*Fel d* I). Environmental control with the cat *in situ*, *Am. Rev. Respir. Dis.*, 143, 1334, 1991.

de Groot, H., Goei, K. G. H., van Swieten, P., and Aalberse, R. C., Affinity purification of a major and minor allergen from dog extract: serological activity of affinity-purified *Can f* I and of *Can f* I-depleted extract, *J. Allergy Clin. Immunol.*, 87, 1956, 1991.

Dreborg, S., Einarson, R., and Longbottom, J. L., The chemistry and standardisation of allergens. In: *Handbook of Experimental Immunology*, Weir, D. M., Ed., 4th ed., Vol. 1, 1986, 10.1.

Donham, K. J., Scallon, L. J., Popendorf, W., Treuhaft, R. W., and Roberts, R. C., Characterization of dusts collected from swine confinement buildings, *Am. Ind. Hyg. Assoc.*, 47, 404, 1986.

Dybendal, T., Vik, H., and Elsayed, S., Dust from carpeted and smooth floors. II. Antigenic and allergenic content of dust vacuumed from carpeted and smooth floors in schools under routine cleaning schedules, *Allergy,* 44, 401, 1989.

Eggleston, P. A., Newill, C. A., Ansari, A. A., Pustelnik, A., Lou, S.-R., Evans, R., III, Marsh, D. G., Longbottom, J. L., and Corn, M., Task-related variation in airborne concentration of laboratory animal allergens: studies with *Rat n* I, *J. Allergy Clin. Immunol.*, 84, 347, 1989.

El-Ansary, E. H., Tee, R. D., Gordon, D. J., and Newman Taylor, A. J., Respiratory allergy to inhaled bat guano, *Lancet,* i, 316, 1987.

Finlayson, J. S., Potter, M., and Runner, C. R., Electrophoretic variation and sex dimorphism of the major urinary protein complex in inbred mice: a new genetic marker, *J. Natl. Cancer Inst.,* 31, 91, 1965.

Ford, A. W., Alterman, L., and Kemeny, D. M., The allergens of dog. I. Identification using crossed radio-immunoelectrophoresis, *Clin. Exp. Allergy,* 19, 183, 1989.

Franke, D., Maasch, H. J., Wahl, R., Schultze-Werninghaus, G., and Bretting, H., Allergens of horse epithelium. I. Physiochemical characterisation of five different horse epithelium raw materials used for allergen extract preparation, *Int. Arch. Allergy Appl. Immunol.*, 92, 309, 1990.

Frankland, A. W., Rat asthma in laboratory workers, *Excerpta Med.*, 123, 1974.

Gelber, L. E., Seltzer, L. H., Bouzoukis, J. K., Pollart, S. M., Chapman, M. D., and Platts-Mills, T. A. E., Sensitization and exposure to indoor allergens as risk factors for asthma among patients presenting to hospital, *Am. Rev. Respir. Dis.,* 147, 573, 1993.

Gergen, P. J., Turkeltaub, P. C. and Kovar, M. G., The prevalence of allergic skin test reactivity to eight common aeroallergens in the US population: results from the second National Health and Nutrition Examination Survey, *J. Allergy Clin. Immunol.*, 80, 669, 1987.

Gordon, S., Tee, R. D., Lowson, D., and Newman-Taylor, A. J., Influence of stock density and cleaning out on static and personal rat urinary aeroallergen (RUA) exposure, *Clin. Exp. Allergy,* 22, 118, 1991 (abstr.).

Herring, S., McGready, S. J., Jones, J. D., and Mansmann, H. C., The maintenance of pets in allergic families. I. A survey of health beliefs, *Ann. Allergy,* 46, 24, 1981.

Iversen, M., Korsgaard, J., Hallas, T., and Dahl, R., Mite allergy and exposure to storage mites and house dust mites in farmers, *Clin. Exp. Allergy,* 20, 211, 1990.

Katila, M.-L., Mantyjarvi, R. A., and Ojanen, T. H., Sensitisation against environmental antigens in swine workers, *Br. J. Ind. Med.*, 38, 334, 1981.

Knysak, D., Animal aeroallergens. In: *Airborne Allergens,* Solomon, W. R., Ed., *Immunol. Allergy Clin. North Am.,* 9, 357, 1989.

Larmi, E., Reijula, K., Hannuksela, M., Pikkarainen, S., and Hassi, J., Skin disorders and prick patch test reactivity in Finnish reindeer herders, *Dermatosen,* 36, 83, 1988.

Lind, P., Norman, P. S., Newton, M., Lowenstein, H., and Schwartz, B., The prevalence of indoor allergens in the Baltimore area: house dust mite and animal-dander antigens measured by immunochemical techniques, *J. Allergy Clin. Immunol.*, 80, 541, 1987.

Longbottom, J. L., Purification and characterisation of allergens from urine of mice and rats. In: *Advances in Allergology and Immunology*, Oehling, A., Ed., Pergammon Press, Oxford, 1980, 483.

Lorusso, J. R., Moffat, S., and Ohman, J. L., Immunologic and biochemical properties of the major mouse urinary allergen (*Mus m* I), *J. Allergy Clin. Immunol.,* 78, 928, 1986.

Lowenstein, H., Ipsen, H., Lind, P., and Matthiesen, F., The physiochemical and biological characteristics of allergens. In: *Allergy: An International Textbook*, Lessof, M. H., Lee, T. H., Kemeny, D. M., Eds., John Wiley & Sons, New York, 1987, 87.

Lowenstein, H., Markussen, B., and Weeke, B., Isolation and partial characterization of three major allergens of horse hair and dandruff, *J. Allergy Clin. Immunol.,* 51, 48, 1976.

Luczynska, C. M., Li, Y., Chapman, M. D., and Platts-Mills, T. A. E., Airborne concentrations and particle size distribution of allergen derived from domestic cats (*Felis domesticus*): measurements using a cascade impactor, liquid impinger and a two-site monoclonal antibody assay for *Fel d* I, *Am. Rev. Respir. Dis.,* 141, 361, 1990.

Makussen, B., Lowenstein, H., and Weeke, B., Allergen extracts of horse hair and dandruff. Quantitative immunoelectrophoretic characterisation of the antigens, *Int. Arch. Allergy Appl. Immunol.*, 51, 25, 1976.

Mansfield, L. E. and Nelson, H. S., Allergens of commercial house dust, *Ann. Allergy,* 48, 205, 1982.

Miller, J. D., Miller, A., Kaminsky, K., Gelber, L., Chapman, M. D., and Platts-Mills, T. A. E., Effect of tannic acid spray on cat allergen levels in carpets, *J. Allergy Clin. Immunol.*, 85, 226, 1990 (abstr.).

Morgenstern, J. P., Griffith, I. J., Brauer, A. W., Rogers, B. L., Bond, J. F., Chapman, M. D., and Kuo, M. C., Amino acid sequence of *Fel d* I, the major allergen of the domestic cat: protein sequence analysis and cDNA cloning. *Proc. Nat. Acad. Sci.,* 88, 9690, 1991.

Newman-Taylor, A. J., Longbottom, J. L., and Pepys, J., Respiratory allergy to urine proteins of rats and mice, *Lancet,* ii, 847, 1977.

Ohman, J. L. and Lorusso, J. R., Cat allergen content of commercial house dust extracts: comparison with dust extracts from cat containing environments, *J. Allergy Clin. Immunol.,* 79, 955, 1987.

Ohman, J. L., Lowell, F. C., and Bloch, K. J., Allergens of mammalian origin. III. Properties of a major feline allergen, *J. Immunol.,* 113, 1668, 1974.

Ohman, J. L., Lowell, F. C., and Bloch, K. J., Allergens of mammalian origin. II. Characterization of allergens extracted from rat, mouse, guinea pig and rabbit pelts, *J. Allergy Clin. Immunol.*, 55, 16, 1975.

Ohman, J. L., Kendall, S., and Lowell, F. C., IgE antibody to cat allergens in an allergic population, *J. Allergy Clin. Immunol.*, 60, 317, 1977.

Platts-Mills, T. A. E., Heymann, P. W., Longbottom, J. L., and Wilkins, S. R., Airborne allergens associated with asthma: particle sizes carrying dust mite and rat allergens measured with a cascade impactor, *J. Allergy Clin. Immunol.,* 77, 850, 1986.

Prahl, P., Allergens in cow hair and dander, *Allergy,* 36, 561, 1981.

Prahl, P. and Lowenstein, H., Quantitative immunoelectrophoretic analysis of extracts from cow hair and dander, *Allergy,* 33, 241, 1978.

Price, J. A. and Longbottom, J. L., Allergy to mice. I. Identification of two major mouse allergens (Ag1 and Ag3) and investigation of their possible origin, *Clin. Allergy,* 17, 43, 1987.

Price, J. A. and Longbottom, J. L., Allergy to rabbits. II. Identification and characterisation of a major rabbit allergen, *Allergy,* 43, 39, 1988a.

Price, J. A. and Longbottom, J. L., ELISA method for measurement of airborne levels of major laboratory animal allergens, *Clin. Allergy,* 18, 95, 1988b.

Price, J. A. and Longbottom, J. L., Allergy to mice. II. Further characterisation of two major mouse allergens (Ag1 and Ag3) and immunohistochemical investigations of their sources, *Clin. Exp. Allergy,* 20, 71, 1990.

Reijula, K., Virtanen, T., Halmepuro, L., Anttonen, H., Matyjarvi, R., and Hassi, J., Detection of airborne reindeer epithelial antigen by enzyme-linked immunosorbent assay inhibition, *Allergy,* 47, 203, 1992.

Salter, J. H., *On Asthma: Its Pathology and Treatment*, 2nd ed., John Churchill & Sons, London, 1968.

Schou, C., Defining allergens of mammalian origin, *Clin. Exp. Allergy,* 23, 7, 1993.

Schou, C., Hansen, G., Littner, T., and Lowenstein, H., Assay for dog allergen, *Can f* I: Investigation of house dust samples and commercial dog extracts, *J. Allergy Clin. Immunol.*, 88, 847, 1991a.

Schou, C., Svendsen, U. G., and Lowenstein, H., Purification and characterisation of the major dog allergen, *Can f* I, *Clin. Exp. Allergy,* 21, 321, 1991b.

Solomon, W. R., Assessing the animal associated environment, *N. Engl. Regul. Allergy Proc.*, 8, 169, 1987.

Swanson, M. C., Agarwal, M. K., Reed, C. E., and Yuninger, J. W., guinea pig derived allergens: clinico-immunologic studies, characterization, airborne quantities and size distribution, *Am. Rev. Respir. Dis.*, 129, 844, 1984.

Swanson, M. C., Campbell, A. R., O'Hollaren, M. T., and Reed, C. E., Role of ventilation rate, air filtration and allergen production rate on determining concentration of rat allergens in the air of animal quarters, *Am. Rev. Respir. Dis.,* 141, 1578, 1990.

Terho, E. O., Husman, K., Vohlonen, I., Rautalahti, M., and Tukianinen, H., Allergy to storage mites or cow dander as a cause of rhinitis among dairy farmers, *Allergy,* 40, 23, 1985.

Twiggs, J. T., Agarwal, M. K., Dahlberg, M. J. E., and Yuninger, J. W., Immunochemical measurement of airborne mouse allergens in a laboratory animal facility, *J. Allergy Clin. Immunol.,* 69, 522, 1982.

Van der Brempt, X., Charpin, D., Haddi, E., de Mata, P., and Vervloet, D., Cat removal and *Fel d* I levels in mattresses, *J. Allergy Clin. Immunol.*, 87, 595, 1991.

Virtanen, T., Two ELISA methods for the measurement of airborne swine epithelial antigens, *J. Immunoassay,* 11, 63, 1990.

Virtanen, T., Louhelainene, K., and Mantyjarvi, R., Enzyme-linked assay (ELISA) inhibition method to estimate the level of airborne bovine epidermal antigen in cowsheds, *Int. Arch. Allergy Appl. Immunol.,* 81, 253, 1986.

Wahn, U, Herold, U., Danielsen, K., and Lowenstein, H., Allergo-prints in horse allergic children, *Allergy,* 37, 335, 1982.

Walls, A. F., Newman-Taylor, A. J., and Longbottom, J. L., Allergy to guinea pigs. I. Allergenic activities of extracts derived from the pelt, saliva, urine and other sources, *Clin. Allergy,* 15, 241, 1984.

Walls, A. F., Newman-Taylor, A. J., and Longbottom, J. L., Allergy to guinea pigs. II. Identification of specific allergens in guinea pig by crossed radio-immunoelectrophoresis and investigation of the possible origin, *Clin. Allergy,* 15, 535, 1985.

Warner, J. A. and Longbottom, J. L., Allergy to rabbits. III. Further identification and characterisation of rabbit allergens, *Allergy,* 46, 481, 1991.

Wentz, P. E., Swanson, M. C., and Reed, C. E., Variability of cat allergen shedding, *J. Allergy Clin. Immunol.,* 85, 94, 1990.

Wood, R. A., Chapman, M. D., Adkinson, F. N., and Eggleston, P. A., The effect of cat removal on allergen content in household dust samples, *J. Allergy Clin. Immunol.,* 83, 730, 1989.

Wood, R. A., Eggleston, P. A., Lind, P., Ingemann, L., Schwartz, B., Graveson, S., Terry, D., Wheeler, B., and Adkinson, N. F., Antigenic analysis of household dust samples, *Am. Rev. Respir. Dis.,* 137, 358, 1988.

Woodfolk, J. A., Luczynska, C. M., de Blay, F., Chapman, M. D., and Platts-Mills, T. A. E., The effect of vacuum cleaners on the concentration and particle size distribution of airborne cat allergen, *J. Allergy Clin. Immunol.,* 91, 829, 1993.

Ylonen, J., Mantyjarvi, A., Taivainen, A., and Virtanen, T., IgG and IgE antibody responses to cow dander and urine in farmers with cow-induced asthma, *Clin. Exp. Allergy,* 22, 83, 1992.

Ylonen, J., Nuutinen, J., Rautiainene, M., Ruoppi, P., and Mantyjarvi, R., Comparative analysis of bovine extract by immunoblotting and ELISA inhibition, *Allergy,* 45, 30, 1990.

9 THE OUTDOOR AEROSOL

Michael L. Muilenberg

CONTENTS

I. INTRODUCTION

A wide variety of particles of biological origin can be found in the atmosphere (bioaerosols). Wind-pollinated flowers depend on air movement to transport pollen to female flower parts of other same-species plants. The fungi

0-87371-724-4/95/$0.00+$.50

rely almost exclusively on air transport for spread to new environments. Although concentrations may be very low, bioaerosols are found the world over, even over the oceans and the arctic (Kramer et al., 1973; Meier, 1935). Bioaerosols are generally concentrated in the lowest portions of the atmosphere, close to their sources. However, fungal spores and bacteria have been recovered at altitudes of over 40,000 ft (Greene et al., 1964).

Historically, the existence of bioaerosols has been recognized for more than 100 years.

> There is no respite to our contact with the floating matter of the air. We not only suffer from its mechanical irritation, but it is a growing belief that a portion of it lies at the root of a class of disorders most deadly to man. (Tyndall, 1884)

This statement by the British physicist John Tyndall (1820 to 1893), with its typical 19th century dramatic flair, was made well over 100 years ago and proved to be surprisingly accurate. Tyndall's statement was based on his work and that of his contemporaries, including Pasteur (see below) and the German naturalist Theodor Schwann (1820 to 1882). Schwann demonstrated in 1837 that meat exposed to heat-purified air would not putrefy, and concluded that putrefaction was caused, not by spontaneous generation, but by something in the air that could be destroyed by high temperatures. Studies intended to better characterize these airborne biological particles were performed by the French chemist Louis Pasteur (1822 to 1895), who also worked to disprove the theory that organisms arise by spontaneous generation. Pasteur filtered air through gun cotton, dissolved the filter, and microscopically analyzed the particles recovered from air. Tyndall (1884) repeated and elaborated upon many of Pasteur's experiments with filtered air. Many of these studies emphasized the importance of bacteria and fungi as contaminants of food and other nutrient sources.

C. H. Blackley did much of the pioneering work on airborne pollen during the last third of the 19th century. He used adhesive coated glass slides, a method still in common use today. Hans Molisch (1856 to 1937), a German botanist and grass allergy sufferer, made references to and expressed concern about the health effects of "aeroplankton", which included pollen, spores, and plant trichomes (hairs), in addition to dust of nonbiologic origin. Another pioneer in the field of airborne biological matter, and the originator of the term "aerobiology", was the plant pathologist Fred C. Meier (1893 to 1938). Meier, working for the U.S. Department of Agriculture, used adhesive-coated glass slides and agar-filled petri dishes suspended from airplanes to study fungal spores and other bioaerosols in an effort to better understand the long-distance transport of plant diseases. Using airplane-borne samplers, he collected spores on numerous flights over the United States, the Arctic, and the Caribbean.

Meier, along with five other passengers and nine crew, were on the first flight of a study to look at long-range dispersal of fungal pathogens when his plane was tragically lost over the Pacific Ocean.

Probably one of the most influential "modern" pioneers of aerobiology was Philip H. Gregory (1907 to 1986), an English phytopathologist who was also interested in the airborne transport of fungi and bacteria and the spread of plant disease. His book *Microbiology of the Atmosphere* is one of the important milestones in the study of aerobiology.

Another modern aerobiologist (to whom this volume is dedicated) is William S. Benninghoff (1918 to 1993). Professor Benninghoff was instrumental in focusing the American scientific community on the principles of aerobiology. In 1964, he stated that "of organic particulate matter in the atmosphere, only bacteria, fungal spores, and spores and pollen of higher plants have been quantitatively sampled and systematically identified." (Benninghoff, 1964). Although these particles are relatively easily sampled and counted, many are difficult or impossible to accurately identify, and we have probably only scratched the surface of the wealth of knowledge that could be gained from research on the outdoor aerosol.

II. PRINCIPLES OF OUTDOOR AEROBIOLOGY

A. Nature of the Particles

The types of particles considered here as bioaerosols cover a very large size range: from viruses, which are as small as a few hundred angstroms (100 Å = 0.01 μm), up to some of the larger pollen grains, which are over 0.1 mm. These larger particles might more properly be called "airborne biological particles" as they are too large to act as true aerosols. However, due to widely accepted usage, the term "bioaerosols" is used here for all organisms, their emanations, and particles of biological origin, smaller than a few hundred micrometers, that are found in the air for extended periods of time and are not airborne via any mechanism of active flying (e.g., small insects).

Bioaerosols originate from diverse sources and can serve a number of different functions. Some bioaerosols are viable organisms and serve as dispersal stages or units (e.g., fungal spores), while others function as agents for the exchange of genetic material (e.g., pollen). Many bioaerosols are not viable but originate from viable organisms (e.g., insect scales) or are metabolic products of organisms (e.g, feces). (Note that the term "viable" means alive and able to grow, while "culturable" is used for viable organisms that can be recovered using artificial culture.) Viable bioaerosols will interact with, and be impacted by, their environment (that is, the air) in different ways than nonviable biological emanations. These different particles will impact human health in very different ways.

B. Biological Consequences of the Airborne State

Airborne organisms are under a variety of stresses (desiccation, radiation, oxygen toxicity, chemical pollutants, etc.), which they may not encounter while not in the airborne state. Bacterial vegetative cells seem to be more susceptible to these stresses than are fungal and plant spores, pollen, and possibly algae. Mancinelli and Shulls (1978), using filter samplers, reported that outdoor culturable bacterial recoveries significantly and directly correlated with total particulates (Pearson correlation coefficient 0.56), inversely correlated with levels of nitric oxide (–0.45) and nitrogen dioxide (0.43), but were not correlated with relative humidity or SO_2. In chamber studies, airborne survival of *Chlamydia pneumoniae* (Theunissen et al., 1993) and *Escherichia coli* (Cox, 1968, 1970) has been shown to be directly related to relative humidity. This is not always a positive correlation (high humidity, higher survival); Cox (1989, p. 487) stated that, in general, Gram negative bacteria (including *E. coli*) seem to be less stable at mid- to high relative humidities than at lower humidity levels. Theunissen et al. (1993) found that the highest percentage of cells often die shortly after aerosolization, a phenomenon not entirely understood but reported by others and thoroughly discussed by Cox (1987). In addition to these humidity stresses, oxygen, especially under desiccating conditions, can be toxic to some bacteria (Hess, 1965; Strange and Cox, 1976; see Chapter 2).

Other factors in outdoor air, such as chemical pollution, radiation, etc., also affect bacterial survival (Cox, 1989). For some time it was noticed that bacteria that were adversely affected when airborne outdoors showed no decrease in culturability when exposed to the same "outdoor" air that was piped into a chamber indoors. Even when factors such as radiation and light were adequately controlled, this effect persisted. This effect, now called the open air factor (OAF), is attributed to ozone-olefin reaction products, which, when brought indoors, readily condense onto tubing or other container surfaces. Most of these effects have not been well studied for fungal spores or other airborne biological particulates.

It should be noted that the above environmental stresses may not necessarily kill an organism but only (temporarily) affect its ability to grow or replicate. Cox (1987) points out that these stresses can affect an organism's ability to replicate (and therefore its culturability), but also its ability to repair itself, which may be of critical importance. Other factors, such as water availability, temperature, nutrient level, and pollutants, can also affect an organism's repair mechanisms after the organism adheres to a surface. These effects contribute to the inaccuracy of culture as a measure of viability. Organisms that can, for example, be counted but not cultured may be dead, damaged, or simply unable to grow under the conditions presented.

There is some speculation that rainfall, or even clouds or fog, may have an effect on the soluble materials in airborne particles. For example, pollen antigens may be leached from the grain by water in the air and become associated with small droplets or with other airborne "rafts" (Solomon et al., 1983; Habenicht et al., 1984). This non-pollen-grain-associated antigen could also result from pollen collecting in standing water. Splash mechanisms could then aerosolize antigen-laden small droplets.

C. Sources and Source Strengths

As mentioned above, bioaerosols originate from many sources, for example, humans (skin scales), animals (dried saliva), plants (pollen and spores), fungi (spores, toxins, volatile organic compounds), bacteria (cells, cell fragments, spores, toxins), and viruses. The strength of the source is defined as the number of particles available for dispersal, and is dependent (among other factors) on the number and "health status" of organisms contributing to the potential aerosol. The health status is, in turn, influenced by nutrient availability (or host density), water availability, temperature, light, substrate pH, pollutants, and other factors.

Many sources exhibit cycles with respect to the availability of particles for release, which may depend on a number of intrinsic (e.g., age, life stage or phase) and extrinsic (e.g., seasons or weather, daylight) factors. Measured cyclic changes in aerosols are related to these to changes in source strength as well as to changes in dispersal influences. In addition to these temporal variations (cycles), the spatial orientation of the source (that is, whether it is a compact "point", a block of many acres or square miles, or a long narrow line source), will affect how bioaerosol recoveries at any one sampling point should be interpreted.

D. Release and Dispersal

Various factors affect the release of different bioaerosols. Some organisms have intrinsic mechanisms that forcibly discharge particles, usually in response to some climatic or other environmental factor. Others require mechanical disturbance for particle release. Local weather factors, including precipitation, wind, changes in humidity, and temperature, all can have a profound effect on both intrinsic and extrinsic release mechanisms. Often the interactions of two or more of these parameters make studies of the individual effects difficult. For example, seeds pods, anther sacs, and spore-bearing structures of lower plants and some fungi are apparently affected by changes in humidity. However, these effects are also influenced by temperature, light, air movement, and other forms of mechanical disturbance. Looking only at correlations with humidity may greatly oversimplify true interactions.

E. Transport and Deposition

The length of time a particle remains airborne depends on a number of factors, including particle density, size (effective aerodynamic diameter), electrical charge of the particle, the nature of the particle surface (e.g., hydrophobicity), ambient air conditions (e.g., humidity, air speed and turbulence, etc.), and the availability of surfaces for impaction (e.g., leaves, building surfaces, etc.). Not surprisingly, particle concentrations, especially during periods of emission, generally are greater close to their source (that is, near ground level in most cases) than at higher altitudes (Ingold, 1971, pp.182–183). Winds and updrafts have a strong influence on the distance particles travel after they are released from their sources. These updrafts have a varying effect on transport depending on the particle's settling velocity (and therefore size). Bacterial-sized organisms (circa 0.5 to 2 μm diameter) have settling velocities in the thousandths of a centimeters per second (cm $\times$ 10^{-3}/s), while larger particles like pollen, large fungal spores, or the smallest seeds, ranging in size from about 20 μm to 100+ μm diameter, settle at the rate of centimeters or tens of centimeters per second (cm/s to cm $\times$ 10^{1}/s). These larger particles require rather strong updrafts, in the range of 1 m/s, to remain airborne for extended periods (Cox, 1989).

At least small numbers of pollen and dry fungal spore aerosols (all of which have earth-bound sources) are able to travel great distances. Proctor and Parker (1942) present a concise summary of the early work on microorganisms recovered at high altitudes. Airplane-mounted samplers have collected fungi and insects many miles above the earth's surface (Greene et al., 1964; Hollinger et al., 1991). A rocket-borne sampler recovered culturable bacteria and fungi at an altitude of between 57 and 77 km (Imshenetsky et al., 1978). Kramer et al. (1973) used ship-borne samplers to recover pollen and spores over the Pacific Ocean, hundreds of kilometers west of North America. Concentrations decreased with distance from land and were very low (3 particles/m^3 of air) at the further distances.

Particles are removed from air by a number of means, including sedimentation, impaction, diffusion to surfaces, and wash-out by raindrops. For larger airborne particles (circa >25 μm), removal by raindrops can be quite efficient. Particles less than about 5 μm are poorly collected by rain (Ingold, 1971, p. 199). Impaction is a major consideration with large particles (circa >10 μm) and is probably of less importance for small particles, except possibly for impaction onto very fine (narrow) substrates (e.g., spider webs). Particle removal efficiency is also related to the type of impaction surface, for example, size, stickiness (dry or wet), electrostatic charge, hydrophobicity, etc. Sedimentation, like impaction, is a predominating consideration only for larger particles (circa >25 μm); sedimentation of smallest particles (less than a few microns) under all but the most still conditions can be more than countered by air turbulence.

When a particle makes initial contact with a surface, it is not only air turbulence or physical disturbance that influence whether it will adhere or be reaerosolized but also moisture, hydrophobic interactions, electrical attraction, and physical configuration. After initial contact, other factors will influence a more permanent attachment. These factors include the surface composition of the particle (e.g., presence of a glycoprotein "mucilage" on some fungal spores) and the ability of the potentially adhering particle and/or landing surface to chemically recognize each other (e.g., a phytopathogenic fungal spore and plant host recognizing each other via lectins and haptens).

F. Patterns of Prevalence

Outdoors (micro vs. macro prevalence). Bioaerosol prevalence patterns outdoors are strongly influenced by climate and weather, often resulting in pronounced seasonal and diurnal cycles. Seasonal climatic changes, especially in temperate and subarctic areas, directly affect the growth cycles of plants, thereby influencing pollen and spore maturation and release cycles. Figure 9.1 displays seasonal variation patterns for levels of ragweed and grass pollen, which respond to different seasonal climatic factors. Seasonal climatic cycles also affect plant senescence and subsequent colonization with saprophytic bacteria and fungi, resulting in seasonal cycles of these microorganisms.

In addition to the effects of weather on an organism's growth and life cycle, weather (especially temperature, humidity, and wind) also can strongly influence the release and takeoff (aerosolization) of small particles. These effects often show up in diurnal cycles, sometimes attributable to large changes in temperature and humidity during the morning and evening hours, which can influence pollen and fungal spore release. Diurnal patterns also result from increases in wind speed and turbulence through the daylight hours, which affects particle lift-off.

The above described factors affect bioaerosol prevalence patterns over relatively large regions or even climatic zones. Other factors will affect airborne concentrations over relatively small areas. Excluding pronounced microclimate or weather differences due to unusual geographic features, these "microprevalence" effects are often due to land use or human activity. Sewage treatment facilities or lagoons and landfill disposal sites can affect surrounding airborne bacterial levels, and composting (municipal or even backyard) can increase downwind fungal spore levels (Crook et al., 1987; Epstein, 1993). Similarly, building construction or destruction can increase fungal and bacterial concentrations, while garden plots (and obviously large-scale agriculture) can affect fungal spore prevalence (Streifel et al., 1983; Burge et al., 1992). Investigators need to be aware during bioaerosol surveys that recoveries will be differentially biased by how close or far the samplers are placed from these sources.

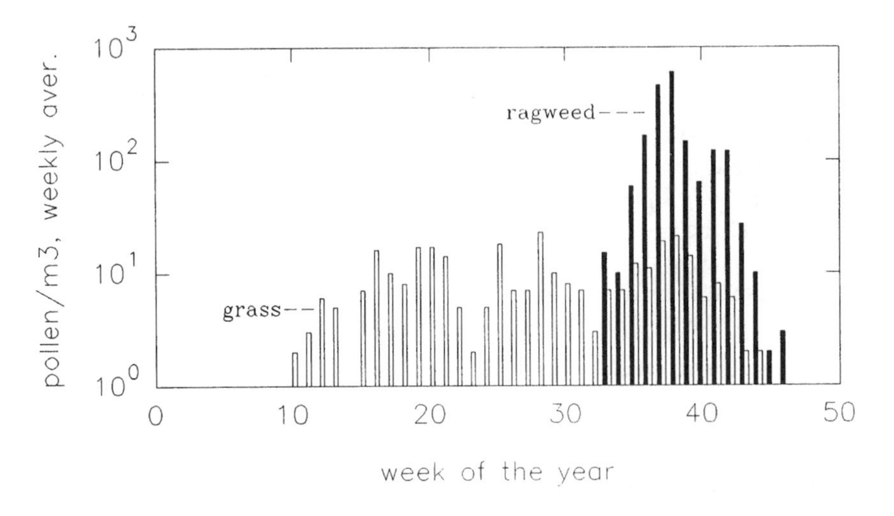

Figure 9.1. Oklahoma City, OK, Rotorod pollen counts, 1992. (Counts courtesy of Dr. Warren Filley, Oklahoma Allergy Clinic.)

Indoor/outdoor relationships. Unless there is an indoor source for specific bioaerosols, concentrations indoors will generally be lower than outdoors. This effect is related to the reasons for occupying enclosures, which are designed to protect us from adverse weather and intrusion by vermin or other unwelcome (sometimes human) visitors. The outdoor aerosol penetrates interiors at rates that are dependent primarily on the nature of ventilation provided to the interior. Indoor/outdoor ratios of specific particle types (of outdoor origin) are highest (tending toward unity) for buildings with "natural" ventilation where windows and doors are opened to allow entry of outdoor air along with the entrained aerosol (Burge, 1994). As the interior space becomes more tightly sealed (usually done to conserve energy required for climate control), the ratio becomes lower and lower. In homes, this sealing usually results in very low air exchange rates (often as low as 0.1 air changes per hour), with equally low penetration of outdoor aerosols. In larger buildings, since more air must be brought in, filtration is used to limit the entrance of aerosols. The effectiveness of such filtration has not been clearly documented, but appears high. The above discussion addresses only the penetration of outdoor aerosols. However, contamination of indoor surfaces can drive this indoor/outdoor ratio well over unity, especially in temperate and subarctic areas during periods with snow cover (Solomon, 1976; Pasanen et al., 1990).

For many enclosures, barriers to penetration of the outdoor aerosol are intentionally reinforced through high-efficiency filtration (see above) and/or the use of air-conditioning, which allows windows and other paths of outdoor air intrusion to be tightly closed. This is often true for homes of people with allergies to outdoor bioaerosols. It is especially important for health care centers, especially where those highly susceptible to infections are housed (see

Streifel et al., 1983). Preventing the common outdoor opportunistic pathogens from entering and growing in buildings has proven only slightly easier than protecting occupants from diseases or organisms carried by other occupants (see Noble and Clayton, 1963; Solomon et al., 1978). However, the effectiveness of air-conditioning, both central systems (Hirsch et al., 1978) and window units (Solomon et al., 1980; Pan et al., 1992), in reducing penetration of outdoor particles into building interiors has been documented. The possibility also exists that these units can become contaminated and serve as an interior source of microbial contamination.

The answer to the question of how to determine if airborne microorganisms in a building are of outdoor origin (indicating penetration) or of indoor origin (often indicating contamination) remains elusive. It is generally assumed that indoor concentrations of bioaerosols will be lower than those outdoors, except possibly for human-source bacteria. Although guidelines for indoor/outdoor ratios have been proposed (ACGIH, 1989), they must be viewed only in a very general way (see Pasanen et al., 1990). Numerous other factors must be taken into account when relating indoor and outdoor bioaerosol levels, not the least being the specific nature of the aerosol of concern; using only generic classification can mask species differences. Unless such care is taken, one is likely to falsely implicate or exonerate buildings with respect to microbial contamination (Holt, 1990).

G. The Human Impact

Probably the most important effects (economically and in terms of numbers of people affected) on the human population attributed to outdoor bioaerosols are the hypersensitivity diseases, especially allergic rhinitis and asthma. The immune system can react, often inappropriately (when there is no real threat to the body), to a variety of antigenic stimuli from pollen, fungi, insects, animal emanations, etc. Infections are also attributable to a number of bioaerosols, although relatively few of these result from outdoor air exposures. In addition to immune responses and infections, toxic reactions can be of concern. Endotoxin, a cell wall component of Gram negative bacteria, has been connected with a variety of health effects. A number of fungi are producers of potent toxins that are often associated with problems upon ingestion of contaminated foods, although disease due to respiratory exposure is also plausible. Discussions of human diseases due to bioaerosols are presented in Chapters 2, 4, 10, and 11.

H. Sampling Considerations

When deciding whether or not a certain sampler is appropriate for a specific type of particle a number of factors must be considered: particle size, particle-specific analytical methods available, approximate concentrations expected, periodicity in abundance (e.g., diurnal cycles), conditions (e.g., weather, airborne dust concentrations) under which sampling will be performed, etc. Samplers generally do not collect all particle sizes with equal efficiency, so that the expected particle size of the aerosol has a large bearing on sampler selection. Familiarity with the types of analytical methods that can be used to identify and quantify the recovered (desired) bioaerosols is essential before selecting a sampler. If the particles are morphologically distinct and of sufficiently large size (e.g., pollen), a sampler that enables one to use a microscope to count and identify particles may be appropriate. If identification of the organism requires growth of specific structures (e.g., most fungi), then sampling in such a manner so as to minimize damage to the organism (i.e., allow its growth in culture) and the use of a nutrient medium that will encourage such growth is required.

Both lower and upper limits of sensitivity depend on the sampling rate, the length of sampling periods, the capacity of the collection medium, and the sensitivity of the analytical method. Some of these are fixed for a specific device; others can be varied to some extent. Cycles of prevalence also influence sampling times and duration. Unless particle concentrations are extremely constant, which is unlikely, attempts should be made to determine which factors may be influencing temporal (and spatial) variations. Regular cycles of bioaerosol prevalence can be quite pronounced in the outdoor environment. Diurnal variation is particularly common, as well as seasonal cycles, especially in temperate areas.

The siting of a sampler will also affect recoveries. Both lateral and vertical profiles will likely show short-term concentration differences (Raynor et al., 1973; Rantio-Lehtimäki et al., 1991), depending on factors such as proximity to sources, air turbulence and velocity, available impaction surfaces (e.g., foliage), and source periodicity. Differences in recoveries between sites close to each other may diminish when averaged over long periods.

When sampling outdoors, temperature and wind speed (or turbulence) can become extreme and can have a pronounced effect on particle collection efficiency. Temperature extremes, as well as humidity, can affect impaction surface "stickiness" and hence, particle retention or bounce. Of course, temperatures approaching 0°C and below preclude using agar as a collection surface unless the sampled air is prewarmed. Wind orientation is critical for suction samplers. Ideally, sampler intake speed should approximate ambient wind speed (isokinetic conditions), a condition that is rarely met when sampling outdoor bioaerosols and that, if ignored, may bias recoveries depending on particle size. The sampler inlet must be constantly oriented into the wind (isodirectional conditions); varying this wind orientation increases the risk of undersampling larger particles. Principles of aerosol and small particle collection that apply to nonbiological particles obviously apply to bioaerosols also, with the added problems of dealing with living organisms (see Willeke and Barron, 1993).

III. THE ORGANISMS AND THEIR AEROSOLS

A. Pollen

The nature of the particle. Pollen grains are relatively large, usually near-spherical particles with an extremely resistant outer wall made primarily of a poorly characterized polymer called sporopollenin and an inner wall of cellulose. Most are spheroidal in the hydrated state, although more exotic shapes (cylindroidal, triangular) also occur. When airborne (and dehydrated), many of the grains collapse inward, often along areas of wall thinning (furrows). Some pollen grains travel in tetrads (cells did not separate after meiosis), or in even larger groups (polyads).

Pollen grains contain the male gametophyte (the male haploid reproductive cells) in addition to a pollen tube cell and are produced by all seed plants (gymnosperms and angiosperms). The pollen grain is not a dispersal unit (as is a fungus spore) but carries genetic material and is "successful" only if it fertilizes a receptive female flower of the same species. Plants are of two general types with respect to the kind of pollen they produce: entomophilous (insect pollinated) and anemophilous (wind pollinated). It should be noted that the vast majority of nonwind pollinators are insects, but other vectors, such as birds and bats, are not uncommon. Because the term "entomophilous" is so commonly used, the term is used here to include all animal-pollinated plants, admitting that "zoophilous" may be more accurate.

Entomophilous plants produce pollen grains that are often sticky due to lipid deposits on the outer wall, which promote adherence to the insect. These grains can be quite large (up to around 250 μm), although most are under 100 μm in diameter. Surface ornamentation, such as spines, heavy ridges, or other elaborate structures, is common among entomophilous pollen. Not surprisingly, pollen grains of entomophilous flowers are not common in the air other than the small number of grains that become airborne due to severe physical disturbance or that become dislodged from flying insects. Even then, these grains tend not to travel far due to size, density, surface ornamentation, or stickiness (which causes clumping). Levels of entomophilous pollen may be relatively high in air collected very close to sources, especially for plants with excerted stamens (i.e., with anthers on long stalks that protrude well beyond other flower parts).

In general, anemophilous (wind-borne) pollen types tend to be small (<50 μm). Some gymnosperm grains, which often have large air-filled "flotation" bladders that allegedly reduce overall density and assist with wind transport, approach 100 μm in diameter. A few grasses, notably corn, have large anemophilous pollen grains (around 100 μm diameter) that fall out of the air within a short distance of the originating plant. On the other hand, nettle pollen (*Urtica* sp., about 15 μm diameter) and three-seeded Mercury (*Acalypha virginica*, about 12 μm diameter) are at the smaller end of the size scale.

Anemophilous pollen is generally smooth or with, at most, low, relatively inconspicuous ornamentation, and the wall is apparently not sticky.

Sources and source strengths. The sources for pollen are flowers, whose appearance can be extremely different depending on whether they produce wind-borne or insect-borne pollen. Flowers of plants producing anemophilous pollen tend to be small and without showy colors or nectaries. There is also a tendency toward separate male and female flowers in these plants. Large, colorful and showy petaled, nectar-producing flowers having functional male and female parts in the same flower ("perfect" flowers) are characteristics that have evolved to attract insect, or other animal, pollinators. Anthers (pollen-producing structures) of anemophilous flowers often have long stalks or "filaments" supporting the pollen sacs (as in silver maples, *Acer saccharinum*) so that when the pollen is shed it is well beyond the flower and other plant parts and is more easily entrained into the wind streams. Alternatively, the entire flower may be on a long stalk (as in sugar maples, *A. saccharum*), or other specialized structures may be present, as in ragweed, where pollen is pushed from the flowers so that it can fall onto leaves below and subsequently be picked up by air currents.

Source strength (i.e., how much pollen is available for dispersal) depends on the stage of the plant life cycle, in addition to a number of weather, moisture, and nutritional factors. Most anemophilous plants produce pollen over a relatively short period. Outside that period, the source strength is virtually zero. Seasonality of flowering and pollen production is regular and predictable (Harvey, 1972; Solomon, 1983) and is dependent on factors such as temperature, day length, precipitation, and water availability. For example, short ragweed (*Ambrosia artemisiifolia*) produces mature pollen only when periods of darkness (nighttime) are sufficiently long (Solomon, 1988), an effect that is readily induced artificially (Dingle et al., 1959). A similar effect, where darkness less than about 10 h per day inhibits flowering, is seen in sunflowers (*Helianthus annuus*) (Dyer et al., 1959). In addition, for ragweed, the overnight temperature must exceed about 10 to 15°C for anther extension to occur, and relative humidity above about 70% inhibits anther opening (Bianchi et al., 1959). Thus, in temperate regions, ragweed pollen is usually not available for dispersal until early August (when days are short enough), and very little additional pollen is produced after the beginning of October (after the first frost).

Pollen production in other kinds of plants can be controlled by different factors. The number of grains available for shedding in trees that form their catkins (male flower clusters) during the prior growing season, such as members of the birch and poplar families (Betulaceae and Salicaceae), is likely influenced by factors that affect the growth and health of the tree the season before flowering. These factors include rainfall, temperature, and adequate nutrients. Prior growing season factors probably have less of an effect on other trees, such as oaks (*Quercus* species) and hickories *(Carya* species), which form their catkins during the months immediately preceding flowering. It is likely that rainfall, temperature, and other weather factors during the spring, when the flowers and pollen are developing, will exert a stronger influence on pollen production in these types than does the prior year's weather.

Wind-pollinated plants are well represented in temperate areas and much less abundant in the tropics (Raven et al., 1976). This scarcity of tropical anemophilous plants may be due to the often dense year-round foliage, which would obstruct wind dispersal. Most wind-pollinated trees in temperate areas shed pollen in early spring before they are in full leaf. Also, populations in the tropics tend to be more complex (i.e., include more species and fewer individuals of each species) than in temperate forests. Gregory (1961, p. 39) claims that, worldwide, entomophilous species probably outnumber anemophilous species by about 10 times.

Factors affecting release and dispersal. Pollen release, like production, is dependent on climatic factors and is seasonal and (within limits) predictable. Factors affecting onset and duration of flowering and pollen shedding vary depending on characteristics of the plant as well as the season in which it flowers. Tree flowers that produce pollen the previous fall generally begin shedding in early spring, probably in response to the arrival of warm, dry weather. This may be as early as January in Texas when mountain cedar (*Juniperus ashei)* begins to shed pollen, and as late as April or May in southern Canada when flowers of the "early flowering trees" (such as alders, *Alnus* spp.) begin to open. Prolonged daytime freezing temperatures during early spring can delay early spring tree pollen shedding by 4 weeks or more. The onset of grass seasons depends, among other parameters, on winter temperatures, cumulated temperatures (especially during the spring), photoperiod, and precipitation (Emberlin et al., 1993).

Daily (or hourly) variation in pollen shedding is directly affected by current weather conditions. Van den Assem (1972) showed that for certain pollen groups, airborne pollen concentrations varied diurnally with temperature (positive correlation) and correlated negatively with relative humidity (temperature obviously having a direct effect on relative humidity). Also, cloudy days (>50% cloud cover) generally resulted in lower pollen concentrations than sunny days. In general, anemophilous pollen is shed during warm dry cloudless periods (Dingle et al., 1959). Ragweed produces mature pollen only when overnight temperatures have been above about 10°C and sheds the pollen when relative humidity is below about 70% (Solomon, 1988). Any sustained rainfall (other than a very light shower) will scrub essentially all pollen from the air (Dingle and Gatz, 1966) while also inhibiting the release of pollen from anthers (due to high relative humidity). These factors make it obvious why wind pollination is not a successful strategy in humid tropical areas with high rainfall.

Patterns of pollen prevalence. The "severity" of a pollen season (i.e., the concentrations of airborne grains encountered) is related to source strengths, release and dispersal factors, and removal factors. As previously mentioned, trees that produce their flowers in the fall (but that do not mature and expand until spring) tend to be the first to shed in the new year. Late-flowering trees (oak, hickory, walnut) shed pollen from late April into June (depending on the area), a period that often overlaps with the grass season. In northern temperate climates, grasses begin flowering in mid to late spring (May) but do not become abundant until early summer (June) (see Figure 9.1). While a few herbaceous, nongrass species (including many "weeds") flower in early summer, for example plantain (*Plantago* species) and sheep sorrel (*Rumex acetosella*), most types reach their flowering peak during mid to late summer (e.g., most ragweeds). These pollen seasonality generalizations apply especially to temperate areas of the United States, but with some modifications can be generalized to other areas as well (see Solomon, 1988; D'Amato et al., 1991). Emberlin et al. (1993) point out how apparent pollen season shifts over a number of years can be attributed to changes in prevalent species. This can be a particular problem among certain groups, for example the grasses, which are often impossible to speciate based on pollen morphology alone.

In addition to whole pollen grains, smaller exine fragments are sometimes encountered, but more importantly, pollen "contents" have also been detected in the air. Suphioglu et al. (1992) reported large numbers ($10^4/m^3$ of air) of small (0.6 to 2.5 μm) starch granules with highest concentrations following rainfall. These granules were mostly from ryegrass (*Lolium perenne*) pollen and were detected using monoclonal antibodies to *Lol p* IX. Other antigenic particles smaller than whole grains have also been found in the air, although they have been poorly characterized (see Human Exposure and Health Effects, below).

The type of pollen dominating the airspora will depend not only on the season, as discussed above, but also on land use, climate, geography, and soil type. Examples of dominant airborne pollen types covering large geographic areas include (1) birches in the northwest United States and Scandinavia, (2) ragweed in the north central United States, (3) pellitory in southern Europe, and (4) mountain cedar in central Texas and into Mexico (see Lewis et al., 1983; D'Amato et al., 1991). Changes in land use, including agriculture, urbanization, and ornamental plantings, can have profound effects on pollen concentrations and types (Solomon and Buell, 1969; Sneller et al., 1993).

Because there are normally few if any pollen sources in homes and even window screens over open windows provide some barrier to outdoor pollen penetration, conventional wisdom has been that pollen is seldom as much of a health threat indoors as out. While particulates do penetrate into interiors, especially of conventionally ventilated homes (no air-conditioning or filtered intake air), indoor concentrations of airborne pollen are generally lower than outside and generally parallel fluctuations in outdoor levels (Solomon et al.,

1980). Concentrations of airborne pollen are generally negligible in buildings with air-conditioning and closed windows. A possible exception to this would be in homes (and, obviously, greenhouses) containing numerous plants with mature flowers that are routinely agitated by watering or other means (Burge et al., 1982). These observations can be broadly applied to automobiles as well (Muilenberg et al., 1991). Recent evidence indicates that there can be significant reservoirs of pollen allergen indoors, indicating penetration from outside. Analyses of house dust samples have shown significant concentrations of grass allergen, both in California homes (Pollart et al., 1988) and in Arizona homes (O'Rourke and Lebowitz, 1984). These samples were collected during or immediately after grass season, and it is not known whether or not the allergens were associated with whole pollen grains. Also, the persistence of these allergens in dust is not known.

Human exposure and health effects. When pollen grains land on a stigma (receptive surface of a female flower), "recognition" chemicals are released. If the pollen grain and flower are compatible (that is, of the same species), pollen tube formation begins. The pollen tube may secrete enzymes that assist in its penetration of the stigma and style on its way to fertilize the ovule. It is thought that these recognition chemicals and/or enzymes may be allergenically important (Knox and Heslop-Harrison, 1971). Antigenic activity has also been associated with pollen exine sporopollenin (Southworth et al., 1988) and with starch granules of certain pollen grains (Knox, 1993). Allergens have been purified and characterized from a number pollens, including those of ragweed (King et al., 1964), birch (Ipsen and Løwenstein, 1983), pellitory (Ayuso et al., 1988), and several grasses (Malley and Harris, 1967; Løwenstein, 1978).

The majority of pollen-associated disease is associated with the upper airway (allergic rhinoconjunctivitis or hay fever), although pollen-related asthma also occurs. Grass pollen exposure has been shown to be a significant risk factor for asthma in a California study (Pollart et al., 1988). Conversely, in New Zealand, Sears et al. (1989) found no significant risk of asthma associated with grass sensitivity, although the selection of the asthma cohort was somewhat different than that of Pollart et al. Disease related to pollen exposure is not necessarily related only to pollen prevalence, and there is evidence that some pollen types carry more allergen activity than others. For example, both oak and pine pollens are extremely abundant seasonally in some parts of the United States, but rates of allergy to these particles appears quite low. On the other hand, grass pollen is usually present in relatively low levels, but is probably among the most important of the outdoor allergens.

Explanations of how particles as large as pollen grains cause asthma are still speculative. The explanation for asthma resulting from exposure to mite fecal particles (which are pollen-sized particles) is that local inflammation results from deposition in the bronchial mucosa, eventually triggering a general airway response (Platts-Mills et al., 1986). Recent studies on particle sizes much smaller than pollen grains (submicronic, i.e., particles less than 1 μm) indicate that antigens of ragweed (*Ambrosia artemisiifolia*; Solomon et al., 1983), ryegrass (*Lolium*; Stewart and Holt, 1985; Suphioglu et al., 1992), and Japanese cedar (*Cryptomeria japonica;* Takahashi et al., 1991) are detectable in outdoor air and may contribute to pollinosis and asthma.

Sampling considerations. Pollen grains, being large and readily identifiable by light microscopy, can be efficiently collected by a variety of sampler types. While settle slides were commonly used until recent years, and still are used by some groups, they are biased toward larger particles, are strongly influenced by air currents, and do not allow the determination of airborne concentrations. Rotating-arm impactors (Rotorod, Sampling Technologies, Inc., Minnetonka, MN) and Hirst-type spore traps (e.g., Burkard Manufacturing Company Ltd., Rickmansworth, England) are usually the samplers of choice for monitoring pollen levels. Because sources can be so intense, samplers located too close to sources are strongly biased and will not reflect average exposures over a larger area. Therefore, it is often suggested that samplers be placed well away from wind-pollinated trees or fields of weeds and grasses. This often means sampling on a rooftop above tree level, which may not directly correlate with ground-level exposure on a day-to-day basis, but should show only small variation compared with ground exposures over a large area when averaged over a number of days (Raynor et al., 1973). Rantio-Lehtimaki et al. (1991) also concluded that ground and elevated sampler recoveries correlated but concentrations at ground level were often significantly higher. This highlights the importance of interpreting recoveries based on sampler and source locations.

B. Other Plant Emanations

Fern spores. Ferns are seedless vascular plants, often with large elaborately compound leaves. Most are not much more than a meter or so tall, but sizes range from very small aquatic ferns (about 1-cm long) to giant tree ferns (over 20-m tall). Although the vast majority of species grow in relatively moist to very wet habitats, some species utilize other habitats as well (see below). Fern spores are asexually produced dispersal units that generally depend on wind for dispersal. Most ferns have only one type of spore (homosporous), unlike some of the club mosses or certainly the seed plants, which are heterosporous (having microspores produced by male flower parts and macrospores produced by the female flower parts). Fern spores develop in structures called sporangia, which are usually clustered into sori. Sori can be located on the underside of leaves or on specialized spore-bearing stalks. Each sporangium contains a group of cells (annulus) with unevenly thick walls, which contract, apparently in response to changes in relative humidity, thereby forcibly discharging the mature spores. Benninghoff (1964) states that fern spores generally range from 20 to 60 μm in diameter and either a straight scar (sulcus) or a trilete scar is evident. These spores also have an outer perispore layer that is frequently pigmented and uniquely ornamented, allowing microscopic differentiation of some genera and species. Each spore germinates to form a very small, free-living, green gametophyte (haploid plant), which produces few-celled antheridia (which produce sperm) and archegonia (which produce eggs). The sperm require water to swim to the archegonia and fertilize the egg, which gives rise to the leafy fern (sporophyte). Because of this dependence on water, ferns are most abundant in damp habitats, including bogs and damp woodlands, although some species are able to colonize well-drained or seasonally dry habitats. Interestingly, ferns are among the first plants to colonize the lava fields on the Hawaiian volcanoes, taking advantage of the abundant rainfall, pockets where water accumulates, and the nutrients in the lava.

Most ferns (in temperate North America) shed their spores during the summer or early fall months, although a few types (e.g., *Osmunda* species) release their spores in the spring. Concentrations of fern spores collected on urban rooftops (where most sampling stations are located) are generally very low: rarely more than a few per cubic meter of air (personal observation; also see Gregory, 1961, p. 110), although much higher concentrations have been reported (Bassett et al., 1978).

Disease due to fern spore exposure, especially in "normal" outdoor concentrations, is rarely reported, although in Thailand, Bunnag et al. (1989) demonstrated that a high percentage of allergic rhinitis patients react to fern spore extracts (intracutaneous test and nasal provocation).

Due to the relatively large size of fern spores, they are efficiently collected using a variety of impactors including rotating-arm impactors and Hirst-type spore traps (this also applies to club moss spores and the larger moss and liverwort spores, see below). Most commonly, fern spores are identified by light microscopy although this often precludes specific, or often even generic, identifications, for which the mature sporophyte is usually required.

Club mosses. The club mosses are a group of primitive vascular plants related to horsetails and ferns. The leaves of these plants are small and contain only one central vein (microphylls), quite unlike the large and often elaborately veined leaves (megaphylls) of ferns and "higher" plants. Like ferns, club moss spores are produced in sporangia, which can be borne on the upper surface of modified leaves or in "cones" on special branches. Club mosses have life cycles similar to ferns, although some (*Selaginella* spp.) are heterosporous, having micro- and macrospores that produce male and female gametophytes, respectively (see ferns, above). While common in the tropics, often in very moist areas or as epiphytes, a number of club mosses also grow in temperate climates, often forming evergreen mats on the forest floor, and a few are desert species.

Spores of club mosses are similar to those of ferns: generally 30 to 45 μm in diameter, often with a trilete scar and commonly with surface sculpturing (reticulate, echinate, grooved, etc.). Most temperate climate species shed spores in late summer and early fall (late July to October). Very few air sampling surveys report club moss spores, although Cua-Lim et al. (1978) reported them as the fifth most abundant type (when compared with pollens) recovered on settle slides in the Philippines. It should be noted that these investigators listed these spores as "spores originated from ferns, *Lycopodium* and *Selaginella*", both of which are genera of club mosses, not ferns. One would assume that samplers located in forested areas, rather than the typically located urban sampler, would recover many more club moss spores (in addition to larger numbers of fern spores).

Salén (1951) reported numerous cases of sensitivity to powders containing *Lycopodium* spores in theater actors (in makeup), pharmacists, and metal workers. It is assumed that these occupational exposures will rarely be encountered today, as the use of *Lycopodium* powders is no longer widespread.

Moss and liverwort spores. Bryophytes are differentiated from ferns and other "higher" plants by their lack of vascular tissue and also the fact that the gametophyte (haploid) stage is dominant whereas in vascular plants (ferns, gymnosperms, and angiosperms) the sporophyte (diploid) is the dominant generation. Like ferns, fertilization is via sperm from antheridia and an egg in an archegonium, and is dependent on abundant water. Mosses and liverworts are plants of very wet environments, and are always small, in part due to lack of the vascular tissue for support of larger structures.

Moss and liverwort spores are generally between about 5 and 30 μm, often have a trilete scar, and may have surface ornamentation. The spores are usually borne in some type of capsule and are shed by a variety of mechanisms. Some liverwort sporangia contain elaters (hygroscopic structures) that are very sensitive to changes in humidity and, through their twisting action, aid in spore dispersal. While not commonly reported, elater fragments have been recovered some distance from possible sources. Spore discharge in the mosses is also largely dependent on changes in humidity. *Sphagnum* spore discharge is atypical in that it uses compressed gas to aerosolize spores. The internal tissue of the capsule shrinks upon maturing, leaving a gas-filled chamber the top of which is blown off when the capsule dries and shrinks, and the forcibly released gas carries the spores into the air.

Gregory (1961, pp. 140, 142) referred to a small number of reports of airborne moss spores indicating very small concentrations at 500 and 1000 m above ground and "numerous" moss spores on a roof top near Hudson Bay, Canada. Gregory also mentioned the work of B. Pettersson in Finland, who recovered liverwort and moss spores from rainwater including some spores of moss species not known in Finland. More comprehensive information on airborne concentrations of moss and liverwort spores and reports of sensitivity is lacking.

Algae. Algae are eukaryotic photosynthetic (containing chlorophyll *a*) organisms composed of single cells or simple multicellular bodies. Although the blue-green algae contain chlorophyll, they are prokaryotic and are more closely related to the bacteria than to plants. In fact, the blue-green algae have a wall structure very similar to Gram negative bacteria. The largest number of eukaryotic algal taxa are aquatic (forming the phytoplankton). However, most viable airborne forms may be from soil or other nonaquatic environments.

Algal cells are commonly encountered in outdoor air, although concentrations never reach those attained by fungi or even pollen. Culturable algae commonly occur in the tens of colonies per cubic meter of air (Tiberg et al., 1984; Brown et al., 1964), but concentrations in the hundreds per cubic meter have been reported (Roy-Ocotla and Carrera, 1993; Gregory et al., 1955). It is likely that most algae recovered from air originate from soil (Schlichting, 1969), although water sources (wastewater treatment plants and naturally occurring sources) can serve as reservoirs and disseminators, especially when bubbling, turbulence, or even splashing from rain drops occurs. The possibility that terrestrial algae are better able to survive the stress of becoming airborne (aquatic algae must survive rapid dehydration immediately after leaving their wet habitat) is discussed by Ehresmann and Hatch (1975). With this dehydration stress in mind, it is interesting that Kramer et al. (1973), using microscopic analysis (where viability is not relevant) of particulate recoveries, made no mention of algal recoveries on a number of air sampling voyages in the Pacific Ocean. Roy-Ocotla and Carrera (1993) state that the commonly recovered taxa reported by different investigators in different parts of the world (Sweden, western Europe, United States, India, Taiwan, and Mexico) are surprisingly very similar. On the other hand, Tiberg (1987) stated that the green algae *Chlorella* and *Chlorococcum* dominate in temperate areas and blue-green algae are more abundant in tropical areas.

Tiberg et al. (1984) found no correlation between airborne algal concentrations and any meteorological factors, although Rosas et al. (1989) found correlations between algal concentrations and wind speed in Mexico City and between levels of algae and vapor pressure in the coastal city of Minatitlán. Gregory et al. (1955) noted diurnal cycles (apparently unrelated to tides) in the recovery of algae at an island sampling site. While there is evidence of sensitivity to algal cells (Bernstein and Safferman, 1973), their role in allergic disease is not clear. This, along with the lack of prevalence data, indicates that more needs to be known before we can analyze the health impact of airborne algae.

Some algal cells can often be enumerated and sometimes identified on spore trap recoveries analyzed by light microscopy, although species identification often requires a culture-plate impactor or all-glass impinger using an inorganic (autotrophic) culture medium.

Seeds. A seed is the undeveloped sporophyte (embryo) along with some nutrients and a protective "coat". Fruits, which consist of one or more seeds enclosed in specialized, modified flower parts, are the dispersal agents of flowering plants. These can be fleshy and attractive to birds and other animals as a food source, such as blueberries (*Vaccinium* species), or spiny or "sticky" with hooked appendages, such as the burdocks (*Arctium* spp.), which are transported on mammal fur, bird feathers, or hikers' clothing. Other seeds have specialized hairs, plumes (dandelion), or wings (maple) that facilitate carriage by air currents. However, very few fruits or seeds are small enough to become truly airborne. The seeds of some orchids are an exception, with many seeds smaller than 0.5 mm, and some *Schizodium* species seeds are as small as 0.17 mm (Kurzweil, 1993). Among the orchids, anemochory (seed dispersal by wind) is common and some seeds may travel great distances (up to 40 km) (Dressler, 1981). However, most seeds probably fall within a few meters of the parent plant, and no reports of human disease related to respiratory exposure to seeds have been reported.

While not well documented, some fruit hairs or tufts have been accused of causing respiratory disease, although only the smallest fragments could be carried much past the anterior nares.

Other plant parts. Many plants are covered with pubescence (consisting of hairs or trichomes), especially when young. The function of these hairs is not completely understood. It has been hypothesized that protection from excessive heat or insects is a likely function of some trichome types. Similarly, glandular hairs on the buds of certain woody plants have a secretory function, possibly producing protective coatings (Esau, 1977). Trichomes are released from plants during maturation and are most commonly recovered from air in the spring months in temperate climates (Ostrov, 1984). Because of the very obvious dense, deciduous pubescence on young leaves and stems of *Quercus* and *Platanus* species, it is speculated that the majority of airborne trichomes originate from these species. Peattie (1953) states in his monograph on western trees that these deciduous woolly hairs of *Platanus* species "for several weeks drift on the atmosphere, setting up an acute inflammation of the mucous membranes of sensitive noses." While the allergenicity of these trichomes has not been conclusively demonstrated, a few studies have shown strong relationship between alleged exposure and either mechanical irritation (Ross and Mitchel, 1974) or allergic symptoms (Zacharin, 1933).

Other parts of vascular plants, besides the readily identifiable pollen, spores, and trichomes, can also become airborne. Most of these particles are impossible to visually identify, and airborne concentrations have never been determined. Benninghoff (1964) described these airborne particles as "fragments of vegetable humus (material of plant origin degraded biologically or autolytically so that characters for determining origin are obscured)." Xylem fragments, which have been recovered on air samples at altitudes of 5000 ft (Vinje and Vinje, 1955), are an example of these airborne plant particles of poorly defined (vascular plant) origin. The importance of any of these particles as contributors to disease is unknown. A notable case of identified plant parts causing disease is the outbreak of asthma associated with soybean dust aerosols in Spain (Anto et al., 1993).

C. Fungi

The fungi are the subject of Chapter 5, and are only considered here from the point of view of their contribution to the outdoor aerosol.

Nature of the particles. Airborne spores are the primary means of dispersal for the fungi. Spores are single or multicellular units surrounded by a rigid (usually chitinous) cell wall. Each spore is capable of reproducing the entire organism (as opposed to pollen grains, which only carry half of the genetic complement of the plant). Fungi can have very complex life cycles, sometimes with up to five morphologically distinct spore types produced during a single cycle. This makes the classification of fungal sources based on airborne spores rather difficult.

Most commonly isolated and identified from air samples are spores of the two largest fungal classes: the Ascomycotina (or Ascomycetes) and the Basidiomycotina (or Basidiomycetes). Mushrooms, puffballs, shelf fungi, rusts, and smuts are members of the Basidiomycetes, a class defined by the production of "basidiospores" as part of the life cycle. Except for the rusts and smuts, basidiospores are the primary means of dispersal for the Basidiomycetes. Basidiospores are sexually produced (following nuclear fusion and meiosis) and are borne on special structures called basidia. Basidiospores are always single-celled, and most have an asymmetrically placed attachment peg (the attachment point to the sterigmata of the basidium). Basidiospores produced by puffballs are an exception to this rule and have a symmetrically placed attachment appendage. Color, size, and shape of basidiospores vary within relatively narrow ranges. For example, colors range from nearly black to colorless, with yellow, gold, and brown pigments prominent. Sizes range from about 2 to 25 μm, with many spores in the 5 to 10 μm range. Shapes are most often ovate, with occasional species producing elongated, angular, nodulose spores. Many basidiospores have a pore at the end opposite the attachment appendage.

The rusts and smuts, on the other hand, produce only a few small, inconspicuous basidiospores, and rely on asexual spore stages for dispersal. These asexual spores are produced within fruiting bodies (smuts) or externally on plant surfaces (rusts). The smut spores are usually more or less globose (although often compressed), usually strongly pigmented, and of a wide variety of shapes, sizes, and surface configurations. These are among the most difficult of the spore types to accurately identify in air samples. The rusts produce two kinds of asexual spores that are commonly recovered from air: one kind is teliospores, which usually are large, often with more than one cell, thick-walled, heavily pigmented, and often with a noticeable attachment appendage. More common are the uredospores, which often have colorless, spiny walls but bright yellow cell contents.

Ascospores are sexual spores produced by Ascomycetes. They are produced inside a special sac called an ascus. Ascospores are more variable morphologically than basidiospores (at least with respect to size and septations), and often are difficult to identify microscopically in air samples. They can be septate or single-celled, tablet-shaped, elongate, or filamentous, or a variety of other shapes. Colors range from colorless through shades of brown to black. A few ascospores are more exotically colored in purple, green, or blue. One characteristic often helpful in identifying ascospores is the lack of attachment scars. Because ascospores are formed from the cytoplasm within the ascus, they do not develop attachment scars or appendages.

Most Ascomycetes also produce asexual spores that are morphologically distinct from the ascospores. In fact, many ascomycete genera have two names, one referring to the sexual stage and one to the asexual stage (e.g., *Eurotium repens* is the sexual stage of *Aspergillus repens*). Sexual/asexual connections remain unknown for many Ascomycetes. Asexual (or "imperfect") spores often have an obvious attachment scar or peg, but vary greatly in other morphological characteristics. Asexual ascomycete spores commonly encountered in outdoor air include those of *Cladosporium, Alternaria, Epicoccum, Penicillium,* and *Aspergillus,* among many others.

Fungal spores of other classes do occasionally become abundant in air. For example, the Zygomycetes are a group of fungi that produce asexual spores inside special sacks borne on long filaments (sporangia). *Mucor* and *Rhizopus* are common zygomycete genera that contribute spores to the outdoor air. A few Oomycetes (fungi that produce motile spores and rely on abundant water for completion of their life cycles) produce airborne propagules that are actually intact sporangia. These sporangia germinate under appropriate conditions to release the motile spores. Included in this group are *Phytophthora infestans*, the "late blight" fungus that caused the Irish potato famine, and *Plasmopara viticola*, the "downy grape mildew" fungus that nearly destroyed the French wine industry in the 1870s.

It should be noted that, in addition to fungus spores, hyphal fragments may be capable of acting as dispersal units, although they probably are not as resistant to environmental stresses (e.g., desiccation, radiation) as are most spores. These pieces of hyphae are commonly encountered in air but are probably not reliably attributable to any particular taxon, although *Cladosporium* has been mentioned as a likely source for some of the fragments (Harvey,

1970). In agricultural situations, where high concentrations of fungal material such as hyphae, spores, and fragments of substrate and soil are aerosolized, the possibility also exists that levels of mycotoxins, in association with these particles, could also be very high.

Sources and source strengths. Fungi are heterotrophic [i.e., obtain their food from preformed organic material that is either nonliving (saprobes) or living (parasites)]. They are single-celled (yeasts) or filamentous (mycelial), eukaryotic organisms that belong to a kingdom separate from both plants and animals. One major food source of fungi is dead plant material. Where there are large amounts of organic debris (with adequate moisture) there will be large numbers of fungi and other microorganisms consuming available nutrients while also producing spores (or other dispersal units).

Deciduous forests or grasslands with large accumulations of vegetable matter serve as natural sources for fungal aerosols. Some of the more notorious sources (i.e., concentrations of dead organic material) attributable to human activity include piles of fresh wood chips, and compost sites (for municipal solid waste, yard waste, and/or sewage sludge) where monocultures of fungi (often *Aspergillus fumigatus*) are encouraged to thrive (Clark et al., 1983; Millner et al., 1977). Agricultural situations, such as fields where plant debris is allowed to die and decompose (Burge et al., 1992), or moldy feed and animal bedding (Gregory and Lacey, 1963), are also good sources for fungal spore aerosols. Some fungal spores are capable of germination after long periods, sometimes years, even under extreme conditions (Benninghoff, 1964), raising the possibility of a change in environment or land use but the persistence of spore sources.

Climate and a variety of environmental factors affect the distribution of fungi on specific sources. Some fungi produce spores (especially sexual spores) in response to specific climatic factors. For example, many Ascomycetes produce ascospores only after the fungus has overwintered in the substrate. These spores are released as the weather begins to warm in the spring, and disappear from the air during the rest of the year. Other fungi (including many of those producing primarily asexual spores) produce spores whenever sufficient water is available, with the stipulation that temperatures are within an appropriate range. Most fungi can produce spores over a range of temperatures from 15 to 30°C. A few, including *Aspergillus fumigatus* can grow and sporulate at higher temperature, up to 45°C, or even in excess of 50°C (e.g., *Thermomyces*).

Plant pathogenic fungi are normally restricted to the range of the host. The "macrocyclic" fungi (those that have multiple spore states in their life cycle) may require the presence of more than one host. For example, wheat stem rust (*Puccinia graminis*) requires both wheat (or other grasses) and common barberry (*Berberis vulgaris*) for completion of its life cycle. An historical method for controlling wheat rust has been to eradicate the local barberry population.

Factors affecting release and dispersal. Factors that affect release and subsequent dispersal of fungus spores can be divided into two interrelated categories: intrinsic and extrinsic. The Basidiomycetes, in particular, have elegant intrinsic mechanisms for spore release and dispersal. Basidia are structures that produce asymmetrically shaped spores on specialized structures (sterigmata) from which spores are forcibly discharged. In addition, basidia are arranged so that when the spores are discharged (such as from mushroom gills) they fall free rather than immediately impacting onto adjacent fungal tissue. Other fungi with well-developed intrinsic mechanisms for spore release include the Ascomycetes, which actively "squirt" sexual spores out of the ascus into the air, and fungi such as *Entomophthora* (an insect pathogen), which release individual spores from aerial strands by a forcible mechanism depending on water relations.

Extrinsic factors that affect spore release include weather parameters (especially humidity, rain, and wind), and other kinds of mechanical disturbance (see below). Relative humidity plays an important role in the release of spores of many fungi. A number of fungi, notably many ascomycetes, release spores in response to high relative humidity, and ascospore concentrations often peak shortly after commencement of light rain showers. Other spore types are released as relative humidity falls (Pedgley, 1982; Leach, 1975). Rain influences spore release both by allowing water absorption and stimulating intrinsic mechanisms, but also mechanically by a mechanism known as rain splash. A dramatic increase in concentrations of certain spores (including some "dry weather" spores) is often observed before and sometimes after rain showers (Hjelmroos, 1993). This effect could be due to splash dispersal or changes in relative humidity affecting intrinsic dispersal mechanisms.

Diurnal fluctuations in outdoor air spora have been noted by a number of investigators (Nussbaum, 1991; Pady et al., 1967). The causes of these fluctuations vary with different spore types and are probably due to interactions between a variety of meteorological and intrinsic parameters. Basidiospores, more so than ascospores, often show distinct diurnal variation, usually with predawn or nighttime peaks (Burge, 1986). Haard and Kramer (1970) attributed this cycle of maximal spore release at night mostly to high relative humidity, and minimal spore release during daylight hours to low relative humidity.

Light–dark cycles also seem to have an effect on spore release. *Hypocrea gelatinosa* (an ascomycete) spore release reaches a peak during dark hours. With artificial continuous dark or light levels, spore release was significantly reduced (Kramer and Pady, 1968). Kramer and Long (1970) were able to induce a pattern of spore release in response to light–dark cycles in *Ganoderma* isolates not originally showing a cyclic spore release pattern. The pattern was not interrupted by continuous light.

Barbetti (1987) found significant correlations between numbers of airborne spores of the plant pathogen *Cercospora* and percentage of clover leaflets infected, but also with weekly rainfall and number of wet days per week. Temperature, humidity, and light levels certainly do not affect spore release in all fungi. For example, Zoberi (1973) found that *Cookeina sulcipes* spore release seems to be dependent on the evaporation of rain water collected in the fruit body; no correlations were found with the previously mentioned weather parameters.

As an interesting, while probably minor, method of long-distance spore dispersal, Evans and Prusso (1969) reported that a number of fungi (*Cladosporium, Penicillium, Fusidium,* and *Alternaria,* among others) were isolated from the feet of 31 birds in the Sierra Nevada. One could speculate that some of these fungal spores are liberated while the birds are airborne, while others are transported long distances and deposited directly by the birds on new substrates.

Patterns of prevalence. Airborne fungal spore concentrations vary seasonally, especially in temperate regions. Few airborne spores are found during periods of snowcover. Following snowmelt, spore concentrations usually increase through the summer months, especially those of "dry-weather" spores (see below), notably *Alternaria, Cladosporium, Epicoccum,* etc. (Figure 9.2). Concentrations of many of these taxa peak during middle to late summer (Mäkinen and Ollikainen, 1972; AAAI, 1990–1993) before decreasing with the arrival of cooler weather. Hjelmroos (1993) noted a decrease in *Cladosporium* and *Alternaria* levels in late summer when daily mean temperature dropped below 15°C.

Fungi can be artificially grouped as "dry-weather fungi" (those releasing spores during dry weather) or "wet-weather fungi." The wet-weather fungi (including many ascomycetes such as *Leptosphaeria* and *Pleospora*) need wet or very humid conditions before spores are released and dispersed (Figure 9.3). Washout of particles, notably pollen and dry-weather fungal spores, from the air during the onset of even a light to moderate rainfall is often quite dramatic (Mäkinen and Ollikainen, 1972).

As mentioned above, diurnal variations are common in fungal spore release, and thus, prevalence. For example, most basidiospores are often most abundant just before dawn, while *Cladosporium* and other dark asexual spores are usually highest in the afternoon.

Human exposure and health effects. Exposure to fungi can cause a variety of diseases, probably the most widespread of which are the hypersensitivity diseases allergic rhinitis and asthma, and, to a much lesser extent, hypersensitivity pneumonitis. These diseases are the subject of Chapter 10. In addition to hypersensitivity disease, fungal emanations can cause infections and toxicoses.

The spores of a small number of "soil fungi" are notorious (although not always serious) causes of respiratory and systemic fungal infections. These include *Cryptococcus neoformans, Coccidioides immitis, Histoplasma*

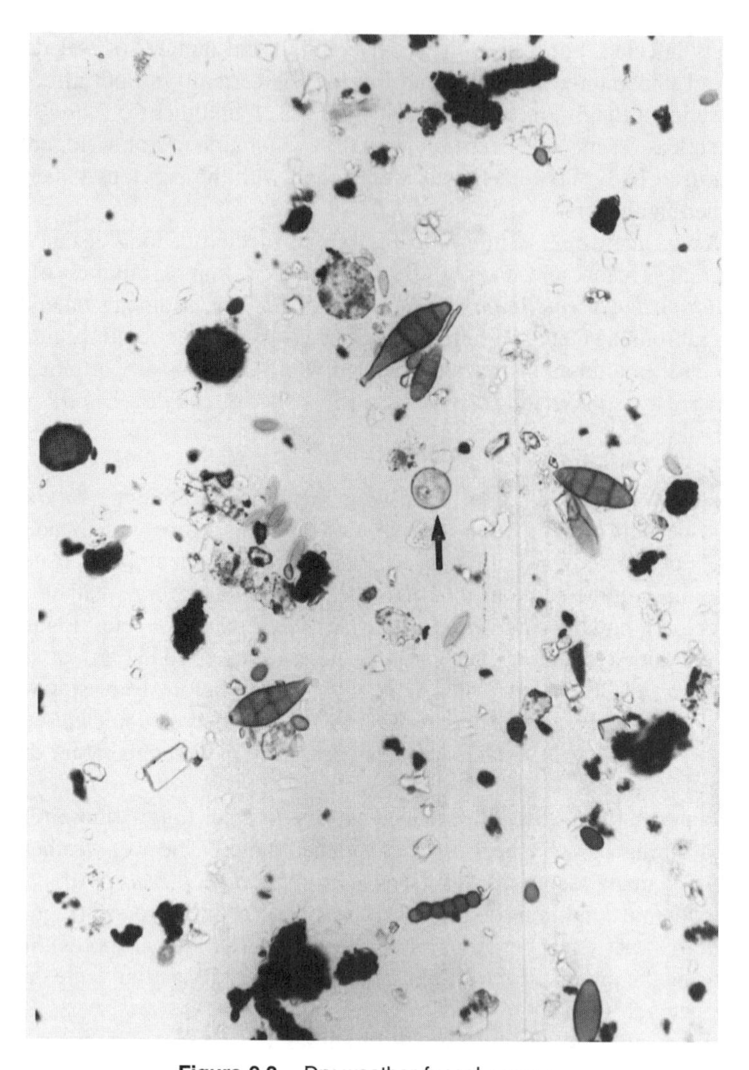

Figure 9.2. Dry-weather fungal spores.

capsulatum, and *Blastomyces dermatitidis*, among others. Often first-time infections with these fungi produce no symptoms, although immunity is developed, and spread of the organism within the host is limited. In the small percentage of patients who do not develop immunity, the disease can be fatal if untreated. *Coccidioides* is most commonly encountered in the arid regions of the southwest United States and in Mexico, where it is found in dry saline soil. *Histoplasma* is often associated with bird droppings or bat guano in contact with soil. *Cryptococcus* is often found in high concentrations in association with dry pigeon or other bird droppings where soil is not present (Ruiz et al., 1981). The small number of *Cryptococcus* infections reported each year

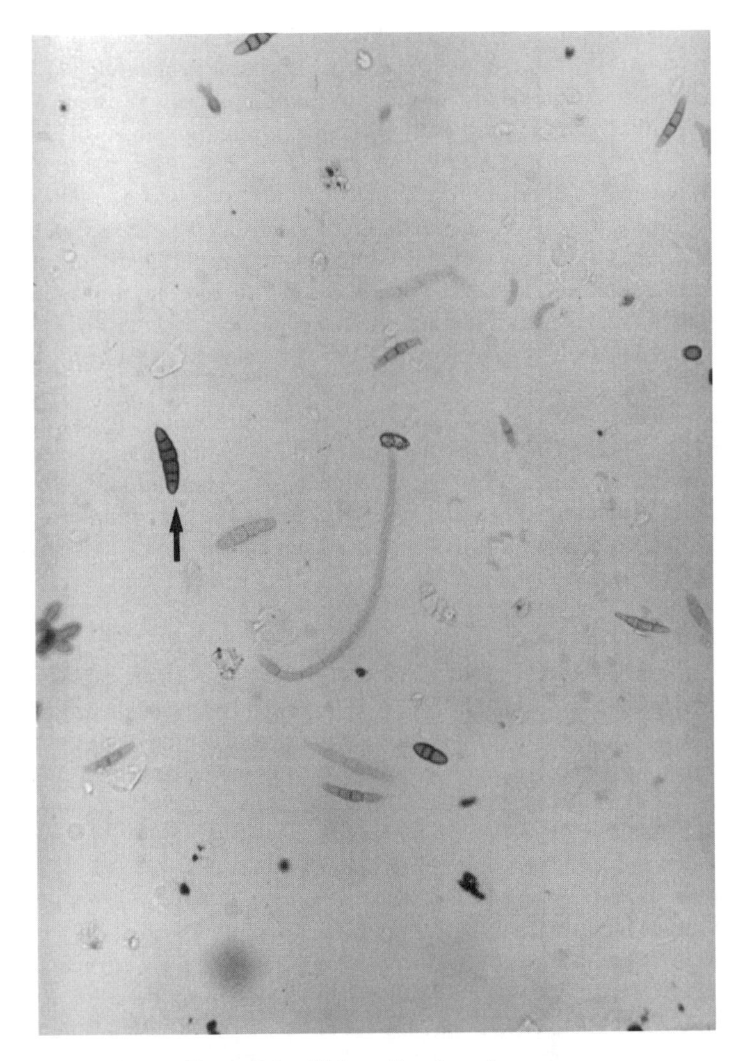

Figure 9.3. Wet-weather fungal spores.

relative to the amount of potential exposures may be due to the relatively large size of the airborne particles (high settling velocity), and the finding that only a small number reach well into the lungs to the alveolar surfaces (Powell et al., 1972), to the opportunistic nature of the infection, which requires some immune system dysfunction, or possibly due to the large dose needed to initiate an infection.

Aspergillus fumigatus as well as some other very common fungal saprophytes can invade severely immunocompromized hosts. This fact has raised concern about the very high levels of *A. fumigatus* that are associated with composting (Millner et al., 1977). Composting facilities should be sited well away from any concentration of highly susceptible people, and individuals who are at risk should avoid exposure to compost aerosols (Epstein, 1993).

In addition to these potentially pathogenic fungi, a number of dermatophytes (including *Geomyces, Chrysosporium, Trichophyton, and Malbranchea* species) have been reported from air, although in very low concentrations (<1 colony-forming unit/m^3 of air) (Della-Franca and Caretta, 1984; Marchisio et al., 1992). Sources of the potential skin colonizers include soil, infected and decomposing feathers or skin, and animal dung. In part due to the very low concentrations encountered, it must be assumed that direct contact with heavily contaminated surfaces is a much more efficient means of dermatophyte disease transmission than the airborne route.

Exposures to high levels of airborne mycotoxins or other fungal metabolites may be assumed in certain environments (e.g., agricultural). These toxins can be associated with airborne spores and hyphae, or can be adsorbed to other particles of respirable size (Sorenson et al., 1987). The risk of airborne exposure to aflatoxin (aflatoxin B_1 is the most potent known natural carcinogen) is discussed by Baxter et al. (1981). Respiratory exposure to other fungal toxins has not been investigated.

Sampling considerations. Documented prevalence patterns for airborne fungal spores depend on the kind of sample collector and on the method of analysis used. The Rotorod sampler, although commonly used especially by the allergy community, is a poor collector for most fungal spores because it does not efficiently collect particles smaller than about 15 μm. Rotorod recoveries will tend to be dominated by large asexual spore types such as *Alternaria*, and large clumps of *Cladosporium* spores. Wind-oriented suction impactors such as the Hirst-type spore traps and related devices commonly used in Europe collect fungal spores much more efficiently than the Rotorod. These devices make it obvious that smaller spores such as basidiospores and ascospores form a greater proportion of the air spora than one would expect based on Rotorod recoveries. Even more efficient are the commonly used Andersen culture plate impactors (Graseby Andersen, Atlanta, GA), efficiently collecting particles well below 1 μm in diameter.

Analyses of air samples for fungi commonly utilize either culture or direct microscopy. When culture is used (i.e., with the Andersen sampler), asexual spores of ascomycetes usually dominate recoveries. These spores germinate and grow to produce new spores and, as spore-bearing structures are usually required for speciation, are readily identifiable. Ascospores are also occasionally produced in culture, but conditions favoring their production appear to be more stringent than for production of asexual spores. Most Basidiomycetes do not grow well in artificial culture and therefore are grossly underestimated if only cultural samplers are used.

Microscopy allows more accurate total counts of fungal spores, but few specific identifications. Some basidiospores (for example those of *Ganoderma* or *Coprinus)* are among the more easily identifiable spores collected on particle (nonculture) samplers. It is probable that the majority of airborne basidiospores are often not even identified as such, except by a few experts, due to their lack of morphological distinctiveness. The very common asexual spores of *Cladosporium, Alternaria, Epicoccum,* etc. are recognizable with light microscopy, and many other spore types can be grouped either generically or as to type by this convenient method.

D. Slime Molds

Plasmodial slime molds (Myxomycetes) differ from true fungi in that, for much of their lives, they have no cell walls and consist of an ameboid mass of multinucleate protoplasm (plasmodium). The slime molds are heterotrophic and animal-like, ingesting yeasts and bacteria for food. However, when food and water become scarce (or sometimes in response to changes in light intensity) the plasmodium differentiates to form stalked, spore-producing structures (sporangia) and spores that are virtually indistinguishable from some fungal spores. The wind-borne spores are the primary dispersal stage and have thick, resistant cell walls. When the spores germinate they become flagellate and "mate" with other flagellated cells to form a zygote, which grows into a new plasmodium.

Myxomycete spores are usually small (commonly 8 to 10 μm), spherical, and are often finely ornamented (spiny, reticulate). Little is known about the airborne numbers of these spores, attributable in part to the difficulty in growing them in culture and also the problems in identifying the spores based on morphology alone (McElhenny and McGovern, 1970). Some slime mold spores are difficult to differentiate from some other fungal spore types, especially certain smut (basidiomycete) spores. Slime mold spores are, in fact, encountered on particulate air samples (personal observation), and have been demonstrated to contain antigens (Giannini et al., 1975). However, their importance in respiratory disease is unknown.

E. Bacteria

Nature of the particles. Bacteria are prokaryotic (lacking an organized nucleus with a nuclear membrane), single-celled organisms usually less than a micrometer or two in smallest diameter and often much smaller. The surface structure of bacteria is more complex than that of animal cells, having a cytoplasmic membrane surrounded by a rigid cell wall. Although bacteria are single-celled organisms, they are commonly grouped into pairs (*Diplococcus*), tetrads (*Micrococcus*), or even long chains (*Bacillus*). The actinomycetes are a group of bacteria that form long, branched chains, which, when viewed microscopically, appear more like very fine fungal hyphae than bacterial cells. The actinomycetes also produce very small (ca. 1 μm in diameter), very resistant spores. Most bacteria are saprobes, decomposing nonliving vegetable and animal materials or effluents. However, a number are opportunistic or obligate parasites.

Sources and source strengths. Bacteria are extremely common in the outdoor environment. Concentrations of 10^9 colony-forming units per gram of fertile soil are not uncommon. They are found in virtually all outdoor environments, including hot springs, Antarctic ice, deep in the ocean as well as in surface water, and on the surfaces of leaves, and are particularly abundant in some soils. Cooling towers of buildings, essentially large aerosol generators, can harbor (and aerosolize) large numbers of bacteria. Under favorable conditions, Actinomycetes produce small, easily respirable spores in great abundance. Gregory and Lacey (1963) reported Actinomycete spore concentrations in the hundreds of millions per gram of "mouldy hay." Actinomycetes are also often considered soil bacteria, where they can be found in very high concentrations (Lacey, 1973), and contribute the distinctive odor of fresh-turned earth.

Factors affecting release and dispersal. As far as is known, bacteria are released from reservoirs by mechanical agitation, including wind, disturbance by animals, human activities such as manipulation of compost, agricultural practices, and, for the human pathogens, coughing, sneezing, singing, etc. No active (intrinsic) release mechanisms are known for bacteria. The background aerosol is probably due primarily to wind release of bacteria from leaf surfaces and possibly from soil where extensive areas of soil are exposed. Under certain standing water conditions, splash mechanisms may also contribute significantly to the bacterial aerosol.

Patterns of prevalence. Concentrations of total culturable bacteria outdoors are commonly in the range of 100 to 1000 CFU/m^3 of air (Bovallius et al., 1978; Jones and Cookson, 1983). Using filter collectors, Mancinelli and Shulls (1978) found concentrations of 13 to almost 2000 CFU/m^3 of air in an urban environment (Boulder, Colorado) with *Micrococcus, Staphylococcus, Aerococcus,* and *Bacillus* species being the most commonly recovered. Bacteria and their by-products (e.g., endotoxin) are found in appreciable levels in outdoor air during some agricultural activities (Siegel et al., 1991), reflecting their role in decomposition of plant materials. Actinomycetes can reach very high concentrations when compost, moist hay, or grain is disturbed. Lacey and Lacey (1964) reported levels above 10^9/m^3 of air in a cowshed when moldy hay was disturbed. Airborne concentrations near such enclosures could be substantially above "ambient" levels, depending, in part, on the tightness of the enclosure. Lloyd (1969) noted increases in levels of actinomycetes with soil disturbance, although levels were not extremely high (up to 6000/m^3 of air) and most of the spores were in association with soil particles and therefore too large to penetrate the lower airways.

Human exposure and health effects. Bacteria are well-known agents of infectious disease, although few of these diseases are regularly transmitted through outdoor air. Infection (or at least increased specific antibodies to *Mycobacterium bovis*, a relative of the organism commonly causing tuberculosis) has been reported upon exposure to infected cattle, elk, rhinoceros, and even seals (Fanning and Edwards, 1991; Dalovisio et al., 1992; Thompson et al., 1993). Legionellosis, including Legionnaires' disease and Pontiac fever, is caused by exposure to *Legionella pneumophila*, usually indoors, but outdoor exposure to high concentrations can also result in disease. Legionellosis is discussed further in Chapter 3.

Some bacteria (particularly the Actinomycetes) cause hypersensitivity pneumonitis in situations where massive aerosols are produced (e.g., farming, compost manipulation, etc.) (Gregory et al., 1963).

Of increasing concern are health effects related to exposure to endotoxin, a cell-wall lipopolysaccharide that forms a part of the wall of Gram negative bacteria. Endotoxin exposure is associated with fever and reduced lung function (airway obstruction) (Castellan et al., 1987) and is further discussed in Chapter 4. Ambient endotoxin levels have not been evaluated.

Sampling considerations. The conventional method of bacterial collection and analysis has been culture (using culture-plate impactors or all-glass impingers) and a series of biochemical tests for identification, although in recent years staining methods for "total" bacteria using filters (acridine orange direct count; Palmgren et al., 1986) as well as for specific types (fluorescent labeled antibodies) have become available. Bacterial vegetative cells can be quite fragile (being susceptible to osmotic shock and physical injury) and can have strict temperature, moisture, and nutritional requirements, mandating care in sampler and culture media selection and incubation conditions. Recoveries from house dust and air are often greater on R2A (a minimal medium) compared to TSA or Nutrient agar (both considered rich media), possibility attributable to the ability of cells to better repair themselves under lower nutrient conditions. While bacteria may average around 0.5 to 1 μm in size, they often are in clumps or "floating" on larger rafts (e.g., skin scales), requiring a sampler that efficiently collects particles from tenths of a micron up to tens of microns to efficiently collect all bacteria.

F. Other Microorganisms

Viruses. Viruses are composed of a core of either DNA or RNA and a protective protein coat and have a size range of about 20 to 300 nm (see Chapter 2). Viruses are by definition pathogens, requiring a living cell for replication. As such one would only expect to encounter significant concentrations in areas with a dense population of hosts, which, with humans, almost always means indoor environments (although the risk of virus transmission at, for example, a Rose Bowl football game has not been examined!).

Animal and plant viruses do spread outdoors. There is strong evidence that foot-and- mouth disease, although usually considered a direct contact disease, can be transported by wind (Daggupaty and Sellers, 1990), especially under cloudy, rainy conditions (see Chapter 2). Fowl pest is another viral livestock disease that may be wind-borne.

It is often assumed that human viruses will not survive well in open air, or that, due to dilution, there will not be sufficient inoculum in outdoor air. A recent outbreak of hantavirus pulmonary syndrome is proving to be an interesting case where outdoor exposures may be playing a part. During the first half of 1993 in the "Four Corners" area of the southwest United States (common border of Utah, Colorado, New Mexico, and Arizona) a number of cases of respiratory illness were diagnosed as hantavirus infections (Hughes et al., 1993). While most exposures seem to be indoors (one was a tentlike enclosure), outdoor exposures have not been ruled out. Disease due to hantavirus is not uncommon in parts of Asia where exposures have been connected to outdoor activities such as plowing agricultural fields. The Asian strains, as well as those reported earlier in the United States, generally do not have a respiratory component but involve renal disease. The new virus, named Four Corners strain, has not been isolated in cell culture. Deer mice (*Peromyscus maniculatus*) seem to be the vectors, and the inoculum is likely from airborne feces and urine (Stone, 1993).

Viruses often require very specific conditions for survival (temperature, humidity, pH), making air sampling and analysis a complex task. Analysis is usually by cell culture or inoculation into susceptible organisms. A number of sampling instruments can be used to collect viruses (culture plate impactors, liquid impingers, filters) but, due to the fastidious nature of these particles, it is advisable to contact a microbiologist experienced in collecting airborne viruses before undertaking any such sampling studies.

Protozoa. Schlichting (1969) listed the types of protozoa recovered from outdoor air by five different investigators in Germany, the United States, and Taiwan. A total of 35 types were reported. In studies in North Carolina, Smith (1973) found 19 species of protozoa with an unidentified zooflagellate, *Monas*, and *Colpoda* being the most frequently encountered. In making correlations with weather parameters, Smith combined algae and protozoa and found positive correlations between wind and numbers of organisms, especially if winds were out of a southerly direction, and an inverse relationship between temperature and levels of algae and protozoa. He did not speculate on possible sources.

Protozoa are known to produce infections (e.g., leishmaniasis, toxoplasmosis, cryptosporidiosis), although these are usually not common in the United States and Canada and unlikely to result from exposure to ambient air. Protozoa are also known to sequester *Legionella pneumophila*. These topics are further discussed in Chapters 3 and 6.

G. Animal Aerosols

Arthropods. Some insects use wind to travel short distances. For example, first-stage gypsy moth caterpillars (*Lymantria dispar*), using abundant body hairs and silken threads, probably can travel hundreds of meters (at most) in light to moderate winds. Depending on one's definition of "airborne", this type of transport may or may not (as a flying squirrel probably would not) be included under aerobiology.

Among the most commonly seen and microscopically identifiable insect parts (the smaller of which we do consider to be truly airborne) on air samples are Lepidoptera (butterfly and moth) wing scales. Usually only a few of these easily identifiable structures are found per cubic meter of air, but numbers may be much higher (dozens per cubic meter) in certain environments (Balyeat et al., 1932). Lepidoptera scale shapes vary, but often resemble a wine glass or snifter with half the stem and base missing. Other insect parts such as "hairs" (for example, the 80 to 100 μm long caddis fly wing hairs), limb fragments, and other appendages are also commonly observed, but their sources have not been identified in most cases. Insect (particularly leafhopper) excretory particles (brochosomes) of about 0.5 μm diameter have been reported in the air at concentrations of 200/m^3 of air (Wiffen and Heard, 1969; Neville, 1970; Smith, 1970). These brochosomes are made of lipoprotein and their health significance is unknown.

While some fragments, especially Lepidoptera wing scales, are likely shed from flying insects, others probably originate from dead insects that are broken up during predation, abraded by other mechanical means (Urbach and Gottlieb, 1940), or broken down due to microbial action. Although the majority of health impacts described for insect fragments have resulted from indoor exposures, high concentrations of insects and their effluents can be found outdoors, often seasonally and in rather specific environments (Osgood, 1957). These insect parts vary over a wide range of sizes, from fractions of a micron to much larger particles, some of which would only become wind-borne in very turbulent air.

Inhalant allergies have been reported from indoor exposures to sewer flies (*Psychoda alternata*) around certain sewage treatment facilities (Gold et al., 1985) and bee moths (*Galleria mellonella*) in fish bait production facilities (Stevenson and Mathews, 1967). Although outdoor concentrations are assumed to be much smaller than those reported for these indoor environments, the possibility of elevated concentrations in air surrounding these facilities or in outdoor environments where high concentrations of similar insects thrive also needs to be considered. In the airborne particle fraction <10 μm in outdoor air, Kino et al. (1987) reported insect-related allergens at nanogram levels per cubic meter of air (levels about equivalent to those reported for *Alternaria* and ragweed antigens). They also reported peak butterfly levels and insect allergen levels in spring and autumn in Japan. While there is very little data available on airborne concentrations of insect particles, Kino and co-workers revealed high rates of skin reactivity to extracts of moth and butterfly wings, caddis fly wing, and chironomid (midge) whole bodies in patients with asthma (Kino and Oshima, 1978; Kino et al., 1987). Gupta et al. (1990) found that 36% of 75 patients had positive responses to intradermal tests with 13 insect extracts. These investigators used RASTs and bronchial provocation tests to rule out the possibility of false positive skin tests. Osgood (1957) claims that caddis fly setae (microscopic hairs) and dried particles from their dead bodies become airborne and are allergenic. He reported large clouds of these insects from about mid-June through September along the Niagara river at the east end of Lake Erie. A significant number (34%, N = 623) of "respiratory allergy" patients showed a moderate or marked positive intradermal skin test to caddis fly extract.

Mammals and birds. Airborne antigens derived from urine, saliva, and epithelium of a variety of mammals and birds have been reported to cause allergic disease in indoor environments (Pope et al., 1993). Sources include dogs, cats, laboratory rats and mice, rabbits, and reindeer, in addition to pigeons, parakeets, chickens, and a number of other mammal and bird species. In general, other than in unusual (usually agricultural) situations, concentrations of these animal-source allergens in outdoor air are nondetectable and probably insignificant. Even in situations where dense populations of animals are encountered outdoors (e.g., commercial reindeer herds) frequency of sensitization is minimal although those exposed to high antigen concentrations indoors (e.g., reindeer leather workers) can become sensitized (Reijula et al., 1992). Mammals and birds as allergen sources are further discussed in Chapters 7 and 8.

IV. RESEARCH DIRECTIONS

Much additional research is needed on factors affecting release and dispersal, as well as disease-causing potential, of outdoor bioaerosols. Nonpropagative plant parts (leaf hairs, small plant parts liberated due to plant decay or abrasion), many microaerosols including microbial toxins, volatile organic compounds produced by living organisms, and viruses have been, in general, poorly studied in outdoor air. Although plant pathogenic fungi have been studied, little is known about concentrations, seasonality, etc. of other fungi, many of which are abundant and impact on human health. This is attributable, in part, to the variety of sampler types and analytical methods employed and the lack of standardization (Muilenberg, 1989). No sampler has yet been devised that will collect desired particles with 100% efficiency and result in species identifications, with information on viability, allergen content, virulence, etc. This is an area that is sorely in need of innovative ideas and additional research.

While much is known about the behavior of aerosols in general, a great deal of work needs to be done on biological particles (bioaerosols) and the effects of becoming airborne: what stresses are involved, how do these stresses affect an organism's infectivity, viability, allergenicity, etc. Also of primary importance is how characteristics of a bioaerosol (as a living organism or product thereof) affect its launching into the air, transport through the air, and deposition on a surface. A few works have gone a long way toward addressing these issues (Cox, 1987; Ingold, 1971; Wolfenbarger, 1975), but the numerous unanswered questions should provide fodder for many interesting research projects.

REFERENCES

AAAI, *Pollen and Spore Report*. American Academy of Allergy and Clinical Immunology, Milwaukee, WI, 1990, 1991, 1992, 1993.

ACGIH, *Guidelines for the Assessment of Bioaerosols in the Indoor Environment*, American Converence of Governmental Industrial Hygienists, Cincinnati, OH, 1989.

Anto, J. M., Sunyer, J., Reed, C. E., Sabria, J., Martinez, F., Morell, F., Codina, R., Rodriguezroisin, R., Rodrigo, M. J., Roca, J., Saez, M., Preventing asthma epidemics due to soybeans by dust-control measures, *N. Engl. J. Med.* 329 (24), 1760, 1993.

Ayuso, R., Polo, F., Carreira, J., Purification of *Par j* I, the major allergen of *Parietaria judaica* pollen, *Mol. Immunol.* 25 (1), 49, 1988.

Balyeat, R. M., Stemen, T. R., Taft, C. E., Comparative pollen, mold, butterfly and moth emanation content of the air, *J. Allergy* 3, 227, 1932.

Barbetti, M. J., Seasonal fluctuations in concentration of airborne conidia of *Cercospora zebrina* and incidence of *Cercospora* disease in subterranean clover, *Trans. Br. Mycol. Soc.* 88(2), 280, 1987.

Bassett, I. J., Crompton, C. W., Parmelee, J. A., *An Atlas of Airborne Pollen Grains and Common Fungus Spores in Canada*, Monograph 18, Canada Department of Agriculture, Ottawa, Canada, 1978.

Baxter, C. S., Wey, H. E., Burg, W. R., A prospective analysis of the potential risk associated with inhalation of aflatoxin-contaminated grain dusts, *Food Cosmet. Toxicol.* 19, 765, 1981.

Benninghoff, W. S., Atmospheric particulate matter of plant origin. In: Sukalo, L. H., Ed., *Atmospheric Biology Conference, Proceedings*, University of Minnesota/NASA, 1964.

Bernstein, I. L., Safferman, R. S., Clinical sensenitivity to green algae demoonstrated by nasal challenge and invitro tests of immediate hypersensitivity, *J. Allergy Clin. Immunol.* 51, 22, 1973.

Bianchi, D. E., Schwemmin, D. J., Wagner, W. H., Jr., Pollen release in the common ragweed (*Ambrosia artemisiifolia*), *Bot. Gazette* 120(4), 235, 1959.

Bovallius, Å., Bucht, B., Roffey, R., Ånäs, P., Three year investigation of the natural airborne bacterial flora of four localities in Sweden, *Appl. Environ. Microbiol.* 35, 847, 1978.

Brown, R. M., Jr., Larson, D. A., Bold, H. C., Airborne algae: their abundance and heterogeneity, *Science* 143, 583, 1964.

Bunnag, C., Dhorranintra, B., Limsuvan, S., Jareoncharsri, P., Ferns and their allergenic importance: skin and nasal provocation tests to fern spore extract in allergic and non-allergic patients, *Ann. Allergy* 62, 554, 1989.

Burge, H. A., Some comments on the aerobiology of fungus spores, *Grana* 25, 143, 1986.

Burge, H. A., Bioaerosols in the residential environment. In: Wathes, C. M., Cox, C. S., Eds., *Bioaerosols: Handbook of Samplers and Sampling*, Lewis Publishers, Chelsea, MI, in press, chap. 20.

Burge, H. A., Muilenberg, M. L., Chapman, J., Crop plants as a source for medically important fungi. In: Andrews, J. H., Hirano, S. S., Eds., *Microbial Ecology of Leaves*. Springer-Verlag, New York, pp. 222–236, 1992.

Burge, H. A., Solomon, W. R., Muilenberg, M. L., Evaluation of indoor plantings as allergen exposure sources, *J. Allergy Clin. Immunol.* 70, 101, 1982.

Castellan, R. M., Olenchock, S. A., Kinsley, K. B., Hankinson, J. L., Inhaled endotoxin and decreased spirometric values, an exposure-response relation for cotton dust, *N. Engl. J. Med.* 317, 605, 1987.

Clark, S. C., Rylander, R., Larsson, L., Levels of gram-negative bacteria, *Aspergillus fumigatus*, dust, and endotoxin at compost plants, *Appl. Environ. Microbiol.* 45, 5, 1501, 1983.

Cox, C. S., The aerosol survival of *Escherichia coli* B in nitrogen, argon and helium atmospheres and the influence of relative humidity, *J. Gen. Microbiol.* 50, 139, 1968.

Cox, C. S., Aerosol survival of *Escherichia coli* B disseminated from the dry state, *Appl. Microbiol.* 19(4), 604, 1970.

Cox, C. S., *The Aerobiological Pathway of Microorganisms*, John Wiley & Sons, Chichester, England, 1987.

Cox, C. S., Airborne bacteria and viruses, *Sci. Prog., Oxford* 73, 469, 1989.

Crook, B., Higgins, S., Lacey, J., Airborne gram negative bacteria associated with the handing of domestic waste. In: Boehm, G., Leuschner, R. M., Eds., *Advances in Aerobiology*, Birkhauser Verlag, Basel, Switzerland, pp. 371–375, 1987.

Cua-Lim, F., Payawal, P. C., Laserna, G., Studies on atmospheric pollens in the Philippines, *Ann. Allergy* 40, 117, 1978.

Daggupaty, S. M., Sellers, R. F., Airborne spread of foot-and-mouth disease in Saskatchewan, Canada, 1951-1952, *Can. J. Vet. Res.* 54(4), 465, 1990.

Dalovisio, J. R., Stetter, M., Mikota-Wells, S., Rhinoceros' rhinorrhea: cause of an outbreak of infection due to airborne *Mycobacterium bovis* in zookeepers, *Clin. Infect. Dis.* 15(4), 598, 1992.

D'Amato, G., Spieksma, F. T. M., Nonini, S., *Allergenic Pollen and Pollenosis in Europe*, Blackwell Scientific Publications, Oxford, U.K., 1991.

Della Franca, P., Caretta, G., Keratinophilic fungi isolated from the air at Pavia, *Mycopathologia* 85(1), 65, 1984.

Dingle, A. N., Gatz, D. F., Air cleansing by convective rains, *J. Appl. Meteorol.* 5, 160, 1966.

Dingle, A. N., Gill, G. C., Wagner, W. H., Jr., Hewson, E. W., The emission, dispersion, and deposition of ragweed pollen, *Adv. Geophys.* 6, 367, 1959.

Dressler, R. L., *The Orchids — Natural History and Classification*, Harvard University Press, Cambridge, MA, 1981.

Dyer, H. J., Skok, J., Scully, N. J., Photoperiodic behavior of sunflower, *Bot. Gazette* 121, 1, 50, 1959.

Ehresman, D. W., Hatch, M. T., Effect of relative humidity on the survival of airborne unicellular algae, *Appl. Microbiol.* 29, 3, 352, 1975.

Emberlin, J., Savage, M., Jones, S., Annual variations in grass pollen seasons in London 1961–1990: trends and forecast models, *Clin. Exp. Allergy* 23, 911, 1993.

Epstein, E., Neighborhood and worker protection for composting facilities: issues and actions, 319-338. In: Hoitink, H. A. J., Keener, H. M., Eds. *Science and Engineering of Composting: Design, Environmental, Microbiological and Utilization Aspects*, Renaissance Publications, Worthington, OH, 1993.

Esau, K., *Anatomy of Seed Plants*, John Wiley & Sons, New York, 1977.

Evans, R. N., Prusso, D. C., Spore dispersal by birds, *Mycologia* 61, 4, 832, 1969.

Fanning, A., Edwards, S., *Mycobacterium bovis* infection in human beings in contact with elk (*Cervus elaphus*) in Alberta, Canada, *Lancet* 338(8777), 1253, 1991.

Giannini, E. H., Northey, W. T., Leathers, C. R., The allergenic significance of certain fungi rarely reported as allergens, *Ann. Allergy* 35, 372, 1975.

Gold, B. L., Mathews, K. P., Burge, H. A., Occupational asthma caused by sewer flies, *Am. Rev. Respir. Dis.* 131, 949, 1985.

Greene, V. W., Pederson, P. D., Lundgren, D. A., Hagberg, C. A., Microbiological exploration of stratosphere, results of six experiments. In: Sukalo, L. H., Ed., *Atmospheric Biology Conference Proceedings*, University of Minnesota/NASA, 1964.

Gregory, P. H., *The Microbiology of the Atmosphere*, London Hill Books Ltd, London, 1961.

Gregory, P. H., Lacey, M. E., Mycological examination of dust from mouldy hay associated with farmer's lung disease, *J. Gen. Microbiol.* 30, 75, 1963.

Gregory, P. H., Lacey, M. P., Festenstein, G. N., Skinner, F. H., Microbial and biochemical changes during the moulding of hay, *J. Gen. Microbiol.* 33, 147, 1963.

Gregory, P. H., Hamilton, E. D., Sreeramulu, T., Occurence of the alga *Gloeocapsa* in the air, *Nature* 176, 1270, 1955.

Gupta, S., Jain, S., Chaudhry, S., Agarwal M.K., Role of insects as inhalant allergens in bronchial asthma with special reference to the clinical characteristics of patients, *Clin. Exp. Allergy* 20, 519, 1990.

Haard, R. T., Kramer, C. L., Periodicity of spore discharge in the Hymenomycetes, *Mycologia* 62(6), 1145, 1970.

Habenicht, H. A., Burge, H. A., Muilenberg, M. L., Solomon, W. R., Allergen carriage by atmospheric aerosol. II. Ragweed-pollen determinants in submicronic atmospheric fractions, *J. Allergy Clin. Immunol.* 74(1), 64, 1984.

Harvey, R., Air-spora studies at Cardiff. III. Hyphal fragments, *Trans. Br. Mycol. Soc.* 54(2), 251, 1970.

Harvey, R., Aerobiological surveys and spore discharge studies at Cardiff, 1942–1972. In: Nilsson, S., Ed., *Scandinavian Aerobiology,* Bulletins from the Ecological Research Committee No. 18, Swedish Natural Science Research Council, pp. 113–130, 1972.

Hess, G. E., Effects of oxygen on aerosolized *Serratia marcescens*, *Appl. Microbiol.* 13, 781, 1965.

Hirsch, D. J., Hirsch, S. R., Kalbfleisch, J. H., Effect of central air-conditioning and meteorological factors on indoor spore counts, *J. Allergy Clin. Immunol.* 62, 22, 1978.

Hjelmroos, M., Relationship between airborne fungal spore presence and weather variables, *Grana* 32, 40, 1993.

Hollinger, S. E., Sivier, K. R., Irwin, M. E., Isard, S. A., A helicopter-mounted isokinetic aerial insect sampler, *J. Econ. Entomol.* 84, 2, 476, 1991.

Holt, G. L., Seasonal indoor/outdoor fungi ratios and indoor bacteria levels in non-compliant office buildings. In: *Indoor Air '90. Preceedings of the 5th International Conference on Indoor Air Quality and Climate,* Vol. 2, Canada Mortgage and Housing Corporation, Toronto, Canada, pp. 33-35, 1990.

Hughes, J. M., Peters, C. J., Cohen, M. L., Mahy, B. W. J., Hantavirus pulmonary syndrome: an emerging infectious disease, *Science* 262, 850, 1993.

Imshenetsky, A. A., Lysenko, S. V., Kazakov, G. A., Upper boundary of the biosphere, *Appl. Environ. Microbiol.* 35, 1, 1978.

Ingold, C. T., *Fungal Spores: Their Liberation and Dispersal,* Clarendon Press, Oxford, 1971.

Ipsen, H., Løwenstein, H., Isolation and immunochemical characterization of the major allergen of birch pollen (*Betula verrucosa*), *J. Allergy Clin. Immunol.* 72(2), 150, 1983.

Jones, B. L., Cookson, J. T., Natural atmospheric microbial conditions in a typical suburban area, *Appl. Environ. Microbiol.* 45(3), 919, 1983.

King, T. P., Norman, P. S., Connell, J. J., Isolation and characterization of allergens from ragweed pollen, II, *Biochemistry* 3, 458, 1964.

Kino, T., Chihara, J., Fukuda, K., Sasaki, Y., Shogaki, Y., Oshima, S., Allergy to insects in Japan III. High frequency of IgE antibody responses to insects (moth, butterfly, caddis fly, and chironomid) in patients with bronchial asthma and immunochemical quantitation of the insect-related airborne particles smaller than 10 microns in diameter, *J. Allergy Clin. Immunol.* 79(6), 857, 1987.

Kino, T., Oshima, S., Allergy to insects in Japan I. The reaginic sensitivity to moth and butterfly in patients with bronchial asthma, *J. Allergy Clin. Immunol.* 61(1), 10, 1978.

Knox, R. B., Grass pollen, thunderstorms and asthma, *Clin. Exper. Allergy* 23, 354, 1993.

Knox, R. B., Heslop-Harrison, J., Pollen-wall proteins: localization of antigenic and allergenic proteins in the pollen-grain walls of *Ambrosia* spp. (ragweeds), *Cytobios* 4(13), 49, 1971.

Kramer, C. L., Long, D. L., An endogenous rhythm of spore discharge in *Ganoderma applanatum, Mycologia* 62(6), 1138, 1970.

Kramer, C. L., Pady, S. M., Spore discharge in *Hypocrea gelatinosa*, *Mycologia* 60(1), 208, 1968.

Kramer, C. L., Wartell, J., Holzapfel, E. P., Surface level trapping of air biota on the Pacific Ocean, *Agric. Meteor.* 12, 49, 1973.

Kurzweil, H., Seed morphology in Southern African orchidoideae (Orchidaceae), *Plant System. Evolut.* 185, 229, 1993.

Lacey, J., Actinomycetes in soils, composts and fodders. In: Skinner, F. A., Sykes, G., Eds., *Actinomycetales: Characteristics and Practical Importance,* Society of Applied Bacteriology Symposium Series No. 2, Academic Press, London, 1973.
Lacey, J., Lacey, M. E., Spore concentrations in the air of farm buildings, *Trans. Br. Mycol. Soc.* 47, 547, 1964.
Leach, C. M., Influence of relative humidity and red-infrared radiation on violent spore release by *Drechslera turcica* and other fungi, *Phytopathology* 65, 1303, 1975.
Lewis, W. H., Vinay, P., Zenger, V. E., *Airborne and Allergenic Pollen of North America.* The Johns Hopkins University Press, Baltimore, MD, 1983.
Lloyd, A. B., Dispersal of *Streptomycetes* in air, *J. Gen. Microbiol.* 57, 35, 1969.
Løwenstein, H., Isolation and partial characterization of three allergens of timothy pollen, *Allergy* 33, 30, 1978.
Mäkinen, Y., Ollikainen, P., Diurnal and seasonal variation s in the airspora composition in Turku, S. Finland. In: Nilsson, S., Ed., *Scandinavian Aerobiology*, Bulletins from the Ecological Research Committee No. 18, Swedish Natural Science Research Council, pp. 143-152, 1972.
Malley, A., Harris, R. L., Biologic properties of a non-precipitating antigen from timothy pollen extracts, *J. Immunol.* 99, 825, 1967.
Mancinelli, R. L., Schulls, W. A., Airborne bacteria in an urban environment, *Appl. Environ. Microbiol.* 35(6), 1095, 1978.
Marchisio, F. V., Cassinilli, C., Tullio, V., Piscozzi, A., Outdoor airborne dermatophytes and related fungi: a survey in Turin, Italy, *Mycoses* 35(9-10), 251, 1992.
McElhenney, T. R., McGovern, J. P., Possible new inhalant allergens, *Ann. Allergy* 28, 467, 1970.
Meier, F., Lindbergh, C. A., Collecting microorganisms from the Arctic atmosphere, *Sci. Monthly* 40, 5, 1935.
Millner, P. D., Marsh, P. B., Snowden, R. B., Parr, J. F., Occurrence of *Aspergillus fumigatus* during composting of sewage sludge, *Appl. Environ. Microbiol.* 34, 6, 765, 1977.
Muilenberg, M. L., Aeroallergen assessment by microscopy and culture, In Solomon, W. R., Ed., *Airborne Allergens. Immunol. Allergy Clin. North Am.* 9(2), 245, 1989.
Muilenberg, M. L., Skellenger, W. S., Burge, H. A., Solomon, W. R., Particle penetration into the automotive interior, I. Influence of vehicle speed and ventilatory mode, *J. Allergy Clin. Immunol.* 87, 581, 1991.
Neville, A. C., "Airborne organism" identified, *Nature* 225, 199, 1970.
Noble, W. C., Clayton, Y. M., Fungi in the air of hospital wards, *J. Gen. Microbiol.* 32, 397, 1963.
Nussbaum, F., Variations in the airborne fungal spore population of the Tuscarawas Valley II, *Mycopathologia* 116, 181, 1991.
O'Rourke, M. K., Lebowitz, M. D., A comparison of regional atmospheric pollen with pollen collected at and near homes, *Grana* 23, 55, 1984.
Osgood, H., Allergy to caddis fly (*Trichoptera*), *J. Allergy* 28(4), 292, 1957.
Ostrov, M. R., Oak leaf hairs as aeroallergens, *Immunol. Allergy Practice* 6(7), 285, 1984.
Pady, S. M., Kramer, C. L., Clary, R., Diurnal periodicity in airborne fungi in an orchard, *J. Allergy* 39(5), 302, 1967.

Palmgren, U., Strom, G., Blomquist, B., Malmberg, P., Collection of airborne microorganisms on Nuclepore filters, estimation and analysis — CAMNEA method, *J. Appl. Bacteriol.* 61, 5, 401, 1986.

Pan, P. M., Burge, H. A., Su, H. J., Spengler, J. D., Central vs room air conditioning for reducing exposure to airborne fungus spores (abstract), *J. Allergy Clin. Immunol.* 89(1), pt. 2, 258, 1992.

Pasanen, A. L., Reponen, T., Kalliokoski, P., Nevalainen, A., Seasonal variation of fungal spore levels in indoor and outdoor air in the subarctic climate. In: *Indoor Air '90. Proceedings of the 5th International Conference on Indoor Air Quality and Climate,* Vol. 2, Canada Mortgage and Housing Corporation, Toronto, Canada, pp. 9-14, 1990.

Peattie, D. C., *A Natural History of Western Trees,* Bonanza Press, New York, 1953.

Pedgley, D. E., *Windborne Pests and Diseases, Meteorology of Airborne Organisms,* Ellis Horwood Ltd., Chichester, England, 1982.

Platts-Mills, T. A. E., Heymann, P. W., Longbottom, J. L., Wilkins, S. R., Airborne allergens associated with asthma: particle sizes carrying dust mite and rat allergens measured with a cascade impactor, *J. Allergy Clin. Immunol.* 77(6), 850, 1986.

Pollart, S. M., Reid, M. J., Fling, J. A., Chapman, M. D., Platts-Mills, T. A., Epidemiology of emergency room asthma in northern California: association with IgE antibody to ryegrass pollen, *J. Allergy Clin. Immunol.* 82(2), 224, 1988.

Pope, A. M., Patterson, R., Burge, H., Eds., *Indoor Allergens: Assessing and Controlling Adverse Health Effects,* National Academy Press, Washington, D.C., 1993.

Powell, K. E., Dahl, B. A., Weeks, R. J., Tosh, F. E., Airborne *Cryptococcus neoformans*: particles from pigeon excreta compatible with alveolar deposition, *J. Infect. Dis.* 125(4), 412, 1972.

Proctor, B. E., Parker, B. W., Microorganisms in the upper air. In: Moulton, F. R., Ed., *Aerobiology,* American Association for the Advancement of Science No. 17, Washington, D.C., 1942.

Rantio-Lehtimäki, A., Koivikko, A., Kupias, R., Mäkinen, Y., Pohjola, A., Significance of sampling height of airborne particles for aerobiological information, *Allergy* 46, 68, 1991.

Raven, P. H., Evert, R. F., Curtis, H., *Biology of Plants,* Worth, New York, p. 378, 1976.

Raynor, G. S., Ogden, E. C., Hayes, J. V., Variation in ragweed pollen concentration to a height of 108 meters, *J. Allergy Clin. Immunol.* 51, 199, 1973.

Reijula, K., Virtanen, T., Halmepuro, L., Anttonen, H., Mäntyjärvi, R., Hassi, J., Detection of airborne reindeer epithelial antigen by enzyme-linked immunosorbent assay inhibition, *Allergy* 47, 203, 1992.

Rosas, I., Roy-Ocotla, G., Mosiño, P., Meteorological effects on variation of airborne algae in Mexico, *Int. J. Biometeorol.* 33, 173, 1989.

Ross, A. F., Mitchell, J. C., Respiratory irritation by leaf hair of the tree *Platanus*, *Ann. Allergy* 32, 94, 1974.

Roy-Ocotla, G., Carrera, J., Aeroalgae: responses to some aerobiological questions, *Grana* 32, 48, 1993.

Ruiz, A., Fromtling, R. A., Bulmer, G. S., Distribution of *Cryptococcus neoformans* in a natural site, *Infect. Immun.* 31, 2, 560, 1981.

Salén, E. G., *Lycopodium* allergy, *Acta Allergol.* 4, 308, 1951.

Schlichting, H. E., Jr., The importance of airborne algae and protozoa, *J. Air Pollut. Control Assoc.* 19(12), 946, 1969.

Sears, M. R., Herbison, G. P., Holdaway, M. D., Hewitt, C. J., Flannery, E. M., Silva, P. A., The relative risks of sensitivity to grass pollen, house dust mite and cat dander in the development of childhood asthma, *Clin. Exp. Allergy* 19, 419, 1989.

Siegel, P. D., Olenchock, S. A., Sorenson, W. G., Lewis, D. M., Bledsoe, T. A., Jay, J. J., Pratt, D. S., Histamine and endotoxin contamination of hay and respirable hay dust, *Scand. J. Work Environ. Health* 17, 4, 276, 1991.

Smith, D. S., "Airborne organism" identified, *Nature* 225, 199, 1970.

Smith, P. E., The effects of some air pollutants and meteorological conditions on airborne algae and protozoa, *J. Air Pollut. Control Assoc.* 23, 10, 876, 1973.

Sneller, M. R., Hayes, H. D., Pinnas, J. L., Pollen changes during five decades of urbanization in Tucson, Arizona, *Ann. Allergy* 71, 519, 1993.

Solomon, A. M., Buell, M. F., Effects of suburbanization upon airborne pollen, *Bull. Torrey Bot. Club* 96, 4, 435, 1969.

Solomon, W. R., A volumetric study of winter fungus prevalence in the air of midwestern homes, *J. Allergy Clin. Immunol.* 57, 1, 46, 1976.

Solomon, W. R., Aerobiology and inhalant allergens I. Pollens and fungi. In: Middleton, E., Jr., Reed, C. E., Ellis, E. F., Eds., *Allergy Principles and Practice*, C. V. Mosby, St. Louis, MO, pp. 1143–1190, 1983.

Solomon, W. R., Common pollen and fungus allergens. In: Bierman, C. W., Pearlman, D. S., Eds., *Allergic Diseases from Infancy to Adulthood*, W. B. Saunders, Philadelphia, p. 141, 1988.

Solomon, W. R., Burge, H. P., Boise, J. R., Airborne *Aspergillus fumigatus* levels outside and within a large clinical center, *J. Allergy Clin. Immunol.* 62, 1, 56, 1978.

Solomon, W. R., Burge, H. A., Boise, J. R., Exclusion of particulate allergens by window air conditioners, *J. Allergy Clin. Immunol.* 65, 4, 305, 1980.

Solomon, W. R., Burge, H. A., Muilenberg, M. L., Allergen carriage by atmospheric aerosol. I. Ragweed pollen determinants in smaller micronic fractions, *J. Allergy Clin. Immunol.* 72(5), 443, 1983.

Sorenson, W. G., Frazer, D. G., Jarvis, B. B., Simpson, J., Robinson, V. A., Trichothecene mycotoxins in aerosolized conidia of *Stachybotrys atra*, *Appl. Environ. Microbiol.* 53(6), 1370-1375, 1987.

Southworth, D., Singh, M. B., Hough, T., Smart, I. J., Taylor, P., Knox, R. B., Antibodies to pollen exines, *Planta* 176, 482, 1988.

Stewart, G. A., Holt, P. G., Submicronic airborne allergens (letter), *Med. J. Aust.* 143(9), 426, 1985.

Stevenson, D. D., Mathews, K. P., Occupational asthma following inhalation of moth particles, *J. Allergy* 39, 274, 1967.

Stone, R., The mouse-pinon nut connection, *Science* 262, 833, 1993.

Strange, R. E., Cox, C. S., Survival of dried and airborne bacteria. In: Bray, T. R. G., Postgate, J. R., Eds., *The Survival of Vegetative Microbes,* Symposia — Society for General Microbiology 26, Cambridge University Press, Cambridge, pp. 111–

Streifel, A. J., Lauer, J. L., Vesley, D., Juni, B., Rhame, F. S., *Aspergillus fumigatus* and other thermotolerant fungi generated by hospital building demolition, *Appl. Environ. Microbiol.* 46, 2, 375, 1983.

Suphioglu, C., Singh, M. B., Taylor, P., Bellomo, R., Holmes, P., Puy, R., Knox, R. B., Mechanism of grass-pollen-induced asthma, *Lancet* 339, 569, 1992.

Takahashi, Y. Sakaguchi, M., Inouye, S., Miyazawa, H., Imaoka, K., Katagiri, S., Existence of exine-free airborne allergen particles of Japanese cedar (*Cryptomeria japonica*) pollen, *Allergy* 46, 588, 1991.

Theunissen, H. J. H., Lemmens-den Toom, N. A., Burggraaf, A., Stolz, E., Michel, M. F., Influence of temperature and relative humidity on the survival of *Chlamydia pneumoniae* in aerosols, *Appl. Environ. Microbiol.* 59(8), 2589, 1993.

Thompson, P. J., Cousins, D. V., Bow, B. L., Collins, D. M., Williamson, B. H., Dagnia, H. T., Seals, seal trainers, and mycobacterium infection, *Am. Rev. Respir. Dis.* 147(1), 164, 1993.

Tiberg, E., Microalgae as aeroplankton and allergens. In: Boehm, G., Leuschner, R. M., Eds., *Advances in Aerobiology,* Birkhauser Verlag, Basel, Switzerland, p. 171–173, 1987.

Tiberg, E., Bergman, B., Wictorin, B., Willén, T., Occurrence of microalgae in indoor and outdoor environments in Sweden. In: Nilsson, S., Raj, B., Eds., *Nordic Aerobiology,* p. 24–29, 1984.

Tyndall, J., *Essays on the Floating-Matter of the Air in Relation to Putrefaction and Infection*, D. Appleton, New York, 1884.

Urbach, E., Gottlieb, P. M., Asthma from insect emanations, *J. Allergy* 12, 485, 1940.

van den Assem, A., Airborne pollen in relation to pollinosis. In: Nilsson, S., Ed., *Scandinavian Aerobiology*, Bulletins from the Ecological Research Committee No. 18, Swedish Natural Science Research Council, 1972.

Vinje, J. M., Vinje, M. M., Preliminary aerial survey of microbiota in the vicinity of Davenport, Iowa, *Am. Midl. Nat.* 54(2), 418, 1955.

Wiffen, R. D., Heard, M. J., Unidentified airborne organism, *Nature* 224, 715, 1969.

Willeke, K., Baron, P., *Aerosol Measurement: Principles, Techniques and Applications*, Van Nostrand Reinhold, New York, 1993.

Wolfenbarger, D. O., *Factors Affecting Dispersal Distances of Small Organisms*, Exposition Press, Hicksville, New York, 1975.

Zacharin, D., Plane tree leaves a cause of seasonal asthma and hay fever (letter), *Med. J. Austr.* 1(15), 467, 1933.

Zoberi, M. H., Influence of water on spore release in *Cookeina sulcipes*, *Mycologia* 65(1), 155, 1973.

10 BIOAEROSOL-INDUCED HYPERSENSITIVITY DISEASES

Cory E. Cookingham and William R. Solomon

CONTENTS

I. INTRODUCTION

Respiratory exposures to airborne allergens (aeroallergens), many of biological origin, account for most allergic asthma and "hay fever," and invariably determine occurrences of typical hypersensitivity pneumonitis. Airborne agents

0-87371-724-4/95/$0.00+$.50

also may induce allergic conjunctivitis and affect atopic eczema but are rarely invoked in urticaria and angioedema, anaphylaxis, and allergic contact dermatitis. In addition, respiratory exposure probably initiates allergic bronchopulmonary mycoses (e.g., bronchopulmonary aspergillosis). This chapter presents an overview of the hypersensitivity diseases as they relate to biological aerosols. For the purposes of this discussion, the following definitions apply:

Hypersensitivity disease: an illness resulting from an abnormal or maladaptive response of the immune system to a substance recognized as "foreign." Allergic contact dermatitis (e.g., poison ivy dermatitis) and common immunoglobulin E (IgE)-mediated conditions such as hay fever and allergic asthma are common examples of "allergic diseases."

Bronchial hyperactivity: an increased tendency to develop airway narrowing in response to nonimmunologic and, often, immunologic stimuli.

Antigen*: an agent that elicits a specific immunologic response, with subsequent development of specific antibody, cell-mediated immunity, or both.

Allergen: an antigen that elicits symptoms of allergic disease in a previously sensitized person.

Atopy: a hereditofamilial predisposition to IgE-mediated allergic disease, especially in response to antigens encountered at mucous membranes.

II. THE HYPERSENSITIVITY RESPONSE

A. Primary/Secondary Response

All hypersensitivity diseases, by definition, involved "immune" responses by the host's body to antigen(s). In order to develop hypersensitivity disease, a person must be exposed to an antigen on at least two appropriately spaced occasions or in a sustained fashion. The immune response is characterized by a primary phase on initial exposure and by secondary (recall) responses on subsequent reintroduction of the specific antigen.

Early in the primary response, antigen is taken up by antigen-presenting cells (notably macrophages), which degrade and display processed antigen on their surfaces in a form recognizable by thymus-derived cells (T lymphocytes). This process can result in functional activation and an increase in number of antibody-secreting cells (B lymphocytes) as well as T lymphocytes that recognize the antigen. Some of the B and T lymphocytes generated become long-lived "memory cells," which "remember" the antigen and proliferate rapidly following reexposure. Individuals with such expanded, antigen-reactive components are said to be "sensitized." When the sensitized host is exposed later to the same antigen, a "secondary" response occurs, with a more rapid expansion

* In other contexts, agents that elicit immunologic responses are defined as "immunogens" while the term "antigens" is reserved for those entities that will react specifically with the products of the immune response. The distinction has validity, since some antigens are not effective immunogens.

of effector cell numbers and/or higher levels of specific antibody, a longer duration of more effective antibody production, and (at times) tissue changes typical of hypersensitivity disease.

B. Kinds of Hypersensitivity Responses

Coombs and Gell (1975) developed a classification of immune responses that are manifest as hypersensitivity diseases. In the Type I or "immediate reaction," B lymphocytes are stimulated to produce antigen-specific antibodies of a particular physical chemical category, namely, immunoglobulin E (IgE). These IgE antibodies bind to tissue mast cells in mucous membranes and connective tissue and to circulating blood basophils via specialized cell surface receptors. When such IgE-coated cells are exposed to a sensitizing antigen, the antigen binds to complementary IgE and results in the release of chemical mediators of inflammation (e.g., histamine, leukotrienes, prostaglandins, etc.) by the affected cells. The effects of these chemical mediators on tissues result in the characteristic changes of disease. Type I reactions are the primary immune mechanism of the so-called atopic diseases: allergic rhinitis, allergic asthma, and gastrointestinal allergy, as well as some forms of urticaria-angioedema and systemic anaphylaxis.

Type II reactions involve primarily immunoglobulin G (IgG) and immunoglobulin M (IgM) antibodies, which circulate in blood and, in disease states, may react with antigenic components on cell surfaces of the host, resulting in tissue damage. These antibodies lack affinity for mast cell/basophil surfaces. Type II reactions are not generally associated with bioaerosol-related hypersensitivity diseases, but may be initiated by viral infection or exposure to environmental toxins.

In the Type III or "Arthus" reactions, circulating IgG and IgM antibodies bind a sensitizing antigen to form blood-borne immune "complexes," which can deposit in tissue (especially in blood vessel walls). These immune complexes activate a series of proteins, known collectively as "complement," which are capable of lysing cells and attracting neutrophils and monocytes (types of phagocytic cells) to the site with further tissue destruction and inflammation. Type III reactions are a probable early component of hypersensitivity pneumonitis and may follow other environmental exposures.

In the Type IV or "delayed" hypersensitivity response, memory T cells reexposed to sensitizing antigen release activating substances (lymphokines) that attract tissue-destructive cells to the site of the T cell–antigen interaction. The tissue inflammation associated with this type of reaction takes several hours to days to become clinically detectable. Type IV reactions are thought to contribute to hypersensitivity pneumonitis and form the basis of allergic contact dermatitis.

C. Genetic Control of the Immune Response

A basic requirement for any immune reaction is that the responding immune system must recognize an antigen as foreign or "nonself." This ability is genetically controlled primarily by the major histocompatibility complex (MHC) on human chromosome six, which encodes human leukocyte antigens (HLA). The HLA are divided into class I (HLA-A, -B, -C) and class II (HLA-DP, -DQ, -DR). The class I HLA are found on all nucleated cells, while the class II HLA occur predominantly on antigen-presenting cells, including B lymphocytes (which also generate antibody-secreting cells) and macrophages. These cells internalize and degrade/modify the antigen, which is displayed later in association with MHC (HLA) molecules on their surfaces. This processed antigen–MHC complex is recognized by specifically reactive T lymphocytes as foreign, facilitating an immune response. The genetic control ("restriction") of the immune response is expressed in whether an individual has MHC molecules that will present specific antigens and whether particular T-cell and antigen receptors (also encoded genetically) are available. The topic of antigen presentation, as well as processing for class I HLA-associated antigens, is reviewed clearly by Brodsky and Guagliardi (1991).

D. Factors Affecting Reactivity to Antigen

Determinants of individual reactivity to antigen include genetic constitution (as discussed above) and health status. Other determinants are the nature of the antigen as well as of associated materials (e.g., bacterial endotoxin), the concentration of the antigen-bearing particles, the size of the particles, and the route, duration, intensity, and frequency of exposure affect response. These factors are subtly interactive. For example, the size of antigen-bearing particles directly affects the decay rate of their airborne concentrations, as well as dosing, at specific respiratory levels, of exposed persons.

The nature of antigen. An antigen is a substance that is specifically recognized by the immune system, with development of specific antibody, specifically reactive cells, or both. Antigenic agents are principally proteins, but polysaccharides, nucleic acids, and lipids may elicit immune responses. Highly reactive small molecules (called haptens) can also induce an immune response after attachment to a larger (protein) molecule. The dinitrophenol (DNP) group is a good example of a hapten. Biogenic antigens are usually complex molecules such as glycoproteins with molecular weights 10,000 to greater than 100,000 Da (Goodman, 1991). The size of an antigen may affect the type of immune response elicited. Stupp et al. (1971) demonstrated this by immunizing guinea pigs with DNP attached to varying numbers of lysine molecules. When only two lysine residues were attached, humoral (circulating antibody) immunity alone was stimulated, while the production of humoral and cellular immunity required at least eight lysine residues conjugated to DNP.

Circumstances of antigen exposure: adjuvants. Antigen exposure, whether as aerosol or by other routes, rarely occurs in the absence of other (nonantigenic) agents; among these is a group of substances known as adjuvants. An adjuvant is a substance that encourages and/or augments an immune response. Examples of naturally occurring adjuvants include endotoxin (lipopolysaccharide present in cell walls of Gram negative bacteria) (see Chapter 4) and complex substances derived from members of the order Actinomycetales (*Mycobacterium, Thermoactinomyces,* etc.). As endotoxin is ubiquitous in the environment, all natural exposures to antigen aerosols probably include endotoxin to some degree. The role of endotoxin in the development of specific hypersensitivity diseases, however, has not been extensively explored. The adjuvant activity derived from mycobacteria is commonly used to stimulate immune responses in animal models of disease. In addition, similar materials derived from actinomycetes may modulate the immune responses to vegetable dusts implicated in hypersensitivity pneumonitis (e.g., in farmer's lung disease).

Nonaerosol routes of exposure. Familiar nonaerosol routes of antigen exposure include ingestion, topical application, and injection into the skin, interstitial tissues, or bloodstream. The ingestion of antigen can result in tolerance (an abrogation of immune response) to the agent, which also can prevent subsequent sensitization (Mowat, 1987). This may have considerable importance, considering the many potential antigens in the human diet, as both foodstuffs and contaminants. In animal models, the quantity of ingested antigen has significantly affected the resulting immune response (Jarrett et al., 1976). Frequently, low doses of antigen have stimulated a response while higher doses induced a state of prolonged unresponsiveness.

Aerosol exposure: the aerosols. Aerosols generally are characterized by (aerodynamic) particle size distribution and concentration (among other factors), and antigen aerosols are no exception. The particle size of airborne antigen determines both the time course of aerosol prevalence within an environment and patterns of respiratory deposition (see below). The effect of particle size on antigen exposure potential is demonstrated by the differing aerosol decay patterns of particles carrying cat and dust mite allergens. Cat allergen is carried on small particles (often <2–3 μm), and much of it remains airborne after disturbance for more than 24 h. On the other hand, most mite allergen particles appear to be effectively much larger (ca., 20 μm), and levels decay rapidly following dispersion (Swanson et al., 1985). Comparable results were reported by Platts-Mills et al. (1986) for dust mite and rat antigens, the latter substantially derived from dried urinary proteins (see Chapters 7 and 8).

Both the airborne concentration and size distribution of antigen-bearing particles vary with the air exchange rate of an enclosure. High ventilation rates tend to lower the total concentration of aerosol and relatively increase mean particle size (Luczynska et al., 1990; DeBlay et al., 1991). Small particles also can travel long distances in moving air as well as adherent to surfaces (e.g., clothing). As a result, allergens often are distributed throughout complex spaces despite discrete and localized sources (Swanson et al., 1984). Cat allergen, for example, has been found in homes and schools where there are no pets (Wood et al., 1989), presumably having been transported on the clothing of students, teachers, and visitors.

Aerosol exposure: respiratory deposition. In the context of hypersensitivity diseases, special importance surrounds the effect of particle size on deposition in the human respiratory tract. Note that the distinction between inhalant and other exposure is blurred here, since considerable antigen deposited within the respiratory tract ultimately is swallowed. Also, there is absorption and blood-borne dissemination of antigen eluted following deposition of material in the respiratory tract.

With regard to aerosol deposition, the respiratory tract may be considered to comprise three compartments: nasopharyngeal, tracheobronchial, and alveolar. Mechanisms of deposition in larger air passages are dominated by inertial impaction, while in small airways (<1 mm) deposition is effected primarily by sedimentation and diffusion. Theoretical and experimental models of aerosol deposition in resting humans reveal that approximately 30% of inhaled 0.2 to 0.5 μm particles are deposited somewhere in the respiratory tract (Stuart, 1973). Total deposition increases above this range with virtually 100% of 10-μm particles captured. For particles less than 0.2 μm diameter, pulmonary deposition also increases, with 50% of 0.1-μm particles deposited, predominantly by diffusion. The Task Group on Lung Dynamics (Morrow, 1966) developed a model to predict regional deposition of particles within the respiratory tract. Here, too, as inhaled particle size increases there is increasing deposition proximally within the respiratory system. Approximately 90% of 10-μm diameter particles are deposited within the nasopharyngeal region, while 60% of retained 0.01 μm particles are located in the ultimate gas exchanging tissues. The hygroscopicity of particles influences their effective size and, therefore, deposition. In addition, electrostatic charge, breathing patterns, exercise, age, and associated diseases also affect deposition (Brain and Valberg, 1979).

Platts-Mills et al. (1986) hypothesized that differences in the clinical asthmatic response to animal-derived and dust mite allergen exposures reflected the characteristic sizes of the respective airborne particles. As noted above, animal allergens typically are associated with smaller particles than dust mite emanations, and this size difference predicts markedly different patterns of respiratory tract deposition. Specifically, airborne animal allergen and responses to exposure should be distributed throughout the respiratory tree, while dust mite allergen should be confined largely to its upper regions. These larger mite particles might be expected to produce a relatively high, local concentration of allergen on a limited area of central airway lining. In a sensitized individual, the resulting Type I immune response might produce an increase in air flow resistance so localized as not to be clinically recognizable. Repeated exposure to dust mite allergen, with consequent inflammation, may be required for bronchial narrowing to produce measurable increases in airway resistance and associated signs and symptoms (e.g., asthma). In contrast, exposure to animal allergen is usually associated with prompt onset of manifest asthma, arguably reflecting a more diffuse increase in airway resistance. However, the primacy of such differences in patterns of airway responsiveness as clinical determinants has not yet been confirmed.

III. CATEGORIES OF HEALTH EFFECTS

A. The Atopic Diseases

"Atopy" is a term coined by Coca and Cooke (1923) to describe hypersensitivity conditions showing a strong familial tendency; the "atopic" illnesses include the well-known "allergies" (i.e., nasal allergy and allergic asthma). These diseases reflect environmental exposure of sensitized subjects, are mediated by IgE, and belong to the Type I category of Coombs and Gell (1975). Atopic dermatitis (infantile eczema) is a strongly linked skin problem for which IgE-based mechanisms are often less obvious.

Risk factors for atopy. Both genetic and environmental factors are implicated in the expression of atopy. In each category, a variety of potential determinants has been suggested, although their relative roles remain to be defined. Whatever the case, a family history of atopy (defined as the presence of allergic rhinitis/asthma in a parent, grandparent, uncle, aunt, or sibling before the onset of the subject's symptoms) is associated with an increased presentation of atopic disease (Smith and Knowler, 1965).

Every individual has several classes of HLA (human leukocyte or "transplantation" antigens), each more or less polymorphic. Since a given HLA gene product may or may not chaperone a specific antigen fragment, HLA constitution determines the breadth of immune responsiveness (vide supra). Zwollo et al. (1991) demonstrated that the immune response to a purified antigen of short ragweed involves particular sequences of HLA-DR, one of the most "influential" HLA gene products. They postulated that differences between responders and nonresponders may be due to variations in the amino acid structure of the HLA-DRB1 chain. Since HLA types are inherited, these observations may help to explain one familial component in the development of atopy.

The absolute magnitude of an individual's IgE responses also appears to reflect hereditable determinants. Meyers et al. (1987) estimated that 36% of the variation in IgE among 278 individuals belonging to 42 families was attributable to genetic factors. However, values for serum IgE alone cannot be used to determine presence or absence of atopy, due to the large overlap in ranges for atopic and ostensibly normal populations (Zetterstrom and Johansson, 1981).

In vitro studies suggest that regulation of IgE production involves the opposing effects of cytokines (chemical mediators released by cells that act on other cells), especially interleukin-4 (IL-4) and interferon gamma (INF-γ). Two principal "helper/promoter" lymphocyte subsets have been identified in mouse models: TH-1 cells (which secrete IL-2 and IFN-γ) and TH-2 cells (which secrete IL-4 and interleukin-5, IL-5). In mouse cell systems, IL-4 augments IgE synthesis by differentiated B lymphocytes, while IFN-γ blocks and/or attenuates this response. TH-2 cells also secrete interleukin-10, which directly inhibits TH-1 cell function, and thus, indirectly, should raise IgE production (Mosmann and Moore, 1991). In allergic individuals, TH-2 cells may be predominant numerically or functionally, promoting IgE synthesis; however, definitive proof of this is lacking. Also recently discovered are circulating low-affinity receptors for IgE known as "soluble CD23," which stimulate IgE-committed B lymphocytes to increase immunoglobulin production (Saxon et al., 1990). The atopic individual may have abnormal regulation of these molecules, favoring an elevated IgE response.

Although genetic determinants appear to be well established, Lubs (1972) demonstrated discordance in the rate of atopic diseases in monozygotic twins, strongly suggesting an additional role for environmental factors. Ratner and Silberman (1952) reviewed the characteristics of 6366 atopic patients and found that only 51.9% had a positive family history of atopy. Further evidence for the role of the environment was presented by Waite et al. (1980). They reported on the 1966 resettlement, in New Zealand, of 1950 Polynesian residents of the island of Tokelau following a devastating hurricane. By 1976, the prevalence of asthma in children 14 years of age or younger was 11% for those still living on Tokelau and 25.3% for Tokelauans living in New Zealand. This difference was attributed to the unlike environments represented by traditional island life styles versus modern urban society. In another study (Dowse et al., 1985) the increasing prevalence of asthma in Papua New Guinea was related credibly to the introduction of blankets, which became reservoirs for house dust mites, into this population.

It is becoming clear that exposure to tobacco smoke favors increased IgE production and is a potential risk factor for allergic sensitization. Maternal (but not paternal) smoking has been reported to elevate blood total IgE and the risk of development of atopic disease by 18 months of age (Magnusson, 1986); the elevation persists in children aged 12 to 16 years (Weiss et al., 1985). Cigarette smoking is also associated with elevated serum IgE in adult smokers (Bahna et al., 1983) and with lower mean levels of total IgG and IgM (Gerrard et al.,

1980). Zetterstrom et al. (1985) found an increased frequency of IgE antibody responses and more pronounced IgE production to aerosolized ovalbumin in rats when exposed to tobacco smoke than in exposed, smoke-free controls.

Viral infection has been shown to affect IgE synthesis in both humans and animal models. Rising viral antibody titers in children up to 4-years-old have been associated with onset of allergic symptoms (Frick et al., 1979). Following this clinical observation, Frick and Brooks (1983) noted that exposure of pups to attenuated canine distemper vaccine increased their specific IgE response to inhaled pollen administered by the investigators.

When antigen is introduced into the skin of subjects who have made a specific IgE response, redness (erythema) and swelling (wheal formation) occur. This is the basis of the skin test approach commonly used to identify atopy or Type I hypersensitivity. Such positive skin reactivity (without reference to family history) has been described as a risk factor for the subsequent development of overt atopic illness in asymptomatic individuals (Hagy and Settipane, 1976). In general, people with symptoms of allergic disease are more likely to have one or more positive skin tests than those with only a family history, or those with neither symptoms nor family history (90, 50, and 9%, respectively, according to Curran and Goldman, 1961).

Early- and late-phase, IgE-mediated reactions. The response of individuals who have made an IgE response to specific allergen challenge potentially involves both early- and late-phase reactions. [Note: These reaction phases should not be confused with the primary (sensitizing) and secondary (eliciting) responses that are always implicit in the immune mechanisms underlying hypersensitivity diseases.] Early ("immediate") reactions occur within minutes of exposure to an allergen and substantially resolve within 60 min. These changes reflect the interaction of allergen with specific IgE on local mast cells, promoting release of proinflammatory mediators (namely, histamine, leukotrienes, prostaglandins) stored in or newly synthesized by these cells. As a result of these agents, small blood vessels become dilated and leaky and, if present, visceral muscle contracts; proinflammatory cells (e.g., eosinophils and neutrophils) also are attracted to the site of antibody–allergen combination. These newly recruited cells and their tissue-reactive products contribute to the "late-phase" reaction, which may be evident by 4 to 6 h after exposure and usually resolves within 36 h. Reactions mediated by IgE can present clinically either as an early- or late-phase reaction or a combined ("dual") response. The late-phase reaction has been documented following pulmonary, cutaneous, and nasal challenges with antigen of sensitized subjects (Lemanski and Kaliner, 1988).

In the lung, both immediate and late IgE responses to inhaled allergens provoke an obstructive respiratory pattern (i.e., narrowed airways and decreased air flow rates) that is potentially reversible with bronchodilator drugs or time. These airway changes are not associated with increased circulating white blood cell count, fever, or change in the lung's diffusing capacity (namely the ability to exchange carbon dioxide and oxygen), and should not be confused with the reaction to antigen challenge in patients with hypersensitivity pneumonitis, in which no role for IgE is evident. However, late IgE-mediated airway responses may initiate long-lived inflammatory changes in the bronchi that promote subsequent airway narrowing in response to specific allergens as well as nonspecific bronchoconstrictors (e.g., cold air).

Allergic rhinoconjunctivitis. Allergic rhinoconjunctivitis (also known as allergic rhinitis or, in its seasonal recurrent form, hay fever) is the most commonly encountered atopic condition and is characterized by paroxysmal sneezing, nasal blockage, rhinorrhea (runny nose), ocular tearing, and pruritus (itching) of the eyes, nose, and throat. Occasionally, constitutional symptoms of fatigue and "grippy" feelings may be present, especially where sleep loss or poor sleep quality results from nasal obstruction. However, when these vague complaints occur alone, they are not easily related to allergy. Symptoms may be perennial, seasonal, or associated with defined episodes of specific exposure. Physical signs may include darkening of the lower eyelids ("allergic shiners"), reddening of the eyes, puffy eyelids, and a pale, swollen nasal mucosa. Stained smears of nasal secretions typically reveal eosinophils (a white blood cell that selectively stains with the aniline dye eosin). Eosinophil numbers also may increase modestly in peripheral blood. Immediate skin reactivity to relevant environmental allergens is often positive. Serum IgE specific for these allergens may also be detected by in vitro tests (e.g., RAST, ELISA, etc.). In one study (Broder et al., 1974b) the disease was shown to go into remission in less than 10% of patients followed into adult life if allergen exposure continues unabated.

Biogenic allergens most commonly associated with allergic rhinitis are derived from plants (especially their airborne pollens), mammals, arthropods, and fungi (see Chapters 5, 7, 8, and 9). The allergen-bearing particles generally range in size from about 2 to 60 μm, although outliers are well recognized. The minimal duration, intensity, or frequency of allergen exposure(s) required to become sensitized and to develop symptoms of allergic rhinitis is unknown and must vary among exposed subjects. Maternowski et al. (1962) found an increased incidence of new ragweed pollinosis in 19% of foreign students within an initial 2 to 5 ragweed seasons in midwestern North America. This incidence was 10 times that observed concurrently among native-born students of comparable age. The peak onset of symptoms occurred during the second and third ragweed seasons experienced by previously unexposed immigrants. This experience strongly suggests that exposure to minute quantities of a potent antigen over relatively brief periods of time can induce symptomatic sensitization. Few studies have addressed the levels of allergen exposure required to produce symptoms in sensitized patients. However, most patients clinically sensitive to grass pollen were found to experience symptoms of allergic rhinitis when the mean daily concentration of grass pollen had reached 50 grains per cubic meter or more (Davies and Smith, 1973). However, threshold levels and their individual variations as well as broader dose-response relationships for this and other allergens remain unknown. A cumulative prevalence of allergic rhinitis approximating 10 to 14% was reported to occur in the unselected total population of Tecumseh, MI (Broder et al., 1974a). The disease apparently begins most commonly at well under 10 years of age, with progressively declining incidence after the early teen years.

Allergic asthma. Asthma is traditionally divided by clinicians into allergic ("extrinsic") and nonallergic (unexplained or "idiopathic") forms. Both types are characterized by cough, wheezing, shortness of breath, and sensations of chest tightness, most typically episodic, but often continuous in severity. The cough may be nonproductive or may raise a scanty, clear, viscous sputum. These features often are precipitated or worsen with exercise. A defining characteristic of asthma is a nonspecific hyperactivity of the airways, such that nonimmunologic stimuli such as cold air, methacholine (a pharmacologic bronchial constrictor), pungent odors, and "inert" dusts also can precipitate airway narrowing in predisposed persons. Physical signs of asthma include prolongation of the expiratory phase of respiration as well as audible expiratory and sometimes inspiratory wheezes. In severe episodes, evidence of impaired respiratory gas exchange such as mental dullness, a decrease in breath sounds over the thorax, and intense breathlessness may be present. Pulmonary function tests reveal an obstructive pattern (i.e., decreased air flow rates without primary change in lung volumes) of ventilatory impairment. The sputum typically contains eosinophils and their products. The chest X-ray reveals lung overinflation due to trapping of air behind narrowed bronchi that narrow further in expiration. Immediate skin reactions often are demonstrable to relevant allergens in allergic asthmatics. Bronchial challenge with natural aerosols or their aqueous extracts may have value in identifying specific responsiveness to suspected allergens, but requires rigid control. Falsely positive bronchial inhalation tests can occur in skin-reactive individuals with allergic rhinitis alone or those with no clinical illness, but merely strong skin reactivity (Townley and Hopp, 1987). These reservations reflect the lack of currently useful challenge approaches that share the time and dose attributes of natural exposure.

Spontaneous asthma typically is a chronic disease with fluctuating symptomatic periods and asymptomatic intervals. During a 4-year follow-up of the Tecumseh community, apparent remissions of asthma occurred in 16% of affected females and 24% of males (Broder et al., 1974b). In another study of 315 asthmatic children, severe persistent asthma was characterized by onset within the first 3 years of life, with an especially high frequency of attacks in the initial year (McNicol and Williams, 1973). Allergic asthma is seen especially during childhood and early adult years, but may present a serious, lifelong health problem. By contrast, idiopathic asthma is recognized especially in early childhood, (when respiratory infections may be the only evident precipitants) and in later adult life. The antigens responsible for allergic rhinitis also produce allergic asthma in some patients. However, levels of most antigens required to cause sensitization and symptomatic impact are not known. An international workshop concluded that 2 μg of *Der p I* (the major allergen of the dust mite, *Dermatophagoides pteronyssimus*) or 100 mites per gram of dust represented a risk for the development of specific IgE antibody and potential asthma (Platts-Mills and de Weck, 1989) (see also Chapter 7). They also noted that 10 μg of *Der p I* or 500 mites per gram of dust was associated commonly with acute asthma and that most mite-sensitive patients would experience symptoms at this level.

Broder et al. (1974a) reported a cumulative prevalence of asthma (without separating allergic and nonallergic forms of the disease) of 4 to 6% in a total community population. The occurrence of asthma seems to vary globally; however, data are not comprehensive, and diagnostic criteria may vary. Regional differences in asthma morbidity rates may be expected in accord with the striking regional variations recognized in potent allergens. For example, asthma is apparently rare in Eskimos (Herxheimer and Schaeffer, 1974). As with allergic rhinitis, the disease is most often observed in children under 10, and declines among teenagers in North America (Smith and Knowler, 1965).

During recent years, asthma-related deaths have been increasing in Australia, Canada, England and Wales, the Netherlands, Sweden, the United States, and West Germany (Jackson et al., 1988). In the United States, asthma deaths increased from 1.2/100,000 in 1979 (NCHS, 1982) to a peak of 2.1/100,000 in 1989 (NCHS, 1990). Over this period, the greatest increase was sustained by the black population, whose mortality considerably exceeds that of whites (3.4 vs. 1.5/100,000 in 1988). Most deaths have occurred in patients over 55 years of age, although rates have also risen for younger persons. Causes of the recent worldwide increase in reported asthma mortality remain enigmatic (Sly, 1989).

Atopic dermatitis. Atopic dermatitis is a skin disorder often associated with elevated serum IgE levels, asthma, allergic rhinitis, and a family history — often biparental — of atopy. There is no single diagnostic lesion; however, acutely involved areas of itchy, red, somewhat swollen, scaling rash — often with small blisters — are characteristic. With longstanding involvement, the skin is thickened and dry. The distribution of affected skin areas varies with age: In infants, the cheeks are preferentially affected; with advancing age, the flexor creases of the arms and legs assume prominence, although whole-body involvement can occur. Useful criteria for the diagnosis of atopic dermatitis have been developed by Hanifin and Rajka (1980). Immediate skin tests to numerous foods and environmental allergens are often positive, IgE is elevated in over 80% of individuals with atopic dermatitis, and in vitro tests may confirm the presence of specific IgE (Chapman et al., 1983). "Patch on scratch" testing (applying allergen to the excoriated skin under an occlusive patch) with environmental allergens may be positive (DeGroot and Young, 1989), but the relationship of these responses to disease causation is not clear.

Although genetic factors are suspected in atopic dermatitis, specific HLA associations have not been identified (Svejgaard et al., 1985). Larsen et al. (1986) found a higher concordance rate of eczema for monozygotic than dizygotic Danish twins, suggesting hereditable determinants. Interestingly, nonatopic individuals have received atopic individuals' bone marrow (for cancer therapy, etc.), and subsequently developed new-onset atopic dermatitis and sensitivity to environmental allergens (Agosti et al., 1988). Although simple IgE-mediated reactions in skin do not account for atopic eczema, a relationship to Type I reactivity seems inescapable. Microscopically, involved skin shows eosinophils and other inflammatory cells. Major basic protein, a toxic product of eosinophils, also is found in biopsies of skin involved by atopic dermatitis (Leiferman et al., 1985), suggesting that these cells contribute to tissue damage.

Observations indirectly implicating airborne allergens in the etiology of atopic dermatitis come from Prausnitz-Küstner (P-K) reactions, a research technique popular before the infectious hazards of blood products became a serious concern. P-K testing involves injecting the serum of an atopic individual into the skin of a nonatopic subject. This (latter) recipient is then challenged (locally or systemically) with a putative allergen and observed for development of a wheal and flare reaction at the passively sensitized site. The basis for a positive P-K response is a Type I or "immediate" reaction. Sulzberger and Vaughan (1934) used the serum from a patient with atopic dermatitis that was exacerbated by the wearing of silk clothing and who was patch test negative to silk, suggesting that a Type IV immune response was not involved. The P-K site developed a wheal and flare after inhalation of silk powder by the recipient. An earlier study by Cohen et al. (1930) had demonstrated a similar phenomenon, that is, production of positive P-K reactions, using serum from ragweed-sensitive individuals and challenging the recipients with ragweed pollen intranasally. In neither of these approaches did serum recipients develop eczematous lesions; however, the responses produced clearly define the possibility of systemic allergen transport following challenge of respiratory mucous membranes. Whether such blood-borne dissemination or direct contact of allergen aerosols with reactive skin promotes flares of eczema is still unclear. However, Clark and Adinoff (1989) evaluated 18 patients who reacted to patch testing using common aeroallergens (pollens, molds, dust mites, and animal "danders") with an eczematous rash. Aeroallergen avoidance resulted in marked improvement or resolution of atopic dermatitis in all patients. Furthermore, a recurrent flare of atopic dermatitis was observed when patients were environmentally rechallenged with incriminated aeroallergens. Although this experience requires confirmation, it accords well with reports of eczema flares in response to specific food allergen challenges and remission with avoidance of the implicated ingestants (Sampson, 1988).

The U.S. Health and Nutrition Examination Survey involving 20,749 people found a prevalence of atopic dermatitis in the United States of 6.9/1000 overall and 19.3/1000 in individuals 1 to 5 years old (Johnson, 1978). Fifty to sixty percent of affected patients apparently develop the disease before age 1 and 90% by age 5 (Rajka, 1975); no racial differences have been seen (Wingert et al., 1968). Among affected children, a majority have resolution of atopic eczema before puberty. However, resurgent involvement in early adult life is well known and may presage lifelong skin disease.

B. Other Diseases Implicating Type I Hypersensitivity

Allergic bronchopulmonary aspergillosis. Allergic bronchopulmonary aspergillosis (ABPA) is a disease of allergic asthmatics in which a fungus (usually *Aspergillus fumigatus,* Af) colonizes the mucous lining of air passages within the lung. Other species of *Aspergillus* and additional fungal genera can rarely cause a similar problem. The characteristics of Af that facilitate disease production include its thermotolerance, ability to utilize the mucus in the asthmatic human lung as a substrate, and production of specific toxins, as well as its small spore size and abundance in the environment, especially in temperate regions. In areas where other thermotolerant aspergilli (e.g., *A. flavus*) predominate, they frequently are implicated in clinically similar disease. Initial establishment of the organism is assumed to result from respiratory exposure to living spores. However, once colonization has occurred, the disease probably can progress without additional external exposure.

Affected patients often have persistent cough, increased amounts of mucoid or pus-containing sputum and worsened asthma; sputum plugs, febrile episodes to 40°C, and variable chest pain may be associated. Physical examination of the chest may be normal; more often, local or diffuse wheezing is present. Diagnostic criteria include asthma, peripheral blood eosinophilia, immediate skin reactivity to an appropriate *Aspergillus* species antigen, serum antibodies (IgG) reactive with *Aspergillus* antigen, strongly elevated total serum IgE, history of pulmonary shadows on chest X-ray (McCarthy et al., 1970), and "central bronchiectasis" denoting irreversible, destructive dilatation of larger airways (Grenier et al., 1986). Pulmonary function tests during an acute episode of ABPA may reveal both decreased air flow rates and lung volume loss as well as a decreased capacity to exchange respiratory gases (Nichols et al., 1979).

Most of the elevated circulating IgE in ABPA patients is not directed against Af (Patterson and Roberts, 1974). However, Af-reactive IgE and IgG levels are significantly greater in patients with ABPA than in those asthmatic patients with positive skin tests to Af but no other manifestations of ABPA (Wang et al., 1978). Results of bronchoalveolar lavage (namely, washing the bronchi and alveoli of a segment of lung and analyzing the recovered fluid) suggest that *Aspergillus*-specific IgE (IgE-Af) and immunoglobulin A (IgA-Af) are produced within the bronchoalveolar compartment, but that elevations in total serum IgE do not arise from this region (Greenberger et al., 1988). Total IgE usually increases prior to a period of worsening of ABPA, but rare exceptions are recorded (Imbeau et al., 1978). Serum IgE will decline to normal with successful therapy of ABPA, usually with use of systemic corticosteroids (Ricketti et al., 1984). Since Af releases a proteinase that is able to induce epithelial cell detachment and might contribute to tissue damage (Robinson et al., 1990), prolonged anti-inflammatory treatment is advocated while the condition is active. Studies by Slavin et al. (1978) in a macaque model suggest that reactions of both IgG-Af and IgE-Af with intrabronchial fungus are necessary for disease production.

The true prevalence of ABPA is not known, but it has appeared to be higher in Great Britain than in North America. Schwartz and Greenberger (1991) found that 28% of their asthmatic patients, drawn largely from Great Lakes states, had positive immediate skin reactivity to *Aspergillus fumigatus* and that 8% of this series met diagnostic criteria for ABPA. If this experience accurately represents the asthmatic population in general, it suggests that ABPA is significantly overlooked by U.S. clinicians; however, true regional prevalence remains obscure. Limited data suggest that outdoor exposure levels of Af are comparable in Great Britain and North America. However, asthmatics affected with ABPA have not been shown to frequent environments imposing excessive risks of Af exposure.

Urticaria and angioedema. Urticaria (hives) and angioedema (similar changes affecting deeper skin layers) may be precipitated by antigen-specific IgE mechanisms, especially involving foods or drugs, but are seldom associated with exposure to bioaerosols. Urticaria is seen as discrete, superficial swellings, often with red haloes and prominent itching, that vary in size from millimeters to several centimeters. In most outbreaks, new hives form as older ones resolve, each lesion lasting no more than 24 h. Angioedema typically produces circumscribed swelling of mucous membranes and fleshy structures such as the eyelids, tongue, and genitalia; it may accompany hives or occur alone. These reactions have been reported (rarely) with exposures to airborne animal allergens, skin contact with penicillin aerosols (Rudzki and Rebandel, 1985), and aerosols of *Hevea brasiliensis* latex. However, neither the true incidence of IgE-mediated urticaria and angioedema nor the role of airborne exposure is defined.

Anaphylaxis. Anaphylaxis is a state in which there is generalized release of mast-cell-derived, proinflammatory mediators resulting in widespread blood vessel dilatation and fluid loss as well as visceral muscle spasm; blood pressure drops, collapse, and death can occur. This condition has been rarely reported in conjunction with exposure to aerosolized penicillin (Rudzki and Rebandel, 1985) and *Hevea brasiliensis* latex. Extensive contact of abraded skin with pollen of flowering grasses has produced similar changes in atopic persons using "alpine slides" (studied by Spitalny et al., 1984). However, the occurrence of anaphylaxis following natural exposure to bioaerosols must be quite low.

C. Hypersensitivity Pneumonitis

Disease definition. Hypersensitivity pneumonitis (HP, also termed "allergic alveolitis") is an inhalant-induced, immunologically mediated inflammation of the alveoli and bronchioles (minute, terminal airways). Of the numerous agents causatively associated with HP, many have been related to occupational activities, with farmer's lung (due to actinomycetes that colonize moist silage) most frequently reported.

Clinical presentation. HP can present characteristically in acute, subacute, and chronic forms or some combination of these. Acute disease follows heavy, discrete respiratory exposure to an offender and comprises sudden onset of breathlessness (essentially without wheezing), chills, and fever as high as 40°C, with muscle aching and "grippy" feelings usually developing 4 to 6 h later. Associated features may include cough, scanty mucoid sputum, and profound loss of appetite. Physical findings often include signs of severe oxygen deficit and abnormal crackling lung sounds ("rales"), but also may be surprisingly normal despite intense subjective symptoms. Pulmonary function tests during the acute phase generally reveal modestly decreased air flow rates and lung volumes (a "restrictive" pattern) and a decrease in gas transfer (using a test agent). Hypoxemia (low blood oxygenation) occurs with modest exercise and even at rest. Blood studies reveal only nonspecific evidence of inflammation. Serum antibodies (of IgG type) reacting with the offending antigen(s) are often present but merely confirm exposure, since the sera of well, similarly exposed persons show comparable reactivity. Serum complement levels are often normal or modestly elevated, while bronchoalveolar lavage fluid complement is increased, suggesting locally distinctive responses in the acute phase of the disease (Yoshizawa et al., 1988). The chest X-ray may be normal or demonstrate diffuse, nodular shadows especially in lower lung fields. Findings on lung biopsies (Reyes et al., 1982) include inflammatory cells (alveolar macrophages, lymphocytes and plasma cells) filling air spaces and infiltrating alveolar walls. Subacute disease presents "granulomas" (dense, discrete collections of mononuclear cells), bronchiolitis obliterans (destructive blockage of the smaller airways), and scarring, which dominates in later stages. The acute form may take as little as hours or as long as several weeks to resolve.

The subacute form of hypersensitivity pneumonitis is seen with repeated episodic exposure and presents as a chronic bronchitis with variable symptoms of recurring cough, breathlessness, weight loss and some degree of general ill health even between acute attacks. An insidious decline in respiratory function may occur despite a relative lack of obtrusive symptoms. Symptomatic pigeon breeders' pulmonary function tests, for example, decrease at four times the rate expected with age (Schmidt et al., 1988); however outcome analyses remain fragmentary. A chronic form of HP is also described with low-level, continuous exposure to offending antigens, most commonly emanations of caged birds. Affected individuals may demonstrate only progressive breathlessness and weight loss, with little else suggesting illness. However, their chest radiographs reveal progressive scarring and tissue loss, and pulmonary function tests document a decline of functioning lung volume. Although removal of identified offenders may slow these functional deficits, all too often these patients may proceed inexorably to invalidism and lung failure (Greenberger et al., 1989).

The diagnosis of hypersensitivity pneumonitis, based on a suggestive exposure history and compatible clinical presentation, often requires extended study of the patient and of environmental suspects. Laboratory challenges using antigenic aerosols can have a diagnostic specificity of 95% and sensitivity of 48 to 85% (Hendrick et al., 1980), but are difficult to control and potentially hazardous. Where occupational or otherwise delimited exposures are the suspected offenders, simple "use tests" have been advocated to effect (uncontrolled but realistic) exposure. In reactive patients so challenged, approximately 4 to 6 h later a restrictive ventilatory deficit develops — often with associated fever, increased white blood cell count, and measurably impaired diffusion of respiratory gases.

Determinants of HP. Beyond the obvious factor of exposure, elements predisposing an individual to develop hypersensitivity pneumonitis are not known. Smoking does appear to decrease the risk of developing of this disease, when studied in the context of several recognized antigens (Warren, 1977) — perhaps by suppressing the activation of alveolar macrophage cells. No HLA associations for these diseases have been conclusively identified (Muers et al., 1982), and many serological findings in exposed people appear unrelated to the presence or absence of symptoms. Specific serum antibodies, principally IgG (Fink et al., 1972), as well as in vitro, antigen-induced proliferation of and secretion by peripheral blood lymphocytes (Fink et al., 1975; Moore et al., 1980), appear unrelated to disease occurrence. On bronchoalveolar lavage, increased total cells recovered, increased lymphocyte activation markers, and increases in the lymphocyte (suppressor T cell) subset (Leatherman et al., 1984) have all been observed in both symptomatic and ostensibly well, exposed individuals. Calvanico et al. (1980) demonstrated that bronchoalveolar lavage fluid IgG levels were significantly elevated in symptomatic, as compared to asymptomatic, pigeon breeders; IgA levels were elevated, and IgM levels revealed no difference, but asymptomatic and symptomatic levels of all serum immunoglobulins overlapped. These findings do little to clarify disease mechanisms but suggest, incidentally, that tissue changes suggestive of HP may occur without overt expression of disease in many exposed persons — especially where avian antigens are implicated.

Extensive studies of humans with HP and their comparably exposed but clinically well peers have offered little fundamental insight into environmental or intrinsic tissue determinants of the disease. However, animal models of HP have been established using active sensitization as well as transfer of serum or cells from sensitized individuals to those previously unexposed. Taken as a group, the resulting reports primarily implicate cell-mediated (Type IV) immunity in disease pathogenesis and suggest that while antibodies may promote some aspects of inflammation following specific challenge, they are, alone, inadequate to produce disease (Bice et al., 1976; Richardson, 1972; Butler et al., 1983; Kopp et al., 1985). In addition, a succession of lymphocyte-mediated activities is suggested, with cells that primarily suppress inflammation ultimately predominating at affected pulmonary sites (Kopp et al., 1985). These observations imply that defects in regulation of the immune or inflammatory responses may ultimately characterize the disease. In addition, some agents of HP have been shown to directly activate proinflammatory enzymatic cascades [specifically, of the classical (Marx and Flaherty, 1976) and alternative (Edwards et al., 1974) complement pathways]. Actinomycetes associated with farmer's lung and humidifier fever have also been shown to exhibit immunologic adjuvant properties, promoting antibody synthesis (Bice et al., 1977) and, possibly, cell-mediated immunity.

Antigens/exposure. The agents associated with HP are numerous and include thermophilic actinomycetes, other bacteria and their products, emanations of birds and mammals, arthropod debris, vegetable dusts, and simple organic chemicals. In addition, instances of HP without a specifically implicated offender have been recorded increasingly. The unifying characteristic of the identified materials is their abundant occurrence as respirable particles (i.e., aerosols) sufficiently small to reach the ultimate conducting airways. Antigen exposure levels necessary for development of clinically evident disease are completely unknown and may vary among affected persons. In many described instances, exposure levels have seemed massive; however, the disease has been described in two wives of pigeon breeders who themselves had no direct contact with the birds (Riley and Saldana, 1973). Presumably, they were exposed to relevant antigens on their husbands' clothing, suggesting the existence of relatively low exposure thresholds for this disease in certain susceptible individuals. In many farm-related cases, by contrast, measured levels of antigen-bearing particles (usually thermophilic actinomycete spores) have approximated 10^{10} spores/m^3 of air (Lacey and Lacey, 1964). Under these conditions, an estimated 750,000 spores per minute could be deposited in the lungs of an exposed person. Although avoidance of the offending aerosols is paramount, and protective respirators have been found beneficial (Muller-Wening and Repp, 1989), compliance with their use is problematic.

Endotoxin(s) from Gram negative bacteria are often environmentally associated with the agents implicated in HP. These substances have a broad range of biological activity and can affect disease processes either as independent, proinflammatory agents or as a nonspecific stimulus to the immune system (Rylander, 1987). Exposure to concentrated endotoxin aerosols alone produces fever and chest tightness in human subjects (Rylander et al., 1989). However, the role of endotoxin in HP remains unknown, although it is a suspected factor in febrile illness occurring with exposure to bacterially contaminated humidifying devices.

Epidemiology. Efforts to derive epidemiological principles relevant to HP are complicated by the numerous etiological agents and lack of specific tests as diagnostic criteria for case finding. Often the diagnosis of HP is suggested only after an extended period of study in which other problems are progressively excluded. Other conditions [e.g., toxic organic dust syndrome (Brinton et al., 1987)] are readily confused with acute HP; chronic disease similarly mimics a variety of similar or associated forms of progressive pulmonary tissue loss. The "healthy worker" effect also increases the difficulty of studying this disease, as those affected by illness frequently change occupation for health reasons. All of these factors impair ascertainment of those truly affected by the disease.

Fink (1987) reviewed available epidemiological studies on HP and derived a prevalence rate of 4 to 30/1000 in the midwestern United States and 25 to 85/1000 in Europe. In related estimates, bird breeder's disease has been described in between 5 and 15% of exposed populations, while a 49% prevalence of HP (namely, bagassosis) had been reported among sugar cane workers in Puerto Rico (Bayonet and Lavergne, 1960). In response to the latter estimate, simple changes were made in processing that have greatly reduced HP among western hemisphere sugar cane workers.

D. Allergic Contact Dermatitis

The disease. Allergic contact dermatitis (ACD) is characterized by a red and intensely itchy rash — often raised — with small blisters. The rash is usually confined to the site of primary exposure to a contact allergen, but may become generalized, perhaps by more delayed inoculation of the sensitizer (e.g., that on clothing or underneath fingernails). ACD secondary to airborne exposure principally affects exposed skin, but covered areas are not entirely exempt. A latent period separates exposure and sensitization. This latent period before sensitization may be as brief as a few days, or, more commonly, may require multiple exposures over extended periods. Diagnosis is made by historical analysis of exposure components and patch testing — in effect, testing the potential of suspected offenders to create the disease in miniature. Caution should be used in applying patch tests as sensitization can occur de novo with this procedure, as well as serious irritation if test materials are not diluted appropriately. Primary irritation can be ruled out as the source of a patch test reaction if several "control" subjects fail to respond to application of a test material.

The antigens. ACD is ascribable commonly to metals, drugs, cosmetics, and plant oils (e.g., that of poison ivy). Airborne exposure from the burning of brush containing poison ivy plants or other members of the cashew family (Anacardiaceae) with resultant rash is not uncommon. ACD has also been attributed to airborne ragweed (*Ambrosia*) pollen in the United States (Arlette and Mitchell, 1981), with related members of the composite family (Compositae) implicated in other areas of the world (Sharma and Kaur, 1989). However, additional plant components (e.g., minute hairs or "indument") may contribute to the airborne allergen burden. The agents responsible for inducing ACD are low-molecular-weight compounds (<1000 Da), functioning as haptens, which must bind to intact proteins to become immunogenic. In both the composite family and poison ivy, responsible haptens are series of small, well-characterized, organic molecules present in the oleoresin. The agents of ragweed that cause sensitization are sesquiterpene lactones (Mitchell and Dupuis, 1971), while quite different sensitizers in the *Rhus* group of plants (poison ivy, poison sumac) are closely related pentadecacatechols.

Risk factors for ACD. Basic factors believed to promote sensitization include high concentration of the sensitizing agent, frequent exposure, and a compromised skin barrier. There is increasing prevalence of ACD with age, but not a consistent sex difference in susceptibility (Goh, 1986). Genetic factors are involved in the susceptibility to ACD as revealed by twin (Lubs, 1971) and family studies (Walker et al., 1967), but examination of an HLA association has produced conflicting results (Hansen et al., 1982; Mozzanica et al., 1990).

At any time point, ACD is reported to occur in 53.9/1000 population in the United States (Johnson, 1978). However, considering that over 95% of persons may be sensitized to *Rhus* antigens, exposure is the arbiter of disease prevalence. The incidence of ACD secondary to airborne agents is not known but appears to be a rather infrequent event. However, the occurrence of an unexplained itchy rash, especially in a work setting, should raise this diagnostic possibility.

IV. HISTORIES AND ENVIRONMENTAL ASSESSMENT

When evaluating patients suspected of having environmentally induced hypersensitivity disease, the importance of a detailed history of the evolution and features of the illness and an objective analysis of exposure risks cannot be overemphasized. The family history also has obvious importance because of the strong genetic component in many hypersensitivity diseases — especially those of the atopic group. A similar pattern of illness in others who share defined exposures is a useful indicator of factors deserving further study. Known exacerbating/alleviating factors should be carefully identified. Similarly, effects of season, job change, vacation, travel, etc. may offer analytic direction. Attention to variation of symptoms throughout the day (a.m. compared to p.m.) and week (especially the work week) should also address severity trends at work and at home, as well as indoors and outdoors. The occupation and hobbies of the patient and other household members — both current and past — should be comprehensively defined. Inquiry should focus on relevant exposures, including animals (at work, school, daycare), and other sources known to produce antigenic aerosols (e.g., humidifiers or activities that disperse natural dusts). Physical findings of disease may be limited but should be sought carefully and quantified where possible. Often, changes in an objective parameter of organ (especially pulmonary) function can be shown to accompany an exposure defined initially by site or specific activity.

Characteristics of enclosed spaces in which the patient works and resides deserve special concern. Attention should be directed especially to the age of the structure(s), duration and type of use patterns, type of construction and foundation, heating/cooling system, floor and wall covering, furnishings, plant and animal materials, dampness, and location of structure(s) vis-à-vis external aerosol sources. Time spent by the patient in different areas of the work/home environment is a key issue, as are sources of moisture incursion — both intentional and incidental.

V. RECOMMENDATIONS FOR RESEARCH

A. Mechanisms of Disease

Specific antibody and T lymphocytes are the immune system's interface with the environment that subserve recognition of antigen. Following their specific interactions, "nonantigen-specific" functions ("effector mechanisms") are initiated, which elicit stereotypic tissue responses. Future research might appropriately target the genes that encode receptors for specific antigen. Once these genes and their products are characterized, disease modification may be possible by two avenues: eliminating cells that have an unwanted antigen-specific T cell receptor ("clonal deletion"), and "repairing" DNA that encodes these molecules, thereby altering the receptors responsible for recognition of, and response to, antigen. Related research is needed to better define the mechanism(s) of inducing immunologic tolerance to antigens both before and after initial sensitization. This strategy is particularly relevant to unavoidable antigens such as ubiquitous pollens, fungus spores, and dust mites. Once the mechanism(s) are understood, tolerance induction to specific antigens might protect susceptible individuals or reverse disease following its initial expression.

While the foregoing approaches are beyond current capabilities, their potential selectivity, with its implication of safety, makes them an especially appealing research and development goal.

Up to the present, virtually all forms of immunomodulation have been antigen-*non*specific, and blockade of proinflammatory mechanisms remains a promising direction for intervention. The development of effective pharmacologic agents that interfere with defined chemical mediators of inflammation will remain a fundamental goal for investigators. The rapidly expanding knowledge base regarding both cytokines and their receptors also may allow for modulation of cell–cell interactions or even control of fundamental processes such as IgE synthesis.

B. Exposure/Dose/Response Relationships

Defining a specific dosage and/or exposure level times duration construct that permits sensitization of predisposed persons or precipitates symptoms in those already sensitized may yet be possible for particular antigens. With such thresholds defined, guidelines might be realistically stated to limit antigen exposure, thereby preventing sensitization and/or disease expression. The current difficulty with this approach is a perception that the antigen exposure necessary for immunological events varies substantially among individuals. In addition, such dose-response constructs would have greatest potential value for antigens from delimited sources that admit practical avoidance strategies and source control on a reasonable scale. The development of innovative avoidance measures that facilitate individual compliance will continue to offer a relatively simple and cost-effective means of limiting the adverse impact of bioaerosols.

C. Antigen Prevalence

The goal of identifying presently unrecognized antigens that promote hypersensitivity disease deserves continuing effort. The list of antigens implicated in Type I-related diseases has expanded quite slowly, and the exposure risks of many situations remain to be assessed. Instances of "air space" disease suggesting HP but without a credible offending agent are well recognized. Difficulties of ascertainment include highly complex exposures and the remarkably low levels of some agents that seem to provoke specific responses in sensitized subjects. A related concern is the possible dispersion of sensitizers in a variety of (native and derived) aerosol sizes, as shown for some pollen allergens (Solomon, 1986). Factors that control the prevalence of indoor antigens especially need to be defined so that preventive construction, use, and maintenance procedures for buildings and their contents can be better designed. In addition, a practical synthesis of the direct effects of physical variables of air quality on human health and comfort is needed, while their (secondary) impact on microbial associations are investigated.

D. Epidemiology

Many public health aspects of the hypersensitivity diseases continue to offer both intellectual opportunity and the promise of helping to reduce the strikingly adverse impact of these conditions on susceptible populations. Specifically with regard to asthma and allergic rhinitis, the "long-term outlook" and natural history should be better examined, and outcomes should be assessed with regard to specific allergens, grades of severity, and the present array of avoidance, drug therapy, and specific (injection) immunotherapy measures. The increased mortality of asthma independently demands further analysis with regard to prognosis determinants as well as means of dealing with already defined risk factors such as mental depression and inadequate patient compliance with treatment measures. The data available on asthma deaths, for example, fail to reveal the extent to which "active" or passive exposure to tobacco smoke may be a factor contributing to mortality. This and other lifestyle-related variables should be examined as asthma death determinants in all age groups. The contrast between approaches that emphasize allergen identification and avoidance and those that do not is an especially intriguing area for outcomes analysis.

Epidemiological studies of HP are at best limited, especially for isolated illness and those cases with mild or modest involvement. Where clinical illness clearly is associated with defined exposures, evidence of additional mild disease or subclinical tissue change should be sought using increasingly sensitive indicators to better describe total rates of involvement. As with asthma, outcome research is needed to define natural history in relation to extremes of exposure experience and various drug treatment options. Predictors of morbidity should be defined such that preemployment evaluation may hope to identify those at significant risk for development of disease in high-exposure occupational settings. Parallel efforts are also essential to develop methods for reducing exposure to acknowledged offenders, once their presence is confirmed.

REFERENCES

Agosti, J., Sprenger, J., Lum, L., Witherspoon, R., Fisher, L., Storb, R., and Henderson, W., Transfer of allergen-specific IgE-mediated hypersensitivity with allogeneic bone marrow transplantation, *N. Engl. J. Med.*, 319, 1623, 1988.

Arlette, J. and Mitchell, J. C., Compositae dermatitis: current aspects, *Contact Derm.*, 7, 129, 1981.

Bahna, S. L., Heiner, D. C., and Myhre, B. A., Immunoglobulin E pattern in cigarette smokers, *Allergy*, 38, 57, 1983.

Bayonet, N. and Lavergne, R., Respiratory disease of bagasse workers; a clinical analysis of 69 cases, *Ind. Med. Surg.*, 25, 519, 1960.

Bice, D. E., Salvaggio, J., and Hoffman, E., Passive transfer of experimental hypersensitivity pneumonitis with lymphoid cells in the rabbit, *J. Allergy Clin. Immunol.*, 58, 250, 1976.

Bice, D. E., McCarron, K., Hoffman, E. O., and Salvaggio, J., Adjuvant properties of *Micropolyspora faeni, Int. Arch. Allergy Appl. Immunol.*, 55, 267, 1977.

Brain, J. D. and Valberg, P. A., State of the art: deposition of aerosol in the respiratory tract, *Am. Rev. Respir. Dis.*, 120, 1325, 1979.

Brinton, W. T., Vastbinder, E. E., Greene, J. W., Marx, J. J., Jr., Hutcheson, R. H., and Schaffner, W., An outbreak of organic dust toxic syndrome in a college fraternity, *J. Am. Med. Assoc.*, 258, 1210, 1987.

Broder, I., Higgins, M. W., Mathews, K. P., and Keller, J. B., Epidemiology of asthma and allergic rhinitis in a total community, Tecumseh, Michigan. III. Second survey of the community, *J. Allergy Clin. Immunol.*, 53, 127, 1974a.

Broder, I., Higgins, M. W., Mathews, K. P., and Keller, J. B., Epidemiology of asthma and allergic rhinitis in a total community, Tecumseh, Michigan. IV. Natural history, *J. Allergy Clin. Immunol.*, 54, 100, 1974b.

Brodsky, F. M. and Guagliardi, L. E., The cell biology of antigen processing and presentation, *Annu. Rev. Immunol.*, 9, 707, 1991.

Butler, J. E., Richerson, H. G., Swanson, P. A., Suelzer, M. T., and Kopp, W. C., Carrier requirement for the development of acute experimental hypersensitivity pneumonitis in the rabbit, *Int. Arch. Allergy Appl. Immunol.*, 71, 74, 1983.

Calvanico, N. J., Ambegaonker, S. P., Schlueter, D. P., and Fink, J. N., Immunoglobulin levels in bronchoalveolar lavage fluid from pigeon breeders, *J. Lab. Clin. Med.*, 96, 129, 1980.

Chapman, M. D., Rowntree, S., Mitchell, E. B., Di Prisco de Fuenmajor, M. C., and Platts- Mills, T. A., Quantitative assessments of IgG and IgE antibodies to inhalant allergens in patients with atopic dermatitis, *J. Allergy Clin. Immunol.*, 72, 27, 1983.

Clark, R. A. and Adinoff, A. D., The relationship between positive aeroallergen patch test reactions and aeroallergen exacerbations of atopic dermatitis, *Clin. Immunol. Immunopathol.*, 53, S132, 1989.

Coca, A. F. and Cooke, R. A., On the classification of the phenomena of hypersensitiveness, *J. Immunol.*, 8, 163, 1923.

Cohen, M. B., Ecker, E. E., Breitbart, J. R., and Rudolph, J. A., The rate of absorption of ragweed pollen material from the nose, *J. Immunol.*, 18, 419, 1930.

Coombs, R. R. A. and Gell, P. G. H., Classification of allergic reactions responsible for clinical hypersensitivity and disease, in *Clinical Aspects of Immunology*, Gell, P., Coombs, R., and Lachmann, P. J., Eds., 3rd ed., Blackwell Scientific Publications, Oxford, 1975, chap. 25.

Curran, W. S. and Goldman, G., The incidence of immediately reacting allergy skin tests in a normal adult population, *Ann. Int. Med.*, 55, 777, 1961.

Davies, R. R. and Smith, L. P., Forecasting the start and severity of the hay fever season, *Clin. Allergy*, 3, 263, 1973.

De Blay, F., Chapman, M. D., and Platts-Mills, T. A. E., Airborne cat allergen (*Fel d* I): environmental control with cat *in situ, Am. Rev. Respir. Dis.*, 143, 1334, 1991.

De Groot, A. C. and Young, E., The role of contact allergy to aeroallergens in atopic dermatitis, *Contact Derm.*, 21, 209, 1989.

Dowse, G. K., Turner, K. J., Stewart, G. A., Alpers, M. P., and Woolcock, A. J., The association between *Dermatophagoides* mites and the increasing prevalence of asthma in the village communities within the Papua New Guinea highlands, *J. Allergy Clin. Immunol.*, 75, 75, 1985.

Edwards, J. H., Baker, J. T., and Davies, B. H., Precipitin test negative farmer's lung — activation of the alternative pathway of complement by mouldy hay dusts, *Clin. Allergy*, 4, 379, 1974.

Fink, J. N., Schlueter, D. P., Sosman, A. J., Unger, G. F., Barboriak, J. J., Rimm, A. A., Arkins, J. A., and Dhaliwal, K. S., Clinical survey of pigeon breeders, *Chest*, 62, 277, 1972.

Fink, J. N., Moore, V. L., and Barboriak, J. J., Cell-mediated hypersensitivity in pigeon breeders, *Int. Arch. Allergy Appl. Immunol.*, 49, 831, 1975.

Fink, J. N., Epidemiological aspects of hypersensitivity pneumonitis, in *Monographs in Allergy, Epidemiology of Allergic Diseases,* Schlumberger, H. D., Ed., Karger, Basel, 1987, 21, 59.

Frick, O. L., German, D. F., and Mills, J., Development of allergy in children, I. Association with virus infections, *J. Allergy Clin. Immunol.*, 63, 228, 1979.

Frick, O. L. and Brooks, D. L., Immunoglobulin E antibodies to pollens augmented in dogs by virus vaccines, *Am. J. Vet. Res.*, 44, 440, 1983.

Gerrard, J. W., Heiner, D. C., Ko, C. G., Mink, J., Meyers, A., and Dosman, J. A., Immunoglobulin levels in smokers and non-smokers, *Ann. Allergy*, 44, 261, 1980.

Goh, C. L., Prevalence of contact allergy by sex, race and age, *Contact Derm.*, 14, 237, 1986.

Goodman, J. W., Immunogenicity and antigenic specificity, in *Basic and Clinical Immunology,* 7th ed., Stites, D. P. and Terr, A. I., Eds., Appleton and Lange, East Norwalk, CT, 1991, chap. 8.

Greenberger, P. A., Smith, L. J., Hsu, C. C., Roberts, M., and Liotta, J. L., Analysis of bronchoalveolar lavage in allergic bronchopulmonary aspergillosis: divergent responses of antigen-specific antibodies and total IgE, *J. Allergy Clin. Immunol.*, 82, 164, 1988.

Greenberger, P. A., Pien, L. C., Patterson, R., Robinson, P., and Roberts, M., End-stage lung and ultimately fatal disease in a bird fancier, *Am. J. Med.*, 86, 199, 1989.

Grenier, P., Maurice, F., Musset, D., Menu, Y., and Nahum, H., Bronchiectasis: assessment by thin section CT, *Radiology*, 161, 95, 1986.

Hagy, G. W. and Settipane, G. A., Risk factors for developing asthma and allergic rhinitis, *J. Allergy Clin. Immunol.*, 58, 330, 1976.

Hanifin, J. M. and Rajka, G., Diagnostic features of atopic dermatitis, *Acta Dermatovenereol. Suppl.* 92, 44, 1980.

Hansen, H. E., Menne, T., and Larsen, S. O., HLA antigens in nickel sensitive females: based on a twin and a patient population, *Tissue Antigens,* 19, 306, 1982.

Hendrick, D. J., Marshall, R., Faux, J. A., and Krall, J. M., Positive alveolar responses to antigen inhalation provocation tests: their validity and recognition, *Thorax,* 35, 415, 1980.

Herxheimer, H. and Schaefer, O., Asthma in Canadian Eskimos, *N. Engl. J. Med.*, 291, 1419, 1974.

Imbeau, S. A., Nichols, D., Flaherty, D., Dickie, H., and Reed, C., Relationship between prednisone therapy, disease activity, and the total serum IgE level in allergic bronchopulmonary aspergillosis, *J. Clin. Allergy Immunol.*, 62, 91, 1978.

Jackson, R., Sears, M. R., Beaglehole, R., and Rea, H. H., International trends in asthma mortality: 1970 to 1985, *Chest*, 94, 914, 1988.

Jarrett, E. E. E., Haig, D. M., McDougall, W., and McNulty, E., Rat IgE production. II. Primary and booster reaginic antibody responses following intradermal or oral immunization, *Immunology,* 30, 671, 1976.

Johnson, M.-L. Skin conditions and related need for medical care among persons 1-74 years. United States, 1971–1974. Vital and Health Statistics: Series 11, Data from the National Health Survey, no. 212, 1978.

Kopp, W. C., Dierks, S. E., Butler, J. E., Updrashta, B. S., and Richerson, H. G., Cyclosporine immunomodulation in a rabbit model of chronic hypersensitivity pneumonitis, *Am. Rev. Respir. Dis.*, 132, 1027, 1985.

Lacey, J. and Lacey, M. B., Spore concentration in the air of farm buildings, *Trans. Br. Mycol. Soc.*, 47, 547, 1964.

Larsen, F. S., Holm, N. V., and Henningsen K., Atopic dermatitis: a genetic-epidemiologic study in a population-based twin sample, *J. Am. Acad. Dermatol.*, 15, 487, 1986.

Leatherman, J. W., Michael, A. F., Schwartz, B. A., and Hoidal, J. R., Lung T cells in hypersensitivity pneumonitis, *Ann. Intern. Med.*, 100, 390, 1984.

Leiferman, K. M., Ackerman, S. J., Sampson, H. A., Haugen, H. S., Venencie, P. Y., and Gleich, G. J. Dermal deposition of eosinophil-granule major basic protein in atopic dermatitis: comparison with onchocerciasis, *N. Engl. J. Med.*, 313, 282, 1985.

Lemanski, R. F. and Kaliner, M., Late-phase IgE-mediated reactions, *J. Clin. Immunol.*, 8, 1, 1988.

Lubs, M.-L., Allergy in 7000 twin pairs, *Acta Allergol.*, 26, 249, 1971.

Lubs, M.-L., Empiric risks for genetic counseling in families with allergy, *J. Pediatr.*, 80, 26, 1972.

Luczynska, C. M., Li, Y., Chapman, M. D., and Platts-Mills, T. A., Airborne concentrations and particle size distribution of allergen derived from domestic cats (*Felis domesticus*), *Am. Rev. Respir. Dis.*, 141, 361, 1990.

Magnusson, C. G., Maternal smoking influences cord serum IgE and IgD levels and increases the risk for subsequent infant allergy, *J. Allergy Clin. Immunol.*, 78, 898, 1986.

Marx, J. J. and Flaherty, D. K., Activation of the complement sequence by extracts of bacteria and fungi associated with hypersensitivity pneumonitis, *J. Allergy Clin. Immunol.*, 57, 328, 1976.

Maternowski, C. J. and Mathews, K. P., The prevalence of ragweed pollinosis in foreign and native students at a midwestern university and its implication concerning methods for determining the inheritance of atopy, *J. Allergy*, 33, 130, 1962.

McCarthy, D. S., Simon, G., and Hargreave, F. E., The radiological appearances in allergic bronchopulmonary aspergillosis, *Clin. Radiol.*, 21, 366, 1970.

McNicol, K. N. and Williams, H. B., Spectrum of asthma in children — I, clinical and physiological components, *Br. Med. J.*, 4, 7, 1973.

Meyers, D. A., Beaty, T. H., Freidhoff, L. R., and Marsh, D. G., Inheritance of total serum IgE (basal levels) in man, *Am. J. Hum. Genet.*, 41, 51, 1987.

Mitchell, J. C. and Dupuis, G., Allergic contact dermatitis from sesquiterpenoids of the Compositae family of plants, *Br. J. Dermatol.*, 84, 139, 1971.

Moore, V. L., Pedersen, G. M., Hauser, W. C., and Fink, J. N., A study of lung lavage materials in patients with hypersensitivity pneumonitis: *in vitro* response to mitogen and antigen in pigeon breeders' disease, *J. Allergy Clin. Immunol.*, 65, 365, 1980.

Morrow, P. E., (Chairman), Task Group on Lung Dynamics, Deposition and retention models for internal dosimetry of the human respiratory tract, *Health Phys.*, 12, 173, 1966.

Mosmann, T. R. and Moore, K. W., The role of IL-10 in crossregulation of TH1 and TH2 responses, *Immunol. Today*, 12, A49, 1991.

Mowat, A. M., The regulation of immune responses to dietary protein antigens, *Immunol. Today*, 8, 93, 1987.

Mozzanica, N., Rizzolo, J., Veneroni, G., Diotti, R., Hepeisen, S., and Finzi, A., HLA-A, B, C and DR antigens in nickel contact sensitivity, *Br. J. Dermatol.*, 122, 309, 1990.

Muers, M. F, Faux, J. A., Ting, A., and Morris, P. J., HLA-A, B, C and HLA-DR antigens in extrinsic allergic alveolitis (budgerigar fancier's lung disease), *Clin. Allergy*, 12, 47, 1982.

Muller-Wening, D. and Repp, H., Investigation of the protective value of breathing masks in farmer's lung using an inhalation provocation test, *Chest,* 95, 100, 1989.

National Center for Health Statistics, Final data from the (1982) Advance Report of Final Mortality Statistics, 1979, *Monthly Vital Statistics Report,* 31(6), Suppl., September 30, 1982.

National Center for Health Statistics, Provisional data from the (1990) Annual Summary of Births, Marriages, Divorces, and Deaths: United States, 1989, *Monthly Vital Statistics Report,* 38(13), August 30, 1990.

Nichols, D., Dopico, G. A., Braun, S., Imbeau, S., Peters, M.E., and Rankin, J., Acute and chronic pulmonary function changes in allergic bronchopulmonary aspergillosis, *Am. J. Med.,* 67, 631, 1979.

Patterson, R. and Roberts, M., IgE and IgG antibodies against *Aspergillus fumigatus* in sera of patients with bronchopulmonary allergic aspergillosis, *Int. Arch. Allergy Appl. Immunol.*, 46, 150, 1974.

Platts-Mills, T. A. and de Weck, A. L. (Chairmen), Dust mite allergens and asthma; a worldwide problem, *J. Allergy Clin. Immunol.*, 83, 416, 1989.

Platts-Mills, T. A., Heymann, P. W., Longbottom, J. L., and Wilkins, S. R., Airborne allergens associated with asthma: particle sizes carrying dust mite and rat allergens measured with a cascade impactor, *J. Allergy Clin. Immunol.*, 77, 850, 1986.

Rajka, G., The clinical aspects of atopic dermatitis, in *Atopic Dermatitis,* Rajka, G., Ed., W. B. Saunders, London, 1975, chap. 2.

Ratner, B. and Silberman, D. E., Allergy: its distribution and the hereditary concept, *Ann. Allergy*, 10, 1, 1952.

Reyes, C. N., Wenzel, F. J., Lawton, B. R., and Emanuel, D. A., The pulmonary pathology of farmer's lung disease, *Chest*, 81, 142, 1982.

Richardson, H. B., Acute experimental hypersensitivity pneumonitis in the guinea pig, *J. Lab. Clin. Med.*, 79, 745, 1972.

Ricketti, A., Greenberger, P. A., and Patterson, R., Serum IgE as an important aid in the management of allergic bronchopulmonary aspergillosis, *J. Allergy Clin. Immunol.*, 74, 68, 1984.

Riley, D. J. and Saldana, M., Pigeon breeder's lung; subacute course and the importance of indirect exposure, *Am. Rev. Respir. Dis.*, 107, 456, 1973.

Robinson, B. W., Venaille, T. J., Mendis, A. H. W., and McAleer, R., Allergens as proteases: an *Aspergillus fumigatus* proteinase directly induces human epithelial cell detachment, *J. Allergy Clin. Immunol.*, 86, 726, 1990.

Rudzki, E. and Rebandel, P., Occupational contact urticaria from penicillin, *Contact Derm.*, 13, 192, 1985.

Rylander, R., Role of endotoxins in the pathogenesis of respiratory disorders, *Eur. J. Respir. Dis. Suppl.,* 154, 136, 1987.

Rylander, R., Bake, B., Fischer, J. J., and Helander, I. M., Pulmonary function and symptoms after inhalation of endotoxin, *Am. Rev. Respir. Dis.*, 140, 981, 1989.

Sampson, H. A., The role of food allergy and mediator release in atopic dermatitis, *J. Allergy Clin. Immunol.*, 81, 635, 1988.

Saxon, A., Ke, Z., Bahatai, L., and Stevens, R. H., Soluble CD23 containing B cell supernatants induce IgE from peripheral blood B-lymphocytes and costimulate with interleukin-4 in induction IgE, *J. Allergy Clin. Immunol.*, 86, 333, 1990.

Schmidt, C. D., Jensen, R. L., Christensen, L. T., Crapo, R. O., and Davis, J. J., Longitudinal pulmonary function changes in pigeon breeders, *Chest,* 93, 359, 1988.

Schwartz, H. J. and Greenberger, P. A., The prevalence of allergic bronchopulmonary aspergillosis in patients with asthma determined by serologic and radiologic criteria in patients at risk, *J. Lab. Clin. Med.*, 117, 138, 1991.

Sharma, C. S. and Kaur, S., Airborne contact dermatitis from Compositae plants in northern India, *Contact Derm.,* 21, 1, 1989.

Slavin, R. G., Fischer, V. W., Levine, E. A., Tasi, C. C., and Winzenburger, P., A primate model of allergic bronchopulmonary aspergillosis, *Int. Arch. Allergy Appl. Immunol.*, 56, 325, 1987.

Sly, R. M., Mortality from asthma, *J. Allergy Clin. Immunol.*, 84, 821, 1989.

Smith, J. M. and Knowler, L. A., Epidemiology of asthma and allergic rhinitis. I. In a rural area, *Am. Rev. Respir. Dis.*, 92, 16, 1965.

Solomon, W. R., Airborne allergens associated with small particle fractions, *Grana*, 25, 85, 1986.

Spitalny, K. C., Farnham, J. E., Witherell, L. E., Vogt, R. L., Fox, R. C., Kaliner, M., and Casale, T. B., Alpine slide anaphylaxis, *N. Engl. J. Med.*, 310, 1034, 1984.

Stuart, B. O., Deposition of inhaled aerosols, *Arch. Int. Med.*, 131, 60, 1973.

Stupp, Y., Paul, W. E., and Benacerraf, B., Structural control of immunogenicity. II. Antibody synthesis and cellular immunity in response to immunization with mono-oligolysines, *Immunology*, 21, 583, 1971.

Sulzberger, M. B. and Vaughan, W. T., Experiments in silk hypersensitivity and the inhalation of allergen in atopic dermatitis (neurodermatitis disseminatus), *J. Allergy*, 5, 554, 1934.

Svejgaard, E., Jakobsen, B., and Svejgaard, A., Studies of HLA-ABC and DR antigens in pure atopic dermatitis and atopic dermatitis combined with allergic respiratory disease, *Acta Dermato-Venereol. Suppl.,* 114, 72, 1985.

Swanson, M. C., Agarwal, M. K., Yunginger, J. W., and Reed, C. E., guinea pig derived allergens: clinicoimmunologic studies, characterization, airborne quantitation, and size distribution, *Am. Rev. Respir. Dis.*, 129, 844, 1984.

Swanson, M. C., Agarwal, M. K., and Reed, C. E., An immunochemical approach to indoor aeroallergen quantitation with a new volumeteric air sampler: studies with mite, roach, cat and guinea pig antigens, *J. Allergy Clin. Immunol.*, 76, 724, 1985.

Townley, R. J. and Hopp, R. J., Inhalation methods for the study of airway responsiveness, *J. Allergy Clin. Immunol.*, 80, 111, 1987.

Waite, D. A., Eyles, E. F., Tonkin, S. L., and O'Donnell, T. V., Asthma prevalence in Tokelauan children in two environments, *Clin. Allergy*, 10, 71, 1980.

Walker, F. B., Smith, P. D., and Maibach, H. I., Genetic factors in human allergic contact dermatitis, *Int. Arch. Allergy Appl. Immunol.*, 32, 453, 1967.

Wang, J. L., Patterson, R., Rosenberg, M., Roberts, M., and Cooper, B. J., Serum IgE and IgG antibody activity against *Aspergillus fumigatus* as a diagnostic aid in allergic bronchopulmonary aspergillosis, *Am. Rev. Respir. Dis.*, 117, 917, 1978.

Warren, C. P., Extrinsic allergic alveolitis: a disease commoner in non-smokers, *Thorax,* 32, 567, 1977.

Weiss, S. T., Tager, I. B., Munoz, A., and Speizer, F. E., The relationship of respiratory infections in early childhood to the occurrence of increased levels of bronchial responsiveness and atopy, *Am. Rev. Respir. Dis.*, 131, 573, 1985.

Wingert, W. A., Friedman, D. B., and Larson, W. R., The demographical and ecological characteristics of a large urban pediatric outpatient population and implications for improving community pediatric care, *Am. J. Public Health*, 58, 859, 1968.

Wood, R. A., Chapman, M. D., Adkinson, N. F., and Eggleston, P. A., The effect of cat removal on the allergen content in household dust samples, *J. Allergy Clin. Immunol.*, 83, 730, 1989.

Yoshizawa, Y., Nomura, A., Ohdama, S., Tanaka, M., Morinari, H., and Hasegawa, S., The significance of complement activation in the pathogenesis of hypersensitivity pneumonitis: sequential changes of complement components and chemotactic activities in bronchoalveolar lavage fluids, *Int. Arch. Allergy Appl. Immunol.*, 87, 417, 1988.

Zetterstrom, O. and Johansson, S. G., IgE concentrations measured by PRIST in serum of healthy adults and patients with respiratory allergy, *Allergy,* 36, 537, 1981.

Zetterstrom, O., Nordvall, S. L., Bjorksten, B., Ahlsted, S., and Stelander, M., Increased IgE antibody responses in rats exposed to tobacco smoke, *J. Allergy Clin. Immunol.*, 75, 594, 1985.

Zwollo, P., Ehrlich-Kautzky, E., Scharf, S. J., Ansari, A. A., Erlich, H. A., and Marsh, D. G., Sequencing of HLA-D in responders and non-responders to short ragweed allergen, *Amb a V*, *Immunogenetics,* 33, 141, 1991.

11 ANALYTICAL METHODS: IMMUNOASSAYS

Martin D. Chapman

CONTENTS

I. INTRODUCTION

Immunoassays have had major impact on many areas of biomedical research. The significance of this technology has been recognized because it has, either directly or indirectly, been the subject of two Nobel Prizes: the first, to Yalow and Berson, for developing the principles of competitive radioimmunoassay in the 1950s (Yalow and Berson, 1960), and the second, to Kohler, Milstein, and Jerne, for their fundamental discoveries that led to the production of monoclonal antibodies (mAb) in the 1970s (Kohler and Milstein, 1975; Jerne, 1974). In essence, the principles expounded by these researchers still form the basis of immunoassay development: that is, that quantitative measurement of a specific ligand (protein, hormone, enzyme, receptor) relies on either using labeled purified ligand (e.g., ^{131}I-labeled insulin in Yalow and Berson's case) or on the use of labeled monospecific antibody to the ligand (for which mAb have many advantages over polyclonal reagents). Once these essentials are in place, immunoassays can be developed using unlabeled ligand to inhibit

0-87371-724-4/95/$0.00+$.50

binding of the label to an antibody, or by directly measuring binding of the labeled antibody to the ligand. Many different assay formats have been used over the past 30 years. Major technological advances have included the use of ^{125}I to label proteins for radioimmunoassay; the use of cyanogen bromide-activated Sepharose and other chemically activated matrices as solid-phase supports; the development of enzyme-linked immunosorbent assays (ELISA); and the use of biotin-streptavidin detection systems (Bolton and Hunter, 1973; Axen et al., 1967; Engvall and Perlmann, 1971; Kemeny, 1991).

Immunoassays have been developed for a wide range of antigenic biological and some nonbiological substances, including proteins, glycoproteins, hormones, peptides, chemical haptens, and drugs. The key advantages of these assays are their sensitivity (in the picogram range), specificity, reproducibility, and the ability to make quantitative measurements of a given antigen in absolute units (e.g., nanograms or micrograms of protein per milliliter). These properties lend themselves to measurements of indoor biological pollutants either in the air or in dust or other reservoir samples. Our research has focussed on the role of sensitization and exposure to indoor antigens (allergens) in causing allergic diseases, such as perennial rhinitis, bronchial asthma, and atopic dermatitis (see also Chapters 7 and 8). Over the past 10 years, we and others have developed immunoassays for major allergens derived from dust mites, cats, cockroaches, rodents, and a few fungi. These assays have been used for a variety of environmental studies, including measurement of defined allergen levels in the houses of patients with asthma; analysis of the particle size and concentration of these allergens in the air; and analysis of the effects of environmental control procedures on levels of these allergens in houses. The immunoassays used in our studies began as double antibody inhibition radioimmunoassays, but our assays have all been converted to mAb-based ELISA formats. This chapter uses our allergen immunoassays to illustrate the general principles of immunoassay development and to provide strategies that may be useful for devising tests for other biological pollutants. For more detailed technical information, the reader is referred to two excellent laboratory manuals on antibodies and ELISA that have recently been produced (Harlow and Lane, 1988; Kemeny, 1991).

II. THE ASSAYS

A. Two-Site mAb-Based ELISA

The principle of the two-site ELISA is outlined in Figure 11.1. Plastic microtiter wells are coated with mAb (e.g., to one of the dust mite allergens) and incubated for 1 to 2 h with dilutions of the unknown allergen mixture (e.g., house dust extract). The mAb binds the specific dust mite allergen in the solution, and bound allergen is detected using a biotinylated second mAb

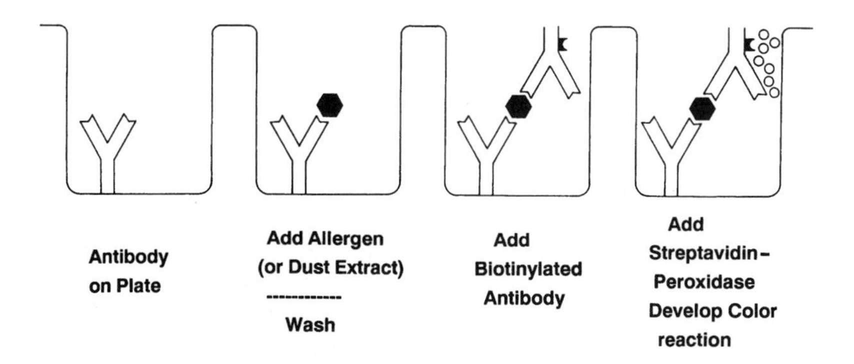

Figure 11.1. Monoclonal antibody-based ELISA for allergens. (From Chapman, M. D., *Allergy and Molecular Biology* (Advances in the Biosciences, Vol. 74), Elsevier, Kidlington, U.K., 1988, 281. With permission.)

directed against a nonoverlapping epitope. The second mAb is detected using streptavidin peroxidase and a suitable chromogenic substrate. In our case, a green color is developed using 2,2-azinobis-(3 ethyl-benzthiazoline) (ABTS), which is stable for several weeks at 4°C, although several other substrates are in common use (Harlow and Lane, 1988). The advantages of the biotin-streptavidin system are that mAb are easily labeled with biotin, and the high affinity of biotin for streptavidin (K_a 10^{-15} l/*M*) increases sensitivity. Streptavidin (from *Streptomyces avidiini*) has four binding sites for biotin and is preferred to avidin from egg white because it does not contain carbohydrate and gives lower assay backgrounds. The allergen concentration of the specific antigen (allergen) in the test extract is proportional to the color development (absorbance at 405 nm), and a sigmoid binding curve is obtained at concentrations of ~0.5 to 250 ng/ml. The ELISA is quantified using doubling dilutions of a reference allergen preparation to form a control curve, from which results for test samples are interpolated.

The development of a two-site ELISA is critically dependent on the production and selection of mAb, which to a large extent determine the sensitivity and specificity of the assay. The advantages of using mAb, as compared to polyclonal antibodies, are that once their specificity has been defined, it remains constant and the clone provides a continuous supply of homogeneous antibody — gram quantities of which can easily be produced. Our standard protocol for mAb production involves immunizing BALB/c mice with 10 to 50 μg allergen in complete Freund's adjuvant (CFA), three times at ~2-week intervals, and boosting the animals intrasplenically 4 days prior to fusion with $SP_2/0$ myeloma cells (Chapman et al., 1984). Animals are selected for fusion based on assessments of their allergen-specific immunoglobulin G (IgG) antibody titers in ELISA or in antigen-binding radioimmunoassay (RIA) using the ^{125}I labeled allergen. These assays are also used to screen the hybrid supernatants, 10 to 14 days after fusion. The allergen-binding RIA (a radioimmuneprecipitation assay) is not commonly used by others. However, it tends to select for high-affinity antibodies, and we have found it consistently useful for identifying mAb with strong binding activity.

The ability of mice to make antibody responses depends on both the immunogenic properties of the allergen (or other antigen) and the genetic and other factors that affect host immuneresponsiveness (Benjamin et al., 1984; Berzofsky and Berkower, 1989). In some cases, BALB/c mice fail to respond, or respond poorly, to a given allergen and other immunization strategies or mouse strains are required. For example, the Group II mite allergens are weak immunogens for BALB/c mice. It has been possible to produce mAb to these allergens by immunizing BALB/b or A/J mice with allergen, using CFA or aluminium hydroxide gel (alum) as adjuvant (Heymann et al., 1989; Ovsyannikova et al., 1994). The A/J strain is useful for mAb production because a BALB/c × A/J F_1 hybrid strain (CAF_1) is available commercially and can be used for ascites production.

The two-site ELISA is dependent on identifying mAb directed against nonoverlapping epitopes on the allergen. In our experience with mAb to mite Group I and Group II allergens and to cat allergen, *Fel d* I, up to 70% of mAb derived from fusions may have the same, or very similar, epitope specificity (Heymann et al., 1989; Ovsyannikova et al., 1994; Chapman et al., 1987; Chapman, 1989; Vailes et al., 1994). Therefore, it is useful to identify mAb directed against nonoverlapping sites at an early stage in the screening process. There are several potential approaches to epitope mapping, using either ^{125}I-labeled antigen or a preexisting labeled mAb, if available. Hybrid supernates can be tested to see whether they inhibit binding of ^{125}I allergen to a solid phase mAb, or for inhibition of the binding of ^{125}I mAb to allergen that is "presented" by another mAb on the solid phase (Figure 11.2) (Chapman, 1989). Antibodies that are thought to be directed against different sites are cloned and purified, and can then be biotinylated and directly tested for binding in the two-site ELISA. It is useful to test different combinations of mAb over a wide range of allergen concentration to establish the combination that shows optimal sensitivity and specificity. Comparisons of RIA binding curves using mAb directed against three nonoverlapping epitopes on *Fel d* I (designated Fd1A-C) are shown in Figure 11.3. The *Fel d* I ELISA was subsequently developed using an Fd1A mAb (clone 6F9) for allergen capture and biotinylated Fd1B (clone 3E4) for detection (Vailes et al., 1994; De Blay et al., 1991a). Whether or not a particular mAb binds to plastic has a major bearing on assay development and is difficult to predict. For both mite- and cat-derived allergens, we obtained mAb that could be used in two-site assays when coupled to cyanogen bromide activated supports but that showed poor binding on plastic. The development of ELISA for these allergens (*Fel d* I and mite Group II allergen) required screening a larger number of mAb from two to three different fusions (Heymann et al., 1989; Ovsyannikova et al., 1994; Chapman, 1989; Vailes et al., 1994; De Blay et al., 1991a).

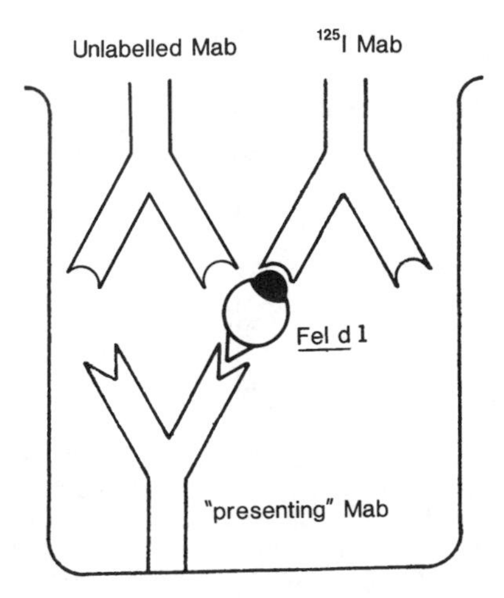

Figure 11.2. Cross-inhibition radioimmunoassay to assess epitope specificity. (From Chapman, M. D., *Allergy and Molecular Biology* (Advances in the Biosciences, Vol. 74), Elsevier, Kidlington, U.K., 1988, 281. With permission.)

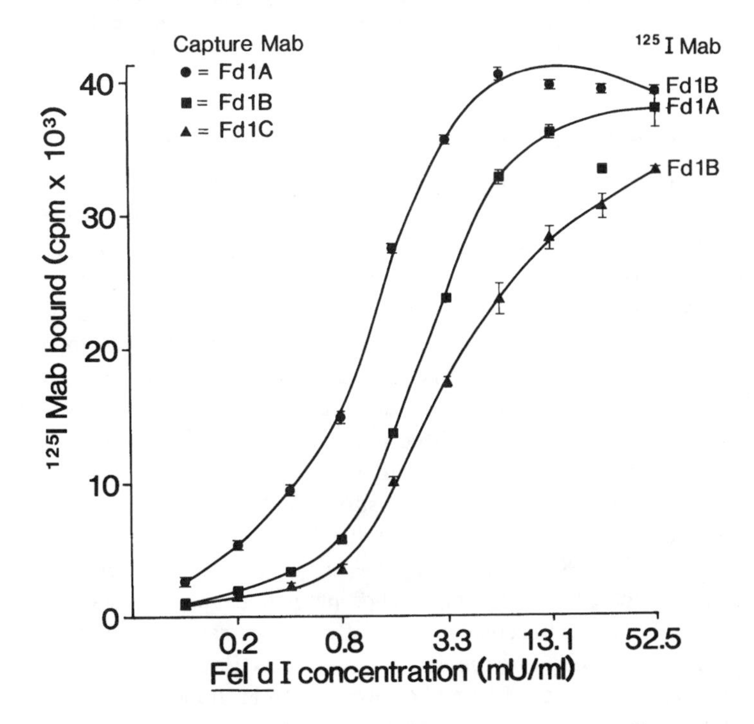

Figure 11.3. Comparisons of different combinations of anti-*Fel d* I mAb in solid-phase RIA. Microtiter wells were coated with three different capture mAb (Fd1A, B, C), and bound allergen was detected using either ^{125}I-Fd1A or -Fd1B. The most sensitive assay used Fd1A mAb (clone 3E4) for antigen capture and an Fd1B mAb (clone 6F9) for detection. This combination was used for the development of an ELISA.

Table 11.1 Two-site immunoassays (ELISA) for common indoor allergens.

Source	Allergen	Solid-phase mAb	Second antibody	Reference standards[a]
Dust mite	*Der p* I	5H8	4C1	NIBSC 82/518 UVA 93/03
	Der f I	6A8	4C1	CBER El-Df UVA 93/02
	Mite group II	7A1	1D8	UVA 92/01
Animal dander				
Cat	*Fel d* I	6F9	3E4	CBER Cat E5
Dog	*Can f* I	Cf-1b	Rabbit IgG ab	NIBSC 84/685
German cockroach	*Bla g* I	10A6	Rabbit IgG ab	UVA 93/04
	Bla g II	TC11	8F4	UVA 93/04
Aspergillus	*Asp f* I	4A6	Rabbit IgG ab	*Asp f* I

Note: MAb combinations shown are for assays developed at the University of Virginia, with the exception of the assay for *Can f* I, which was developed with mAb provided by Dr. R. Aalberse (University of Amsterdam) and polyclonal IgG ab provided by Dr. Carsten Schou (ALK Laboratories, Horsholm, Denmark) (DeGroot et al., 1991; Schou et al., 1991a). Similar assays for mite, cat, and cockroach allergens have been developed by other investigators (Lind, 1986; Yasueda et al., 1990; Lombardero et al., 1988, Schou et al., 1991b).

[a] Developed by national or international standardization agencies, such as the National Institute of Biological Standards and Control (UK) or Center for Biologics Evaluation and Research (USA) (Platts-Mills et al., 1992). The UVA standards have been substandardized against known international references, where available.

B. Immunoassays for Characterized Indoor Allergens

Immunoassays that are currently in use for measuring characterized indoor allergens, including mAb and polyclonal antibody based assays, are described in Table 11.1 (Heymann et al., 1989; Ovsyannikova et al., 1994; De Blay et al., 1991a; Luczynska et al., 1989; Lind, 1986; Yasueda et al., 1990; Chapman et al., 1988; DeGroot et al., 1988, 1991; Lombaradero et al., 1988; Schou et al., 1991a; Pollart et al., 1991a, 1991b; Arruda et al., 1992). Most of these assays use two mAb directed against nonoverlapping epitopes. However, for some allergens (derived from cockroach, dog, and *Aspergillus*), only mAb directed against a single epitope have been produced, and in those cases polyclonal rabbit antibodies are used for detection. Several other groups have produced allergen immunoassays of similar sensitivity and specificity using either mAb or polyclonal antibodies in ELISA (Table 11.1) or using radioimmunoassays (Yasueda et al., 1990). In most cases, the assays are quantitated using reference standards prepared as part of national or international allergen standardization programs, for which the content of specific allergen has been determined (e.g., the *Dermatophagoides pteronyssimus* standard, NIBSC 82/518, contains 12.5 μg/ml *Der p* I) (Platts-Mills and Chapman, 1991). Important features of the ELISA assays for monitoring environmental exposure are that they are technically straightforward and can be used to test large numbers of samples (several hundreds per week). This makes them extremely useful for epidemiologic and clinical studies. On the other hand, it should be noted that the single-allergen specificity of mAb allows detection only of the specific allergen. In other words, a *Der p* I mAb assay cannot be used to measure total dust mite allergen. This is especially problematic when evaluating exposures to agents such as fungi where a single allergen may not be dominant in all strains.

Exposure to allergens produced by arthropods and mammals is usually assessed by measuring allergen levels in three to four dust samples collected from appropriate "reservoirs" within the home. For dust mite allergens, these reservoirs include mattresses, bedding, carpets, and soft furnishings (e.g., sofas), and for other allergens from other sources (e.g., cockroach) sites such as kitchens may also need to be sampled (Pollart et al., 1991a; Schou et al., 1991b; Tovey et al., 1981; Platts-Mills et al., 1992). It is, perhaps, stating the obvious to say that the gross composition of house dust varies considerably from site to site within a house and from houses in different geographic areas. Other potential sources of variability are the sampling conditions used (e.g., the area to be sampled, the collection device and flow rates). Nonetheless, "standardized" sampling techniques and extraction procedures have evolved over the past 5 to 10 years, and aqueous extracts of these samples are used in allergen analyses. In our protocols, dust is collected from a 1-m^2 area for 2 min using a modified hand-held vacuum cleaner. The dust is sieved through an 0.3-mm mesh screen, and 100 mg of this "fine dust" is extracted in 2 ml borate-buffered saline (Tovey et al., 1981). The results of allergen assays are expressed as micrograms or units of specific allergen per gram of fine dust. Similar procedures have been used by other laboratories and have provided epidemiologic data on allergen content of dust from many parts of the world. Recommendations for assessing mite allergen exposure were recently reviewed as part of the second international mite workshop, and similar considerations apply to the other indoor allergens (Platts-Mills et al., 1992). Levels of allergen in dust that are likely to result in sensitization for IgE antibody responses have been proposed: for dust mite, 2 μg Group I allergen/g of dust; for cat, 8 μg *Fel d*I/g; and for cockroach, 2 units *Bla g* I/g. Levels of >10 μg mite Group I/g allergen have also been proposed as thresholds for causing exacerbations of symptoms in patients with asthma. These levels have been proposed on the basis of case control studies, comparisons of seasonal variations in allergen levels, studies on patients presenting to emergency rooms with asthma, and a prospective study on the development of childhood asthma (reviewed in Platts-Mills et al., 1992; see also Chapters 7 and 8).

Immunoassays have also been used to compare the form and concentration of specific allergens in the air. These studies have involved using a variety of air samplers, including a cascade impactor collecting particles on filters, liquid impingers, and Andersen culture-plate cascade impactor, and assaying the allergen collected on filters, in fluid in the chambers of the impinger, or extracted from agarose gel (De Blay et al., 1991a; Tovey et al., 1981; Platts-Mills et al., 1992; Van Metre et al., 1986; Luczynska et al., 1990; Sakaguchi et al., 1990; De Blay et al., 1991b; Swanson et al., 1989; Wentz et al., 1990).

Although measurement of airborne allergen might appear to be a more useful index of allergen exposure than reservoir dust, airborne measurements are technically more involved than those of dust (Platts-Mills et al., 1992). Mite allergens become airborne on 10 to 40 μm particles (primarily mite feces) following disturbance of dust within a house (e.g., vacuum cleaning, bedmaking, etc.). However, these particles settle rapidly and remain airborne for only 20 to 30 min. In contrast, airborne cat allergen includes both large (10 to 40 μm) and small (2 to 5 μm) particles, and the smaller particles, which may comprise 10 to 60% of the total, can remain airborne for several hours (De Blay et al., 1991a; Luczynska et al., 1990). Differences in the aerodynamic properties of the allergen-containing particles are thought to explain the reasons why cat allergic patients often develop immediate symptoms on entering a house with a cat, whereas there is no clear temporal relationship between exposure and symptoms for dust mite. Cat allergen has proved to be an excellent model for airborne studies, both experimentally and in patients' houses. Dust samples containing large amounts of *Fel d* I (up to 3000 μg/g) can be used for artificial disturbance, and it has been possible to define conditions of exposure in experimental rooms and in houses. These studies have made it possible to estimate the quantity of allergen shed by cats; the levels of exposure that precipitate asthma attacks; and the effects of air exchange rates, furnishings, air filtration, and vacuum cleaners on airborne levels (De Blay et al., 1991a; Van Metre et al., 1986; Luczynska et al., 1990; De Blay et al., 1991b; Swanson et al., 1989; Wentz et al., 1990). As a result of these studies, environmental control procedures that reduce airborne levels of cat allergen have been developed, and these procedures may reduce the symptoms of cat allergic patients, while allowing them to keep their cats (De Blay et al., 1991a).

C. Alternative Methods for Assessing Allergen Exposure

Most extracts of allergen-containing materials (e.g., pollen, fungus materials, environmental samples) are a heterogeneous mixture of proteins and other macromolecules (many of which are nonallergenic). In cases where neither purified allergens nor monospecific antibodies are available for measuring specific allergens, alternative immunoassay techniques can be used for comparing the "total" allergen content of extracts. These include using the crude allergen-containing mixture to inhibit IgE antibody binding to extract coupled to a solid phase in the radioallergosorbent test (RAST) or ELISA, or using the extract to inhibit immunoblots developed with IgE or IgG antibodies.

RAST. When the RAST assay for measuring allergen-specific IgE antibodies was first developed by Wide and colleagues in 1967, it soon became apparent that inhibiting this assay using dilutions of crude allergen extracts could be used as a measure of allergenic potency (Wide et al., 1967; Gleich et al., 1974). The method was superior to other techniques in use at that time (weight per volume or protein nitrogen units) because IgE antibody binding to the solid phase was only inhibited by the relevant allergens and linear dose-response curves were obtained (Gleich et al., 1974). In the RAST inhibition assay, macromolecules (including allergens) in crude extracts are coupled to cyanogen bromide-activated Sepharose or paper discs. Serum containing IgE antibodies to allergens in the extract is mixed with dilutions of a test extract and the mixture is incubated with the solid-phase allergens. Allergens in the test extract that match those on the solid phase inhibit the binding of specific serum antibodies. The results are expressed either as the dilution required to cause 50% inhibition, or in arbitrary units relative to a reference allergen preparation. IgE antibody binding is detected using ^{125}I anti-IgE as tracer or, in more recent modifications, enzyme-labeled anti-IgE is used (ELISA inhibition). RAST inhibition cannot be used to measure absolute quantities of allergen (in nanograms or micrograms), because binding of IgE antibodies is inhibited by multiple allergens in the extract. The results are also dependent on the compositions of both the extract coupled to the solid phase and the allergic serum pool used as source of IgE antibodies. These assays are most effective when the extracts being tested have composition similar to that used to bind to the solid phase. For example, mite body extracts should be compared with other mite body extracts and not with whole mite culture extracts, as these different kinds of source materials show both qualitative and quantitative differences in allergen content. For example, mite culture extracts contain much higher ratios of Group I to Group II allergens than body extracts (Heymann et al., 1989). Bearing these factors in mind, RAST inhibition is most useful for comparing the potency of allergen extracts (e.g., by researchers or manufacturers). For environmental studies, the technique is limited by the complexity of most environmental samples and by the availability of adequate supplies of sera containing appropriate IgE antibodies.

Immunoblot inhibition. Inhibition of antibody binding on immunoblots ("immunoprint inhibition") has recently been developed as a method of identifying allergens following isoelectric focusing (IEF) and for studying the antigenic relationships between them. The crude allergen extract is separated by IEF and tranferred to cyanogen bromide-activated nitrocellulose membrane. Binding of IgE (or IgG) antibodies to the separated bands is inhibited by allergen extract (or partially purified allergen), and the immunoblots are developed using either radioactive or enzyme-labeled second antibodies. This technique has been extensively used by Lehrer and colleagues to study antigenic relationships between basidiomycete allergens, and has identified a major cross-reactive allergen, pI 9.3 (DeZubiria et al., 1990). Immunoprint inhibition is being used to analyze allergens from fungi, foods, and pollens, and has the advantage (compared to RAST inhibition) of identifying individual allergens, which can then be isolated (e.g., by electroelution) and used for amino acid sequence analysis or mAb production.

Chemical assays as indicators of allergen sources. An alternative chemical test for assessing exposure to dust mite allergens is the guanine assay, developed by Bischoff and Bronswijk (Bronswijk et al., 1989). This test relies on the fact that dust mites are the major source of guanine in house dust extracts (spiders excepted!). The chemical test for guanine correlates with mite counts and assays for Group I allergens, and is marketed as a semiquantitative "dipstick" test that can be performed in physicians' offices or in the patients' houses (Van der Brempt et al., 1991; Hoyet et al., 1991; Quoix et al., 1993). It is a colorimetric test, and color development from yellow to dark red is visually assessed on a four-point scale (guanine classes 0, 1, 2, and 3). A quantitative guanine assay has also been developed, allowing guanine levels to be expressed as micrograms or milligrams per gram dust. Recently, the validity of the guanine assay has been questioned on the grounds that the test also measures xanthine (a plant constituent), and that guanine levels do not tightly correlate with those predicted from measured concentrations of specific allergens (Hallas et al., 1993). However, most published studies suggest that guanine classes 2 or 3, or guanine concentrations of >3 mg/g dust, correlate with >10 μg Group I mite allergen/g. Conversely, guanine class 0, or <0.9 mg/g, usually represents <2 μg Group I allergen/g (Quoix et al., 1993; Hallas et al., 1993; Chapman, 1993). The guanine test appears to be a useful test for semiquantitative assessment of mite allergen exposure and can be performed without sophisticated laboratory equipment.

III. CONCLUSIONS

Immunoassays have many advantages for the specific measurement of foreign proteins in the environment and in other biologic specimens. Inhaled allergens are among the most common proteins to which humans are exposed to indoors, and mAb-based immunoassays are now established as the method of choice for assessing environmental exposure to the characterized allergens. The strategies and applications involved in using immunoassays to assess allergen exposure provide a model for investigating other indoor biologic pollutants, both in the domestic environment and in the workplace. Allergen immunoassays are now being used to monitor the efficacy of regimes for avoidance of some allergens. For example, acaricides and chemicals that denature allergens, both in the laboratory and in patients' houses, are being tested for their ability to reduce mite allergen concentrations in dust (Woodfolk et al., 1993). Domestic cleaning appliances (e.g., vacuum cleaners and air filtration units) are being examined for their effects on levels of airborne cat allergen (Woodfolk et al., 1994). However, many appliances are marketed specifically at patients with allergic disease, with minimal data on efficacy in preventing symptoms, and usually no data demonstrating reductions in allergen levels. Indeed, some cleaners may generate small airborne allergen particles that would be expected to cause symptom exacerbations (Luczynska et al., 1990; Woodfolk et al., 1993). Immunoassays provide the means for researchers, and others interested in indoor air quality, to assess the efficiency of the various air cleaners on the defined allergens and provide objective methods for manufacturers to improve their products. At present, the ELISA tests are still essentially "laboratory based", and the widespread application of these measurements will require simpler home- or office-based tests. Such tests are currently being developed and should enable doctors, or patients themselves, to assess levels of some allergens in the home on a semiquantitative basis. Once these tests become available, it will be possible to assess whether home monitoring of these allergen levels improves patient compliance with avoidance procedures.

The use of immunoassays to measure allergen exposure is successful only when the offending allergen sources are recognized and the specific allergens have been immunochemically well defined. There remain many allergens both indoors and out that cannot currently be measured using immunoassays, including most of the common fungi. In addition, the application of this technology to other environmental pollutants is likely to be biased toward protein or glycoprotein antigens or haptens, to which antibodies can be made. It remains critical to establish the biologic importance of the specific ligand to be measured.

REFERENCES

Arruda, L. K., Mann, B. J., and Chapman, M. D., Selective expression of a major allergen and cytotoxin, *Asp f* I, in *Aspergillus fumigatus:* implications for the pathogenesis of *Aspergillus* related diseases, *J. Immunol.*, 149, 3354, 1992.

Axen, R., Porath, J., and Ernback, S., Chemical coupling of peptides and proteins to polysaccharides by means of cyanogen halides, *Nature* (*London*), 214, 1302, 1967.

Benjamin, D. C., Berzofsky, J. A., East, I. J., Gurd, F. R. N., Hannum, C., Leach, S. J., Margoliash, E., Michael, J. G., Miller, A., Prager, E. M., Reichlin, M., Sercarz, E. E., Smith-Gill, S. J., Todd, P. E., and Wilson, A. C., The antigenic structure of proteins: a reappraisal, *Annu. Rev. Immunol.*, 2, 67, 1984.

Berzofsky, J. A. and Berkower, I. J., Immunogenicity and antigen structure, in *Fundamental Immunology*, Paul, W. E., Ed., Raven Press, New York, 1989, 169.

Bolton, A. E. and Hunter, W. M., The labeling of proteins to high specific activities by conjugation to a ^{125}I-containing acylating agent. Application to the radioimmunoassay, *Biochem. J.*, 133, 529, 1973.

Bronswijk, J. E. M., Bischoff, E., Schirmacher, W. and Kniest, F. M., Evaluating mite (acari) allergenicity of house dust by guanine quantification, *J. Med. Entomol.*, 26, 55, 1989.

Chapman, M. D., Guanine — an adequate index of mite exposure?, *Allergy*, 48(5), 301, 1993.

Chapman, M. D., Monoclonal antibodies as structural probes for mite, cat and cockroach allergens, in *Allergy and Molecular Biology* (Advances in Biosciences, Vol. 74), El Shami, S. and Merrett, T., Eds., Elsevier, Kidlington, U.K., 1988, 281..

Chapman, M. D., Aalberse, R. C., Brown, M. J., and Platts-Mills, T. A. E., Monoclonal antibodies to the major feline allergen *Fel d* I. II: Single step affinity purification of *Fel d* I, N-terminal sequence analysis and development of a sensitive two site immunoassay to assess *Fel d* I exposure, *J. Immunol.*, 140, 812, 1988.

Chapman, M. D., Heymann, P. W., and Platts-Mills, T. A. E., Epitope mapping of two major inhalant allergens, *Der p* I and *Der f* I, from mites of the genus *Dermatophagoides*, *J. Immunol.*, 139, 1479, 1987.

Chapman, M. D., Sutherland, W. M., and Platts-Mills, T. A. E., Recognition of two *Dermatophagoides pteronyssimus* specific epitopes on Antigen P1 using monoclonal antibodies: binding to each epitope can be inhibited by sera from mite allergic patients, *J. Immunol.*, 133, 2488, 1984.

De Blay, F., Chapman, M. D., and Platts-Mills, T. A. E., Airborne cat allergen *Fel d* I: environmental control with the cat *in situ*, *Am. Rev. Respir. Dis.*, 143, 1334, 1991a.

De Blay, F., Heymann, P. W., Chapman, M. D., and Platts-Mills, T. A. E., Airborne dust mite allergens: comparison of Group II mite allergens with Group I mite allergen and cat-allergen, *Fel d* I, *J. Allergy Clin. Immunol.*, 88, 919, 1991b.

DeGroot, H., Goei, K. G. H., van Swieten, P., and Aalberse, R. C., Affinity purification of a major and minor allergen from dog extract: serologic activity of affinity purified *Can f* I and of *Can f* I depleted extract, *J. Allergy Clin. Immunol.*, 87, 1056, 1991.

DeGroot, H., van Swieten, P., van Leeuwen, J., Lind, P., and Aalberse, R. C., Monoclonal antibodies to the major feline allergen *Fel d* I. I. Biologic activity of affinity purified *Fel d* I and of *Fel d* I depleted extracts, *J. Allergy Clin. Immunol.*, 82, 778, 1988.

DeZubiria, A., Horner, W. E., and Lehrer, S. B., Evidence for cross-reactive allergens among basidiomycetes: immunoprint inhibition studies. *J. Allergy Clin. Immunol.*, 86, 26, 1990.

Engvall, E. and Perlmann, P., Enzyme linked immunosorbent assay (ELISA): quantitative assay of IgG, *Immunochemistry*, 8, 871, 1971.

Gleich, G. J., Larson, J. B., Jones, R. T., and Baer, H., Measurement of the potency of allergy extracts by their inhibitory capacities in the radioallergosorbent test, *J. Allergy Clin. Immunol.*, 53, 158, 1974.

Hallas, T. E., Xue, Y., and Schou, C., Does guanine concentration in house-dust samples reflect house dust mite exposure? *Allergy*, 48(5):303, 1993.

Harlow, E. and Lane, D., *Antibodies — A Laboratory Manual*, Cold Spring Harbor Laboratory Press, Cold Spring Harbor, NY, 1988.

Heymann, P. W., Chapman, M. D., Aalberse, R. C., Fox, J. W., and Platts-Mills, T. A. E., Antigenic and structural analyses of Group II allergens (*Der p* II and *Der f* II) from house dust mites (*Dermatophagoides* spp.), *J. Allergy Clin. Immunol.*, 83, 1055, 1989.

Hoyet, C., Bessot, J. C., Le Mao, J., Quoix, E., and Pauli, G., Comparison between *Der p* I and *Der f* I content determination and guanine measurement in 239 house dust samples, *J. Allergy Clin. Immunol.*, 88, 678, 1991.

Jerne, N. K., Toward a network theory of the immune system, *Ann. Immunol.* (Paris), 125C, 373, 1974.

Kemeny, D. M., *A Practical Guide to ELISA*, Pergamon Press, Oxford, U.K., 1991.

Kohler, G. and Milstein, C., Continuous cultures of fused cells secreting/antibody of pre-defined specificity, *Nature,* 256, 495, 1975.

Lind, P., Enzyme linked immunosorbent assay for determinantion of major excrement allergens of house dust mite species *D. pteronyssimus, D. farinae* and *D. microceras*, *Allergy*, 41, 442, 1986.

Lombardero, M., Carreira, J., and Duffort, O., Monoclonal antibody based radioimmunoassay for the quantitation of the major cat allergen (*Fel d* I or Cat-1), *J. Immunol. Methods*, 108, 71, 1988.

Luczynska, C. M., Li, Y., Chapman, M. D., and Platts-Mills, T. A. E., Airborne concentrations and particle size distribution of allergen derived from domestic cats (*Felis domesticus*): Measurements using cascade impactor, liquid impinger and a two site monoclonal antibody assay for *Fel d* I, *Am. Rev. Respir. Dis.*, 141, 361, 1990.

Luczynska, C. M., Arruda, L. K., Platts-Mills, T. A. E., Miller, J. D., Lopez, M., and Chapman, M. D., A two-site monoclonal antibody ELISA for the quantification of the major *Dermatophagoides* spp. allergens, *Der p* I and *Der f* I, *J. Immunol. Methods*, 118, 227, 1989.

Ovsyannikova, I. G., Vailes, L. D., Li, Y., Heymann, P. W., and Chapman, M. D., Monoclonal antibodies to Group II Dermatophagoides spp. allergens: murine immune response, epitope analysis, and development of a two-site ELISA, *J. Allergy Clin. Immunol.,* 94, 537, 1994.

Platts-Mills, T. A. E., Thomas, W. R., Aalberse, R. C., Vervloet, D., and Chapman, M. D., Dust mite allergens and asthma: report of a second international workshop, *J. Allergy Clin. Immunol.*, 89, 1046, 1992.

Platts-Mills, T. A. E., and Chapman, M. D., Allergen standardization, *J. Allergy Clin. Immunol.*, 87, 621, 1991.

Pollart, S. M., Mullins, D. E., Vailes, L. D., Sutherland, W. M., and Chapman, M. D., Identification, quantitation and purification of cockroach allergens using monoclonal antibodies, *J. Allergy Clin. Immunol.*, 87, 511, 1991a.

Pollart, S. M., Smith, T. F., Morris, E. C., Platts-Mills, T. A. E., and Chapman, M. D. Environmental exposure to cockroach allergens: analysis using a monoclonal antibody based enzyme immunoassay, *J. Allergy Clin. Immunol.*, 87, 505, 1991b.

Quoix, E., Le Mao, J., Hoyet, C., and Pauli, G., Prediction of mite allergen levels using guanine measurements in house-dust samples, *Allergy*, 48(5):306, 1993.

Sakaguchi, M., Inouye, S., Yasueda, H., Tatehisa, I., Yoshizawa, S., and Shida, T., Measurement of allergens associated with house dust mite allergy. II. Concentrations of airborne mite allergens (*Der* I and *Der* II) in the house, *Int. Arch. Allergy Appl. Immunol.*, 90, 190, 1990.

Schou, C., Hansen, G. N., Lintner, T., and Lowenstein, H., Assay for the major dog allergen, *Can f* I: investigation of house dust samples and commercial dog extracts, *J. Allergy Clin. Immunol.*, 88, 847, 1991a.

Schou, C., Fernandez-Caldas, E., Lockey, R. F., and Lowenstein, H., Environmental assay for cockroach allergens, *J. Allergy Clin. Immunol.*, 87, 828, 1991b.

Swanson, M. C., Campbell, A. R., Klauck, M. J., and Reed, C. E., Correlations between levels of mite and cat allergens in settled and airborne dust, *J. Allergy Clin. Immunol.*, 83, 776, 1989.

Tovey, E. R., Chapman, M. D., Wells, C. W, and Platts-Mills, T. A. E., The distribution of dust mite allergen in the houses of patients with asthma, *Am. Rev. Respir. Dis.*, 124, 630, 1981.

Vailes, L. D., Li, Y., Bao, Y., DeGroot, H., Aalberse, R. C., and Chapman, M. D., Fine specificity analyses of B cell epitopes on *Felis domesticus* allergen I (*Fel d* I): effect of reduction and alkylation or deglycosylation on *Fel d* I structure and antibody binding, *J. Allergy Clin. Immunol.,* 93(1), 22–33, 1994.

Van der Brempt, X., Haddi, E., Michel-Nyugen, A., Fayon, J. P., Soler, M., Charpin, D., and Vervloet, D., Comparison of the Acarex test with monoclonal antibodies for the quantification of mite allergens, *J. Allergy Clin. Immunol.*, 87, 130, 1991.

Van Metre, T. E., Marsh, D. G., Adkinson, N. F., Fish, J. E., Kagey-Sobotka, A., Norman, P. S., Radden, E. B., and Rosenberg, G. L., Dose of cat (*Felis domesticus*) allergen 1 (*Fel d* I) that induces asthma, *J. Allergy Clin. Immunol.*, 78, 62, 1986.

Wentz, P. E., Swanson, M. C., and Reed, C. E., Variability of cat allergen shedding, *J. Allergy Clin. Immunol.*, 85, 94, 1990.

Wide, L., Bennich, H., and Johansson, S. G. O., Diagnosis of allergy by an *in vitro* test for allergen antibodies, *Lancet,* ii, 1105, 1967.

Woodfolk, J. A., Luczynska, C. M., de Blay, F., Chapman, M. D., and Platts-Mills, T. A. E., The effect of vacuum cleaners on the concentration and particle size distribution of airborne cat allergen, *J. Allergy Clin. Immunol.*, 91(4):829, 1993.

Woodfolk, J. A., Hayden, M. L., Miller, J. D., Rose, G., Chapman, M. D., and Platts-Mills, T. A. E., Chemical treatment of carpets to reduce allergen: a detailed study of the effects of tannic acid on indoor allergens, *J. Allergy Clin. Immunol.*, 94, 19, 1994.

Yalow, R. S. and Berson, S. A., Immunoassay of endogenous plasma insulin in man, *J. Clin. Invest.*, 39, 1157, 1960.

Yasueda, H., Mita, H., Yui, Y., and Shida, T., Measurement of allergens associated with house dust mite allergy I. Development of sensitive radioimmunoassays for the two groups of *Dermatophagoides* mite allergens, *Der* I and *Der* II, *Int. Arch. Allergy Appl. Immunol.*, 90, 182, 1990.

12 SAMPLING AND ANALYSIS OF BIOLOGICAL VOLATILE ORGANIC COMPOUNDS

Stuart A. Batterman

CONTENTS

0-87371-724-4/95/$0.00+$.50

I. INTRODUCTION

A. Sources and Nature of Biogenic VOCs

Both fungi and bacteria produce complicated VOC mixtures that include alcohols, aldehydes, ketones, aromatics, amines, terpenes, chlorinated hydrocarbons, and sulfur-containing compounds. The VOCs discussed in this chapter arise from fungi and bacteria. It should be noted that there are many other biogenic VOC emission sources that emit a spectrum of VOCs resembling those emitted by fungi and bacteria (e.g., algae, pine trees, etc.) Some compounds emitted by fungi and bacteria are listed in Table 12.1.

Non-biogenic indoor VOC sources include paints, varnishes, pesticides, insecticides, personal use products (e.g., hair sprays, cosmetics), binders and resins in carpets, furniture and building materials (Wallace, 1991a). Outdoor sources and manufacturing processes also may provide significant contributions. These sources produce many VOCs, some of which are also listed in Table 12.1.

While the identification of many of the biogenic VOCs remains tentative, it is clear that the composition of biogenic emissions varies significantly from other kinds of VOCs found in residential, commercial, and industrial settings. These differences in composition suggest the possibility of using biogenic VOCs as tracers for biological contamination, as discussed later. However, ambient concentrations of biogenic VOCs are probably very low (although in most cases the levels have not been quantified) and VOC concentrations from typical indoor sources clearly exceed biogenic levels measured to date by orders of magnitude.

B. Mechanisms of Production

The mechanisms of VOC production by fungi and bacteria are poorly understood. Most fungi obtain energy by aerobic respiration, producing CO_2 and water. A few use fermentation, producing alcohols (usually ethanol) or lactic acid. Both fungi and bacteria digest their food outside the cell by excreting enzymes into the environment, which are then used to metabolize hydrocarbons including aromatic forms, during which time a variety of metabolites are produced. These metabolites include soluble sugars that are used as food, complex secondary metabolites which may be toxic to mammalian and other cells, and VOCs. The highest production of VOCs appears to precede toxin production as well as the peak spore production period (Abramson et al., 1983) although factors controlling the composition and concentration of emissions are unknown.

Table 12.1. VOCs common in indoor environments.

	Also recovered from microorganisms?
1,1,1-Trichlorethane	No
1,2,Dichloroethane	No
1,2-Dibromoethane (ethylene dibromide)	No
Acetone	Yes
Benzene	No
Carbon tetrachloride	No
Chloroform	No
Formaldehyde	No
m, *o*, and *p*-Xylene	No
Methyl ethyl ketone	Yes
Methylene chloride	No
p-Dichlorobenzene	No
Styrene	No
Terpenes (limonene and a-pinene)	Yes
Tetrachloroethylene	No
Toluene	Yes
Trichlorethylene	No
Undecane	No
Vinylidene chloride	No

From Shah and Singh, 1988; Wallace, 1991a and 1991b.

C. Health and Comfort Effects

Many of the biogenic VOCs are polar compounds that have distinct odors and low odor thresholds, and have been reported to produce discomfort in building occupants (McJilton et al., 1990). For example, the musty or dank odor associated with damp basements is largely caused by VOCs associated with fungi (Burge and Hodgson, 1988), as is the familiar odor of moldy bread (Harris et al., 1986). Odor thresholds for some biogenic VOCs are very low, as little as 10 ppt for dimethyl trisulfide or even 1 ppt for 1,5-octadien-3-ol (Harris et al., 1986).

Little evidence exists to date that links biogenic VOCs found in contaminated indoor environments to specific health effects. Such cause-effect relationships have not been examined critically, due in part to the difficulty of estimating exposures, problems associated with quantifying trace VOCs in the complex mixtures typically found indoors, adequate identification and management of potential confounding factors in epidemiologic investigations, and the difficulty of evaluating symptoms associated with few objective findings and uncertain etiology that are typically found in building-related health problems (e.g., Samet et al., 1988; Kreiss, 1989; Otto, 1990). These problems are not easily resolved. In view of these problems, and because health effects caused by other fungal emissions are well known, minimizing indoor biological contamination is usually recommended and should minimize levels of biogenic VOCs.

II. VOC SAMPLING

A. Sorbent Sampling

Sorbent sampling involves the collection of VOCs on adsorbents that attract and retain the compounds for later analysis. Active sorbent sampling is a well-developed technique in which sample air is drawn via a vacuum pump through a small tube filled with one or several adsorbents onto which the VOCs adsorb. The method permits the extraction of low concentration VOCs from large volumes of air. For example, atmospheric sampling in remote locations (e.g., midocean or polar sites) pass very large volumes of air (1000 m^3) through sorbents, permitting measurement of VOCs in the parts per trillion range. Most industrial hygiene applications require only several hundred liters of air to detect high (ppm) VOC levels; as little as several liters have been used in some indoor air studies. Because the sampling equipment is small and portable, sorbent tube sampling is well suited to field studies of biogenic VOCs (as performed by McJilton et al., 1990).

Passive sorbent samplers are a second well-developed collection method. In these matchbook-sized devices, a flat sorbent is supported via a screen a small distance from the open sampling end. VOCs diffuse through the samplers' internal dead air space and are collected on the sorbent. Since diffusion rates are known (a function of temperature and compound), quantitative results can be obtained. Passive samplers do not require a pump, which is a major advantage. However, sampling rates are very low. These samplers are commonly used in occupational settings where high VOC concentrations are expected. They are usually clipped to a worker's clothing and provide a measure of personal exposure. Passive samplers have been used in environmental applications with low concentrations that require or benefit from the long sampling periods (days or weeks) required to collect sufficient material for analysis, including indoor air studies (Seifert and Abraham, 1983; Weschler et al., 1988). No application for biogenic VOCs has yet been described.

Common sorbents used for both active and passive VOC sampling include activated charcoal and carbon black (e.g., Carbosieve®* and Carbotrap®*) silica gel, polymers such as Tenax®**, Chromosorb®*, and Poropak®*, and resins such as XAD-4®* and polyurethane foam. A relatively new sorbent called Eu-sorb appears effective for VOCs such as ketones and aldehydes (Williams and Sievers, 1984) and thus may be useful for biogenic compounds.

Sorbent selectivity is based on boiling point, polarity, and other VOC properties, and thus must be matched to the VOCs of interest. For example, Tenax GC® and Tenax TA®, both 2,6-diphenylparaphenylene polymers, capture 5-carbon and higher non-polar aromatics (TA may be somewhat more stable). Carbosieve®, molecular sieve, and silica gel capture lighter VOCs (2 to 4 carbons). VOCs collected on resins and polymers may have limited stability;

* Registered trademark of SKC, Eightfour, Pennsylvania.

** Registered trademark of Alltech Associates Inc., Deerfield, Illinois.

also, thermal desorption may cause the sorbent to degrade thermally and introduce artifacts. Artifacts that have been identified include aliphatic, alicyclic, and aromatic hydrocarbons. These compounds may persist despite careful cleanup and conditioning (MacLeod and Ames, 1986) and may interfere in the analysis of low-level biogenic compounds.

The selection and use of sorbents for biogenic VOCs in field studies must address breakthrough and competitive sorption issues. Breakthrough is the saturation and diffusion of the sorbent with VOCs (or other compounds) that causes erroneously low concentration results. Several studies (e.g., Rothweiler et al., 1991) have indicated that the capacity of Tenax® decreases when the same tube is used repeatedly (>5 times), thus increasing the likelihood of breakthrough. A back-up sorbent (following the primary sorbent) can be used to determine whether breakthrough has occurred.

Competitive sorption occurs between nonpolar and polar compounds (and within these classes), and high concentrations of nonpolar compounds may tend to displace polar compounds (Goller, 1985). Thus, in settings where relatively high levels of nonpolar VOCs are expected, an effective strategy might be to utilize charcoal as the primary sorbent and Tenax® or silica gel as a backup sorbent. While most studies of biogenic VOCs have used Tenax®, a combination of sorbents (e.g., Carbosieve®, molecular sieve, and Tenax TA®) used in series may capture a wider range of VOCs than any single sorbent. This approach may permit detection of biogenic VOCs in the presence of the nonbiogenic aliphatic, aromatic, and chlorinated VOCs found in buildings and other settings.

Given the low concentrations of biogenic VOCs, great attention must be given to recovery and artifact issues inherent in the use of sorbents. This includes the careful use of blanks and standards. It is likely that some of the biogenic VOCs reported in the next section represent only artifacts and not true metabolites.

B. Whole Air Sampling

Whole air sampling with cryogenic preconcentration can capture a wide range of VOCs. These samples are collected in stainless steel canisters or sampling bags constructed of Teflon®, Tedlar®, Mylar®, aluminum, and/or other materials. Because sampling bags may be difficult to clean and may introduce artifacts, their use for fungal VOCs is not recommended. Summa®* canisters can provide lower background levels and enhanced VOC stability than sample bags. The electropolished stainless steel canisters are typically 6 liters in volume (1 to 20 liter canisters are available) and are equipped with tightly sealing bellows valves. Canister sampling is simple. Sample air is collected in the previously cleaned and evacuated canister by vacuum alone or with the assistance of a small pump.

* Registered trademark of Scientific Instrumentation Specialists Inc., Moscow, ID; Andersen Samplers Inc., Atlanta, GA.

Whole air sampling provides considerable flexibility for a wide range of VOCs that can be sampled without concerns of sorbent breakthrough, selectivity, displacement, and recovery at low detection limits (~0.1 ppb). Multiple samples for GC-MS analysis can be extracted from a single canister, as its volume greatly exceeds the sample volume (~0.3 liter). Multiple samples permit fine-tuning of GC-MS parameters, decreased uncertainties (Gholson et al., 1990), and quality assurance measurements. In contrast, sorbent samples can be desorbed only a single time.

On the other hand, whole air sampling has several disadvantages. Some alcohols and reactive VOCs (containing oxygen, nitrogen and sulfur) may be poorly preserved in the canister due to reaction and deposition on canister walls. Whole air sampling equipment is bulky and expensive. The typical 6 liter canister is about 0.3 meters high and weighs several kilograms.

Direct cryogenic sampling is an approach that represents the preconcentration stage of whole air sampling. Here, sample air is passed directly over a cryotrap to condense the VOCs. The trap is maintained at cryogenic temperatures until analysis. Since there is no sampling container, there are no wall losses, and the approach is suitable for very reactive compounds. However, water condensation may be a problem, and the approach is relatively expensive and is limited to laboratory studies (Pleil, 1991; Rivers et al., 1992).

III. VOC ANALYSIS

A. Transfer from Sorbents

Following collection, VOCs are transferred to the analytical instrument (e.g., GC-MS), taking care to avoid interferences or artifacts due to contaminants, solvents, impurities, moisture, decomposition, etc. The procedures selected for biogenic VOCs depend on the sampling strategy employed and the target compounds.

Either chemical or thermal techniques can be used to desorb the VOCs from the sorbent and introduce them to the GC. In chemical desorption, a small quantity of high purity methanol, carbon disulfide, diethyl ether, or other compound is injected directly into the sorbent. In thermal desorption, the sorbent is heated to release the VOCs. Both techniques have been used for polymer sorbents (e.g., Tenax®), but chemical desorption is typically used with charcoal and silica gel media. Most fungal VOC studies have employed Tenax® with thermal desorption, which is faster than chemical desorption and avoids many interferences. Further, it generally offers better sensitivity since the sample is not diluted (Westendorf, 1985). On the other hand, thermal desorption requires more expensive equipment to control temperatures and gas flows.

B. Whole Air Analysis

For whole air analysis using canisters, a portion of the canister contents is removed and passed through a cryogenic trap cooled by liquid nitrogen from –196 to –90°C. VOCs condense in the cold trap. The sample flow is then terminated and the GC carrier gas (e.g., nitrogen or helium) is passed through the trap which is rapidly heated to 100 to 150°C, which is above the boiling point of the VOCs. The condensed VOCs are thus concentrated into a small gas volume and introduced into the GC. This procedure introduces few artifacts and allows high sensitivity. Its disadvantages include the condensation of water in the cryotrap, which, in sufficient concentration, will interfere with the elution and separation of compounds in the GC column. Also, ice formation may clog the trap. Thus, samples must be dried using a selectively permeable polymer membrane (e.g., Nafion®*) that removes water (Pleil et al., 1987). Unfortunately, this membrane may remove some polar VOCs and low molecular weight species. The drying step may be avoided by gradually heating the trap, assuming that water is volatilized before the target compounds (Pankow, 1991). Precise temperature control is needed to achieve high repeatability in whole air analyses.

C. Gas Chromatography

Typically, high resolution gas chromatography is used to separate biogenic VOCs as well as provide some information on identity. The VOC sample is introduced into the (nitrogen or helium) carrier gas flow which then passes through a packed or capillary column. VOCs are separated in the column by elution time, which is a function of column-VOC interactions that include polarity, molecular size, and other physical-chemical properties.

To improve the separation of compounds, a small cryogenic trap may be used at the GC column inlet through which the GC carrier gas flows. When cooled, this trap captures the VOCs as they are introduced from the sample. This trap is then rapidly heated above the boiling point of the VOCs, which flow with the carrier gas into the column for separation. This method aids in separation of compounds on the column, producing very clean and well-defined peak shapes. The method exceeds the performance of sampling loops, inlet purging, and other techniques designed to facilitate chromatogram interpretation, especially if capillary columns are used.

Once the sample has passed through the column, several methods of detection can be used.

D. Mass Spectroscopy

In most cases, retention time alone does not provide the selectivity required for the complex biogenic mixtures, and analysis of eluates by mass

* Registered trademark of PermaPure Products, Inc., Toms River, New Jersey.

spectroscopy is usually necessary. Selective mass spectrometer (MS) detectors provide the greatest certainty in identifying VOCs, and most studies of biogenic VOCs have used GC-MS analyses. The MS indicates the molecular weight of fragments produced upon ionization of the VOC as it elutes through the GC column. Often, mass spectra are distinctive and provide sufficient information for identification. Libraries containing spectra of over 100,000 compounds are available from the National Institute of Standards and Technology. The use of retention time and mass spectra together is considered the definitive method for compound identification.

The development of mass spectra in MS requires scanning of mass ranges and makes MS much less sensitive than that of nonselective detectors such as flame ionization detectors (see below). Sensitivity can be increased using selective ion mode of operation if only specific ions are of concern. For biogenic VOCs, MS sensitivity of roughly nanogram levels can be obtained. Sensitivity can be enhanced using sensitive MS technologies such as ion trap configurations and greater sample preconcentration.

With the use of selective MS detectors, misidentification may occur with cluttered chromatograms (i.e., the mixture problem) and if retention times and spectra are not experimentally determined. In several published studies, the identification of VOCs remains tentative for these reasons.

Dual mass spectrometers (MS-MS) instruments may provide enhanced identification of VOCs. MS-MS using direct atmospheric pressure inlets do not require sample conditioning and represents the most sophisticated technique available for VOC analysis. To date, the limited availability and high cost of this technology has precluded its routine use.

E. Other Detectors

Flame ionization detectors are universal hydrocarbon sensors that respond to compounds based on carbon content. These detectors are very sensitive, with a dynamic range of 4 or 5 orders of magnitude. Equally sensitive and somewhat selective, photoionization detectors (PIDs) respond to compounds based on ionization potential. PIDs give strong responses to aromatics, unsaturated compounds, and sulfur-containing compounds, but weaker response to alkenes, ketones, aldehydes, mercaptans, etc.

F. Other Sampling and Analysis Methods

Several other sampling and analysis techniques are potentially applicable to biogenic VOCs. Impinger sampling can be used to collect reactive gases and vapors. For example, aldehydes and ketones are efficiently collected in dinitrophenylhydrazine (DNPH) solution by passing sampling air through the solution, which is then extracted and analyzed using liquid chromatography.

IV. STUDYING EMISSIONS IN THE LABORATORY

A. Sample Generation

Biogenic compounds can be identified from cultures as well as from contaminated materials brought into the laboratory using either dynamic or static headspace sampling approaches. In these two laboratory methods, fungi or bacteria either purposely grown on some substrate (standard culture media, building materials, etc.) or already growing on field-collected material are placed into a clean and air-tight chamber. Ideally, the chamber is made of deactivated glass or electropolished stainless steel. The organisms are allowed to grow in the chamber, and chamber air containing the VOCs is sampled during active growth using adsorbents, canisters, or other means.

In dynamic sampling, the chamber is continuously purged using a clean gas [e.g., air (when active growth must continue) or nitrogen]. VOCs emitted in the exhaust stream are captured using a sorbent tube. Sampling may continue for minutes to hours, depending on the application and sensitivity required. Dynamic purging with sorbents has been used in VOC studies of food samples (Harris et al., 1986; Kirk, 1987 and 1988; Bartley and Schwede, 1988; Narain et al., 1990; Borjesson et al., 1990).

In equilibrium headspace sampling, the chamber air (headspace) is stagnant, and a small portion of the air is removed for analysis. This approach has yielded reproducible results in studies of food (Olafsdottir et al., 1985), fungi (Batterman et al., 1991) and bacteria (Pleil, 1991; Rivers et al., 1992).

Static headspace sampling approaches are very simple and cost-effective; dynamic approaches provide a higher degree of flexibility, experimental control, and possibly realism. However, dynamic sampling has several advantages. Chamber humidity and temperature are readily regulated. Air in the chamber has a short residence time which helps to minimize wall losses and flow paths over the substrate could be designed to avoid contact with chamber walls, further minimizing wall losses. Finally, dynamic purging maintains a low VOC concentration in the chamber and thus should provide higher and more consistent results for emissions that are diffusion-limited. The continuous purging produces low concentrations, however, that may require the use of sorbent tube sampling or nonselective detectors. The dynamic headspace approach closely resembles the standardized chamber testing protocol being used to investigate VOC emissions of household products and materials (ASTM, 1991). This protocol is designed to provide reproducible and quantitative results, but is not well suited to biogenic VOCs given their polarity and low concentrations.

Samples may also be generated directly from fungi, bacteria, and/or the substrate to determine VOCs as well as semi- and non-volatile components that may be present in both liquid and solid phases. In this case, sample collection involves homogenizing a bulk (e.g., fungus) sample, extracting the organics using a suitable prepurified solvent (e.g., methylene chloride or methanol), concentrating the solvent using physical (e.g., evaporation) or chemical (e.g., sequential extractions) means, and injecting the extract into a GC-MS system for identification and quantification of compounds. Other approaches that have been used to extract VOCs from food samples include vacuum distillation (e.g., Kaminski et al., 1974) and steam distillation (e.g., Lin and Jeon, 1985). Finally, supercritical fluids may be used to extract organics from complex samples for GC or GC/MS analysis, as performed for flavor and fragrance analysis (Hawthorne et al., 1988). The relationship of the VOCs found by these methods to VOCs that are emitted into the air has not been investigated.

B. Sample Collection

Both whole air and direct cryogenic sampling approaches are amenable to laboratory chamber studies, especially static headspace approaches where concentrations are relatively high. Recovery issues due to the attenuation of polar VOC on canister walls (as well as losses due to the use of air dryers in the VOC analysis) could be addressed by quality assurance and calibrations tasks that are a normal part of laboratory analyses.

V. VOCs ASSOCIATED WITH MICROORGANISMS

Research regarding VOCs emitted by fungi and bacteria is sparse. Most of this information is found in food-related literature.

A. Organisms Studied and VOCs Recovered

Only a few fungi have been studied with respect to VOCs. These are *Aspergillus flavus, A. clavatus, A. niger, Penicillium caseicolum, P. camemberti, P. chrysogenum, P. aurantiogriseum, Botrytis cinerea, Hydnellum suaveolens, Cladosporium sphaerospermum,* and others that were not specifically identified. Predominant among the VOCs identified in these studies were 3-methyl-1-butanol, 3-octanol or 1-octan-3-ol, 3-octanone, and 1-octen-3-ol, each identified in at least five different studies. These and other VOCs recovered from fungi and bacteria are listed in Table 12.2.

One study reports VOCs recovered from the actinomycetes *Streptomyces griseus* and *Streptomyces odorifer* (Harris et al., 1986). These organisms produced a series of compounds that were not identified in fungi also included in the study (dimethyl trisulfide, geosmin, heneicosane, docasane, tricosane, tetracosane, pentacosane, hexacosane, and heptacosane), as well as 3-methyl-1-butanol, 4-decanol, and 2-methylisoborneol, which were recovered from both fungal and actinomycete cultures.

The literature on bacterial VOCs is extremely limited. In general, identifications for the strains used are not provided. VOCs recovered from bacteria

Table 12.2. VOCs recovered from fungi and bacteria.

Recovered in more than one study

Compound	Study
1-Octen-3-ol	1,2,3,4,5,12
1-Propanol	4,7
2-Heptanone	7,11
2-Octen-1-ol	2,12
3-Methyl-1-butanol	1,2,4,5,11,12,11
3-Octanol (1-octan-3-ol)	1,2,3,5,12
3-Octanone	1,2,3,5,12
Acetone	7,9
Benzaldehyde	2,12
Dimethyl disulfide	7,9
Dimethyl trisulfide	7,9
Ethanol	7,9
p-Anisaldehyde (4-methoxybenzaldehyde)	6,12
Hexanol	2,12
Indole	9,12
Methanol	7,9
n-Hexanal	7,12
Nonanone, nonan-3-one	2,12
Phenol	7,9
Trimethyl amine	7,9

Recovered in one study from headspace analysis or ambient air

Compound	Study	Compound	Study
1,2,Pentadiene	7	Cresol	9
1,5-Octadien-3-ol	5	Damascenone	5
1-Butoxy-2-propanol	10	Dimethyl trisulfide	5
2 Hexanone	11	Docasane	5
2-Butanol	7	Ethanol	7
2-Methoxy-3-isopropylpyrazine	5	Geosmin	5
2-Methyl propionic acid	10	Heneicosane	5
2-Methyl-1-propanol	4	Heptacosane	5
2-Methylisoborneol	5	Hexacosane	5
2-Pentanone	7	Limonene	7
2-Propanol	7	Methanol	7
2h-1-Benzopyran-2-one (coumarin)	6	Methyl ethyl ketone	9
3,3-Dimethyl-1-octene	8	Methyl mercaptan	9
3,3-Dimethyl-2-oxetanone	8	Methyl-tert-butyl ether	7
3-Ethyl-2,4-dimethyl pentane	8	Pentacosane	5
3-Methylfuran	4	Tetracosane	5
3-Octen-2-ol	4	Toluene	7
4-Decanol	5	Tricosane	5
Acetone	7	Trimethyl hexene	5
Chloranisoles	5		

Recovered only from fungus mat or substrate extraction (1,2,or 12)

1,2-Dimethyloxybenzene	Diemthoxyethane
1-Decen-3-ol	Dimethylpyrazine
1-Methoxy-3-methylbenzene	Ethyl benzoate
1-Octanol	Heptanol
1-Octenol	Hexan-2-ol
2,3,5-Trimethylpyrazine	Hexan-2-one
2,5,-Dimethylpyrazine	Hexan-3-ol
2-Ethyl-5-methylpyrazine	Hexanol
2-Ethyl-5-methylpyridine	Isobutyl alcohol
2-Methoxyacetophenone	Nonanone, nonan-3-one
2-Methylbutanol	Octan-2-one

Table 12.2 (continued). VOCs recovered from fungi and bacteria.

Recovered only from fungus mat or substrate extraction (1,2,or 12)	
2-Methylphenol	Octane
2-Methylpyridine	Octyl acetate
2-Octen-1-ol	*p*-Ethlylphenol
2-Pentylfuran	*p*-Methoxybenzaldehyde
3 & 4-Methylphenol	Pentan-2,4-dione
3,5-Dimethyl-1,2,4-trithiolane	Pentan-2-ol
Acetophenone	Penten-4-ol
Alpha-bergmotene	Phenethyl alcohol
Amyl acetate	Phenylacetaldehyde
Benzaldehyde	Phenylethanol
Benzothiazole	Propylbenzene
Butyl alcohol	Pyridine
cis-2-Octen-1-ol	Tetramethylpyrazine
Decalin	Thialdin
Undecan-2-one	

Note: Numbers refer to the following literature citations: 1. Kaminski et al., 1972; 2. Kaminski et al., 1974; 3. Karahadian et al., 1985; 4. Borjesson et al., 1990; 5. Harris et al., 1986; 6. Wood et al., 1988; 7. Pleil, 1991; 8. Batterman et al., 1991; 9. Rivers et al., 1992; 10. McJilton et al., 1990; 11. Miller et al., 1988; 12. Seifert et al., 1992.

are listed in Table 12.2. Bacteria and fungi appear to produce several VOCs in common (acetone, 2-butanol, dimethyl trisulfide, methanol, and 1-propanol) (Pleil, 1991).

B. Methods of Analysis

All of the microbial VOC studies in the literature utilized GC-MS analysis. Three of the studies reviewed used extraction directly from fungus material to obtain VOCs (Kaminski, 1972 and 1974; Seifert and King, 1982; Wood et al., 1988), while all others used headspace analysis. Only one fungus (*Aspergillus flavus)* was included in both kinds of studies. Table 12.3 compares VOCs recovered by the two methods for this fungus. Only two compounds (3-octanol, 1-octen-3-ol) were recovered by both investigators. Although culture methods differed, both used non-defined media based on wheat (wheat flour or whole wheat bread). Although inconclusive, this comparison emphasizes the intuitive idea that direct extraction techniques might liberate VOCs not normally present in air.

C. Inter-Strain and Inter-Species Variability

Seifert and King (1982) were able to distinguish between three strains of *Aspergillus clavatus* by their VOC composition as measured by vacuum steam extraction directly from the fungus.

Batterman et al. (1991) compared VOC recoveries from *Penicillium chrysogenum* and *Cladosporium sphaerospermum*, using canister headspace

Table 12.3. Comparison of two studies using different sampling methods to study *Aspergillus flavus* VOCs.

	Kaminsky (1972) direct extraction	Harris et al. (1986) headspace analysis
3-Octanol	X	X
1-Octen-3-ol	X	X
4-Decanol		X
2-Methylisoborneol		X
1-5-Octadien-3-ol		X
Trimethyl hexene		X
Damascenone		X
3 Methyl-butanol	X	
3-Octanone	X	
1-Octenol	X	
cis-2-Octen-1-ol	X	

samples and GC-MS analysis. He classified recoveries as fungal tracers (compounds found in inoculated samples but not in controls), enhanced control emissions (compounds that were higher in inoculated samples than in controls), suppressed control emissions (compounds that were lower in inoculated samples than in controls), and unaltered control emissions (compounds found in both controls) and inoculated samples without significant difference). Fungal tracers identified included 3,3-dimethyl-2-oxetanone for *P. chrysogenum*, and 3,3-dimethyl-1-octene and ethyl-2,4-dimethylpentane for *C. sphaerospermum.*

Harris et al. (1986) compared *P. roqueforti, A. flavus, A. niger,* and *Botrytis cinerea* as well as two streptomyces (actinomycete) species with respect to headspace VOCs (Table 12.4). Effects of strain on VOC production in bacteria were reported by Rivers et al. (1992).

D. Effect of Substrate and Other Environmental Variables

Borjesson et al. (1990) found some differences in the VOC composition derived from headspace analysis of *Penicillium aurantiogriseum* grown on six agar substrates, as well as on coarse meal without agar. For example, the production of 8-carbon alcohols and 3-methyl-1-butanol was higher and the terpenes lower when the fungus was cultured on an oat grain medium rather than an agar medium. However, the VOCs produced on the various cereal-based substrates were generally similar. Rivers et al. (1992) report recovery differences related to substrate for bacteria.

Effects of substrates and environmental variables such as temperature, humidity, light, and the presence of other contaminants are poorly understood. In addition, emissions might differ significantly between environmental and laboratory studies due to the large number of species and substrates found in environmental settings. In particular, the emissions from fungi or bacteria

Table 12.4. Comparison of headspace VOCs from four fungi and two actinomycetes.

	P. roqueforti	*A. flavus*	*A. niger*	*B. cinerea*	*S. griseus*	*S. odorifer*
4-Decanol	X	X	X		X	X
2-Methylisorborneol	X	X			X	
3-Octanol	X					
1-Octen-3-ol	X		X			
1,5-Octadien-3-ol	X					
Trimethyl hexene	X					
Damascenone	X					
3-Methyl-1-butanol					X	X
Dimethyl trisulfide					X	X
Geosmin					X	X
Heneicosane					X	
Docasane					X	
Tricosane					X	
Tetracosane					X	
Pentacosane					X	
Hexacosane					X	
Heptacosane					X	

Based on Harris, N. D., Karahadian, C., and Lindsay, R. C., *J. Food Prot.,* 49, 964, 1986.

growing on heterogeneous substrates, such as coatings and printed paper, might differ from those reported here. Fungi and bacteria might release halogenated or possibly metallic compounds as a result of co-metabolic degradation of the substrate (in which the degrading organisms receive no benefit). These processes have been observed in other contexts (e.g., a sequence of partially dechlorinated by-products are found in the degradation of halogenated substances present in soils and groundwater at waste sites.

E. Conclusions with Respect to Biochemistry

Extracted VOCs were estimated in one study to comprise 5 to 15 ppm of the dry weight of the fungal mat (Seifert and King, 1982). A negative correlation between 1-octen-3-ol and phenolic content led Seifert and King (1982) to suggest that methyphenols may act as antioxidants that inhibit the formation of the alcohol. Rivers et al. (1992) reported that growth rate and metabolic activity influenced VOC production in bacteria.

In response to suggestions that the VOC composition changes with growth phase (Batterman, 1991), Rivers et al. (1992) proposed that metabolic changes occur as microorganisms (bacteria in this case) successively use different substrates: first simple sugars (producing ethanol as a by-product of glucose fermentation), then peptones and fatty acids (producing methyl mercaptan and other sulfur-containing compounds resulting from amino acid degradation). However, little information is available on the underlying metabolic processes, life cycle, and substrate interactions that affect or cause these emissions. Abramson et al. (1983), Borjesson et al. (1990), and others have indicated that the dominant fungal VOCs (low-weight alcohols) were always the same regardless of the substrate (based on tests using various kinds of grains). However, terpene production was most pronounced on artificial substrates, possibly as a result of a shortage of certain nutrients.

F. Odors and Microbial VOCs

Maximum odor production coincided with conidia maturity (Seifert and King, 1982). Harris et al. (1986) report recovery of some fungal volatiles in levels that exceeded odor thresholds. Using both experimental results and a literature review, these authors also characterized fungal odors into the following aroma categories:

Old musty paper-like, mildew-like mustiness due to a variety of unknown compounds, possibly including 1-octen-3-ol, produced by fungi including *A. niger, A. flavus,* and *P. roquefortii.*
Fresh mushroom-like mustiness due to 1-octen-3-ol, 8-carbon alcohols, and ketones, produced by *P. roquefortii, P. caseicolum,* and *A. niger.*
Obnoxious, heavy, cat-like mustiness due to unidentified VOC(s), produced by *A. flavus.*
Jute sack-like mustiness due to chloranisoles and chlorophenols (of industrial origin), produced by some *Penicillium* and *Aspergillus* species.
Snow pea pod-like mustiness due to 2-methoxy-3-isopropylpyrazine, produced by *P. caseicolum.*

Specific studies related to bacterial odors and VOCs have not been reported.

G. Relationship to Indoor Air

Borjesson et al. (1990) studied emission rates of both VOCs and CO_2 emitted by *P. aurantiogriseum.* By 4.5 days after inoculation, VOC emission rates ranged from 5 to 24 ng/hour (propanols and butanols), 10 to 170 ng/hour (sesquiterpenes), and 2,700 to 18,000 ng/hour for ethanol (probably released from the substrate). These rates apply to an inoculated area about the size of a petri dish.

In an indoor air study using Tenax® GC, thermal desorption, and GC-MS Miller et al., (1988) detected 3-methyl-1-butanol, 2-hexanone, and 2-heptanone, three compounds of possible fungal origin during a survey of 52 Canadian homes, most of which had been reputedly associated with health problems. The three compounds were found in 44, 89, and 89%, respectively, of the houses tested. In many cases, concentrations of these VOCs had little correlation with other measures of fungal activity (e.g., CFU concentrations). Only a few air samples were taken in each house.

In a laboratory study using direct cryogenic sampling and GC-MS analysis, Pleil (1991) identified VOC emissions from unidentified bacteria isolated from a residential air filter and grown in a trypticase soy growth medium.

In a field study, McJilton et al. (1990) found 1-butoxy-2-propanol and 2-methyl propionic acid in air samples in several buildings, and the same compounds in a pure culture of a bacterium isolated from the heating, ventilating, and air conditioning unit. The "red" bacteria were found in condensate water of a heat pump. The VOCs were dispersed throughout three buildings, and appeared to be related to odor and symptom complaints of building occupants.

Batterman et al. (1991) provided some evidence that fungi might modify "background" emissions of substrates (e.g., rates of VOCs outgassed from building surfaces may be altered by the presence of fungi). Mechanisms that could increase emissions include (1) opening of micro-channels by fungal hyphae penetrating into the substrate, (2) uptake and subsequent volatilization of the substrate material, and (3) fungal production and release of the same chemical. On the other hand, suppression of emissions might occur as (1) fungi presenting a diffusion barrier, (2) fungi providing a larger surface area increasing vapor deposition and thus decreasing airborne concentrations, and (3) the compound depleted by the fungus. All of this suggests that fungi (and possibly bacteria) would have adverse effects on the integrity and performance of coatings and finishes applied to building and furniture surfaces with the intention of minimizing outgassing and diffusion of VOCs, radon, and water vapor.

Most of the biogenic VOCs identified are not emitted by furnishings, building materials, cleaners, or other materials and products used in buildings. Seifert and King (1982), Borjesson et al. (1990), and others have noted that VOC emissions correspond well to fungal growth (e.g., as measured via fungal CO_2 production). If the biogenic compounds are unique, grab samples of indoor air suitably analyzed for VOC composition might provide the ability to predict and/or confirm biological growth. The VOC tracers would provide a diagnostic indicator that could be useful and advantageous in many applications. Results would be obtained quickly without culturing. VOC concentrations (or emission rates) might provide an indication of the metabolic status of the microbial contaminants. If VOC emission patterns were unique for the fungal or bacterial species and the substrate, then air monitoring could indicate this information and the possible location of contamination. On the other hand, air samples that were free of biogenic VOC "tracers" would indicate that large areas were free of active microbial growth. The selection of sampling sites would not be critical given the rapid dispersion of VOCs in air. Finally, as Borjesson et al. (1990) noted, tracer VOCs could be detected before any visible signs of microbial growth were apparent.

Several factors may limit the application of tracers of biological contamination. First, a reliable database or library is needed that contains VOC compositions and concentrations for combinations of species and substrates. Second, the analysis must account for the time evolution and physiological factors that may alter the emission rate and composition of biogenic VOCs (Rivers et al., 1992). Third, fungal and bacterial contamination may occur simultaneously (Borjesson et al., 1990) and there is likely to be a natural or "background" level of these VOC tracers present. These factors might tend to confound the use of tracers, although multivariate methods have been successfully used in similar situations [e.g., receptor modeling that apportions sources of ambient air pollutants (Koutrakis and Spengler, 1987)]. The low correlation found between fungal measurements and VOC measurements in one field study led Miller et al. (1988) to suggest that fungal VOCs may detect only the most serious cases of fungal contamination. However, this also points to the need for sampling techniques that account for spatial and temporal variability expected in the field measurements, and for the use of alternate VOCs or sets of VOCs that better correspond to biological activity. As sorption of the largely polar VOCs on interior surfaces will decrease concentrations below those obtained in laboratory studies with high loading rates, improved collection and analysis method are needed. While the practicality of using tracers has not been established, the technique might provide building investigators with a flexible diagnostic tool.

VI. SUMMARY AND RESEARCH DIRECTIONS

Taken together, the studies in the food, flavor, environmental, odor, and indoor air quality literature indicate that fungi and bacteria produce a variety of VOCs, including many polar and odoriferous compounds. These VOCs are certainly present in indoor air, but they appear to be minor constituents among many other VOCs found at much higher concentrations. Biogenic emissions appear to vary by species and strain, substrate, and metabolic activity. The most common compounds (those identified in five or more studies) include 3-methyl-1-butanol, 3-octanol (1-octan-3-ol), 3-octanone, and 1-octen-3-ol. Numerous other VOCs, including sulfur-containing compounds have been recovered from fungi and bacteria.

The available studies have several limitations. The studies are largely qualitative due in part to the significant technical challenges in sampling and identifying trace level polar VOCs. The sampling and analysis protocols used generally favored non-polar compounds for a variety of reasons that include the use of sorbents that have higher affinity for non-polar compounds, the use of permeable membrane dryers that also remove some polar VOCs, and the loss of polar VOCs on the surfaces of culture chambers. Also, all studies used a high loading rate (i.e., a large area of fungal growth compared to chamber volume) and a low flow rate near the fungi, two factors that tend to decrease emission rates that are diffusion-dependent. The different sampling and analysis methods employed in the studies tend to limit the relevance of interstudy comparisons.

Available studies are also limited with respect to the species and substrates used. All of the laboratory studies used artificial (agar or grain) substrates and controlled environments, and most used pure cultures of fungi or bacteria. Only a few studies have attempted to identify fungal or bacterial VOCs in nonlaboratory settings, and only one (McJilton et al., 1990) attempted to confirm the VOCs at the probable source location.

These limitations in the available literature make the use of VOCs as tracers currently speculative. The deficiencies in the available studies also indicate a range of research opportunities. Improved experimental protocols are needed, including the formulation of standardized procedures for the confirmation of compounds and odors, the quantification of emission rates, and the assessment of effects of temperature, humidity, species, substrate, and possible control measures. A mechanistic understanding of the VOC emission process is lacking. Simultaneous examination of VOCs released from single organisms and substrates as well as from mixed populations would be helpful. Finally, studies are needed to address potential health and comfort impacts of biogenic VOCs, and to establish the relationship between laboratory chamber measurements and actual environmental concentrations. Such studies might also address the use of tracers of biological contamination in buildings.

REFERENCES

Abramson, D., Sinha, R. M., and Mills, J. T., Mycotoxins and odor formation in barley stored at 16 and 20% moisture in Manitoba, *Cereal Chem.*, 60, 350, 1983.

ASTM, Standard guide for small-scale environmental chamber determinations of organic emissions from indoor materials/products, American Society for Testing and Materials, Philadelphia, PA, 5116, 1991.

Bartley, J. P. and Schwede, A., Volatile flavor components in the headspace of the Australian or Bowen Mando, *J. Food Sci.*, 52, 353, 1988.

Batterman, S., Bartoletta, N., and Burge, H., Fungal volatiles of potential relevance to indoor air quality, Paper 91-62.9, presented at the Annual Meeting of the Air and Waste Management Association, Vancouver, Canada, June 1991.

Borjesson, T., Stollman, U., and Schnurer, J., Volatile metabolites and other indicators of *Penicillium aurantiogriseum* growth on different substrates, *Appl. Environ. Microbiol.*, 56, 3705, 1990.

Burge, H. A. and Hodgson, M., Health risks of indoor pollutants, *ASHRAE J.*, July, 1988.

Gholson A. R., Storm, J. F., Jayanty, R. K., Fuerst, R. G., Logan, T. J., and Madgett, M. R., Evaluation of canisters for the collection and storage of air toxics, *Anal. Chem.*, 62, 1899, 1990.

Goller, J. W., Displacement of polar by non-polar organic vapors in sampling systems, *Am. Ind. Hyg. Assoc. J.*, 46, 170, 1985.

Harris, N. D., Karahadian, C., and Lindsay, R. C., Musty aroma compounds produced by selected modes and actinomycetes on agar and whole wheat bread, *J. Food Prot.*, 49, 964, 1986.

Hawthorne, S. B., Kriefer, M. S., and Miller, D. J., Analysis of flavor and fragrance compounds using supercritical fluid extraction coupled with gas chromatography, *Anal. Chem.*, 60, 472, 1988.

Kaminski, E., Stawicki, S., and Wasowicz, E., Volatile flavor compounds produced by molds of *Aspergillus, Penicillium*, and *Fungi imperfecti, Appl. Microbiol.*, 27, 1001, 1974.

Kaminski, E., Libbey, L. M., Stawicki, S., and Wasowicz, E., Identification of the predominant volatile compounds produced by *Aspergillus flavus, Appl. Microbiol.*, 24, 721, 1972.

Kirk, B. D., Analysis of cereal products by dynamic headspace/gas chromatography, presented at the Pittsburgh Conference on Analytical Chemistry and Applied Spectroscopy, New Orleans, LA, February 1988.

Kirk, B. D., Volatile organic profiles of processed foods, presented at the Pittsburgh Conference on Analytical Chemistry and Applied Spectroscopy, Atlantic City, NJ, March 9 to 13, 1987.

Koutrakis, P. and Spengler, J. D., Source apportionment of ambient particles in Steubenville, OH using specific rotation factor analysis, *Atmos. Env.*, 21, 1511, 1987.

Kreiss K., The epidemiology of building-related complaints and illness, *Occup. Med. State Art Rev.*, 4, 575, 1989.

Lin, J. C and Jeon, I. J., Headspace gas sampling/GC method for the quantitative analysis of volatile compounds in cheese, *J. Food. Sci.*, 50, 843, 1985.

MacLeod, G. and Ames, J. M., Comparative assessment of the artifact background on thermal desorption of Tenax GC and Tenax TA, *J. Chromatog.*, 355, 393, 1986.

McJilton, C. E., Reynolds, S. J., Streifel, S. K., and Pearson, R. L., Bacteria and indoor odor problems — three case studies, *Am. Ind. Hyg. Assoc.*, 51, 545, 1990.

Miller, J. D., LaFlanme, A. M., Sobel, Y., LaFontaine, P., and Greenhalph, R., Fungi and fungal products in some Canadian homes, *Int. Biodeterioration*, 24, 103, 1988.

Narain, N., Hsieh, T., and Johnson, C. E., Dynamic headspace concentration and gas chromatography of volatile flavor components in peaches, *J. Food Sci.*, 55, 1303, 1990.

Olafsdottir, G., Steinke, J. A., and Linday, R. C., Quantitative performance of a simple Tenax-GC adsorption method for use in the analysis of aroma volatiles, *J. Food Sci.*, 50, 1431, 1985.

Otto, D., Mollhave, L., Rose, G., Hudnell, H. K., and House, D., Neurobehavioral and sensory irritant effects of controlled exposure to a complex mixture of volatile organic compounds, *Neurotox. Teratol.*, 12, 1, 1990.

Pankow, J. F., Technique for removing water from moist headpace and purge gases containing volatile organic compounds, application in the purge with whole column cryotrapping (P/WCC) method, *Environ. Sci. Technol.*, 25, 123, 1991.

Pleil, J. D., Oliver, K. D., and McClenny, W. A., Enhanced performance of Nafion dryers in removing water from air samples prior to gas chromatographic analysis, *J. Air Pollut. Control. Assoc.*, 37, 244, 1987.

Pleil, J. D., Demonstration of a valveless injection system for whole air analysis of polar VOCs, in *Proc. 1991 Int. Symp. Measurement of Toxic and Related Air Pollutants*, Air and Waste Management Association, Pittsburgh, PA, 1991.

Rivers, J. C., Pleil, J. D., and Wiener, R. W., Detection and characterization of volatile organic compounds produced by indoor air bacteria, *J. Exp. Anal. Environ. Epidemiol.*, Suppl. 1, 177, 1992.

Rothweiler, H., Wagner, P. A., and Schlatter, C., Comparison of Tenax TA and Carbotrap for sampling and analysis of volatile organic compounds in air, *Atmos. Environ.*, 25B, 231, 1991.

Samet, J. M., Marbury, M. C., and Spengler, J. D., Health effects and sources of indoor air pollution (Parts 1 and 2), *Am. Rev. Respir. Dis.*, 136, 1486, 1987; 137, 221, 1988.

Seifert, B., and Abraham, H. J., Use of passive samplers for the determination of gasous organic substances in indoor air at low concentration levels, *Int. J. Environ. Anal. Chem.*, 13, 237, 1983.

Seifert, R. M. and King, A. D., Jr., Identification of some volatile constituents of *Aspergillus clavatus, J. Agric. Food Chem.*, 30, 786, 1982.

Wallace, L., Volatile organic compounds, in *Indoor Air Pollution: A Health Perspective,* Samet, J. M. and Spengler, J. D., Eds., Johns Hopkins University Press, Baltimore, MD, 1991a.

Weschler, C. J., Shields, H. C., and Ranier, R., Concentrations of volatile organic compounds at a building with health and comfort complaints, *Am. Ind. Hyg. Assoc. J.*, 51, 2161, 1988.

Westendorf, R. G., Thermal desorption of adsorbent traps, *Am. Lab.*, Dec. 1985.

Williams, E. J. and Sievers, R. E., Synthesis and characterization of a new sorbent for use in the determination of volatile, complex-forming organic compounds in air, *Anal. Chem.*, 56, 2523, 1984.

Wood, W. F., DeShazer, D. A., and Largent, D. L., The identity and metabolic fate of volatiles responsible for the odor of *Hydnellum suaveolens, Mycologia*, 80, 252, 1988.

13 BIOSTATISTICS AND BIOAEROSOLS

Lynn Eudey, H. Jenny Su, and Harriet A. Burge

CONTENTS

0-87371-724-4/95/$0.00+$.50

I. INTRODUCTION

Jerrold Zar, in his book *Biostatistical Analysis* (1984) quotes Mark Twain as saying "There is something fascinating about science. One gets such a wholesale return of conjecture out of a trifling investment of fact." Statistics is the science of making sure that such conjecture is accurate and meaningful.

Statistical analysis has become a routine part of nearly all research efforts. However, problem-solving investigations directed at bioaerosols often fail to take into account the statistical principles that allow conclusions to be drawn from a well-designed research study. For example, a few samples may be collected in a building, analyzed, and conclusions drawn from the resulting data without consideration of the representativeness of the sample, or the variability inherent in the underlying population. While it is often not possible to conduct problem-solving investigations at a level where formal statistical analysis is possible, it is necessary to consider statistical principles in drawing conclusions from even the smallest study. To this end, and also in the hope of making some useful statistical methods available to investigators new to the field of bioaerosol research, we present this chapter.

Statistics involves methods for collecting data from a sample in order to draw inferences about a well-defined population (in other words, collecting a few facts about which we can make many conjectures). These methods include the design of the experiment or survey that defines the sample; the collection of sample data from the population; the organization, description, and summarization of the data; and the process of drawing inferences from the data. Inherent in the inferential step is a measure of the reliability of the inference, which is based on the variability of the data. In addition, statistics encompasses the exploration and modeling of the data to gain insight about the structure of the population. The major steps to a statistical analysis can be written as follows:

1. Formulate a research question.
2. Design the experiment (experimental design).
3. Collect and organize the data (data management).
4. Describe and summarize the data (descriptive statistics).
5. Perform an inferential analysis (inferential statistics).
6. Address the research question.

This chapter presents a brief overview of some procedures used within descriptive statistics and inferential statistics. We have not covered aspects of experimental design and sampling that are essential in planning a well-designed statistical study. A poorly designed sampling scheme can yield a worthless answer from an eloquently performed inferential analysis. The key is to carefully consider the research question (or investigational problem) in light of the kind of information needed to draw conclusions. Further information on experimental design can be found in *Statistical Design and Analysis of Experiments* (John, 1971). A very readable text on the basic concepts behind experimental design and inferential statistics can be found in *Statistics*, Freedman et al., (1991). A good text on some of the methods used in health sciences can be found in *Biostatistics: A Foundation for Analysis in the Health Sciences,* W. W. Daniel (1991).

II. SOME DEFINITIONS

A. Populations, Variables, Samples, Data

A population (statistically speaking) is the universe of all possible measurements of a variable. Note that a variable is a measurable characteristic of the elements of a population (bivariate means there are two and multivariate more than two variables under consideration for each element). For example, all of the possible measurements of the variable "total colony forming units of fungi per cubic meters of air (CFU/m^3)" under specific conditions constitute a population. Unfortunately, because the number of measurements that can realistically be performed is less than "all", one must rely on a sample, a set of some measurements to describe the population of all measurements. Therefore, we rely on the hope that the measurements (data) resulting from a sampling experiment constitute a sample that is representative of the population. It is intuitively obvious that the larger the sample, the more likely it is to be representative of the population. Note also that many different samples (of the same size and with the same sampling design) can usually be collected from a single population. (The word "sample" will be used in this chapter in a statistical context, to mean a subset of the well-defined population.)

B. Parameters and Statistics

A parameter is a numerical measurement of the population while a statistic is a numerical measurement of a sample which has been selected from this population. Usually, statistics are represented by Latin letters while parameters use Greek letters. Thus, the population mean [μ] is a parameter, while the sample mean [$\bar{x}$] is a statistic. One can usually calculate the statistic (e.g., sample mean), and from that statistic infer the parameter (e.g., population mean). Note here that each sample taken from a population has its own characteristic statistics. For example, the mean of one sample will be different to some degree from the mean of any other sample drawn from the same population. For this reason, a statistic is in itself a variable.

C. Random Sampling

Most statistical principles assume that samples are randomly selected from a population, and that selection of the sample does not change the population. Bioaerosol sampling is never truly random because both collectors and analytical methods are biased for particular aerosol characteristics. The act of collecting bioaerosol samples also changes the population. For example, each sampling event permanently removes some individuals from the population. For populations that are small relative to the sampling rate, this removal changes the population during sampling and, for enclosed populations, results in underestimates of concentrations. Recognizing and minimizing these biases is an important part of experimental design. In spite of these problems, we are usually forced to do the best we can with experimental design, and consider our samples close to random.

III. DESCRIPTIVE STATISTICS

Descriptive statistics allows a data set (or data sets) to be summarized and presented in ways that provide insight into its distribution, and often some indication of relationships between data sets (i.e., between samples). The methods presented in this section are as they apply to sample data. They can also be used, after a few minor changes, for population data.

Most data of interest in bioaerosol research can be placed on a ratio scale. A ratio scale has a constant interval between measurements (i.e., the difference between 9 and 10 is the same as the difference between 100 and 101), and the scale always has a zero point. Nominal scales are also used in bioaerosol data. Nominal simply means name, and applies to the names of, for example, fungus taxa that are identified in bioaerosol samples. Statistical methods appropriate for ratio scale (quantitative) data are usually not appropriate for nominal (qualitative) data.

A. Tabular and Graphical Methods of Presenting Qualitative Data

Frequency distribution tables. Qualitative data (i.e., data that is represented by categories rather than numbers) can be tabulated by category and the respective frequency of each category. The categories chosen must be exhaustive of all possibilities for the variable and must not overlap. For example, two questions were asked in 100 households: "Does the house have a gas stove?" and "Is there a resident smoker?". Data resulting from answers to these questions can be put into a table (Table 13.1).

Table 13.1. Frequency distribution based on responses from 100 households.

	Frequency	Cumulative frequency
Gas stove, no smoker	23	23
Gas stove, smoker	35	58
No gas stove, no smoker	17	75
No gas stove, smoker	25	100

Using frequency tables to compare samples taken from different populations requires either that the size be the same for each sample, or that the scale be relative frequency. Relative frequency is the frequency of a category divided by the total sample size. For example, the question "Is there mold growth in your house?" was asked of 200 households in City No. 1 and 150 households in City No. 2. These data are tabulated in Table 13.2. Note that comparison between the two cities for comparison of mold growth is appropriate in the relative frequency scale but not in the frequency scale.

Table 13.2. Survey of mold growth in households.

	City #1			City #2		
	Frequency	Relative frequency	Percent	Frequency	Relative frequency	Percent
Mold growth	50	.25	25	50	.33	33
No mold growth	150	.75	75	100	.67	67
Total	200	1.00	100	150	1.00	100

Bar charts (or graphs) and pie charts. For bar charts, either of the frequency scales (frequency or relative frequency) can be presented on one axis and the exhaustive categories of interest on the other. The data from Table 13.1 is presented as a bar chart in Figure 13.1.

For a pie chart, the categories are represented by an appropriately sized slice of a circle. Each category of a variable is identified with a relative portion of the circle which is the same as the relative frequency for that category.

Multivariate data. A frequency distribution table can be extended to include more than one category by constructing a table that has a column categorization for one of the variables and a row categorization for another of the variables. This is called a bivariate frequency distribution for two variables because it represents a joint distribution of the two variables. Table 13.3 contains the same data as Table 13.1 presented as a bivariate frequency distribution.

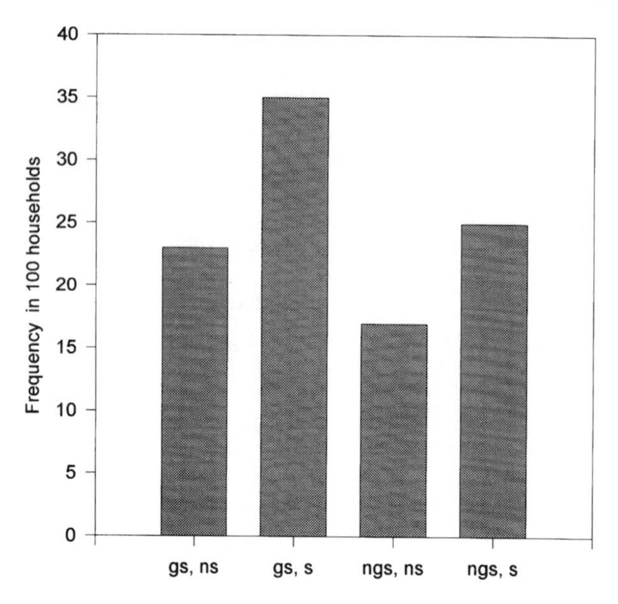

Figure 13.1. Bar graph of frequencies from Table 13.1

Table 13.3. Bivariate frequency distribution based on responses from 100 households.

	Smoker		No smoker		
	Frequency	**Relative frequency**	**Frequency**	**Relative frequency**	**Total**
Gas stove	35	.35	23	.23	58
No gas stove	25	.25	17	.17	42
Total	60	.60	40	.40	100

Note that tables representing the joint distribution of more than two variables are rather cumbersome because they either collapse the higher dimension variables (third on up) or they present a series of bivariate tables.

B. Tabular and Graphical Methods of Presenting Quantitative Data

Frequency distribution tables. Quantitative data can also be organized into a table. If the data are discrete and do not take on many possible values, then the methods described for qualitative data can be applied directly. If the data are continuous, or discrete with many possible values, then the data can be divided into non-overlapping, exhaustive intervals. The number of intervals is not important, but too few will group the data too much and too many will cancel out the advantages of grouping the data.

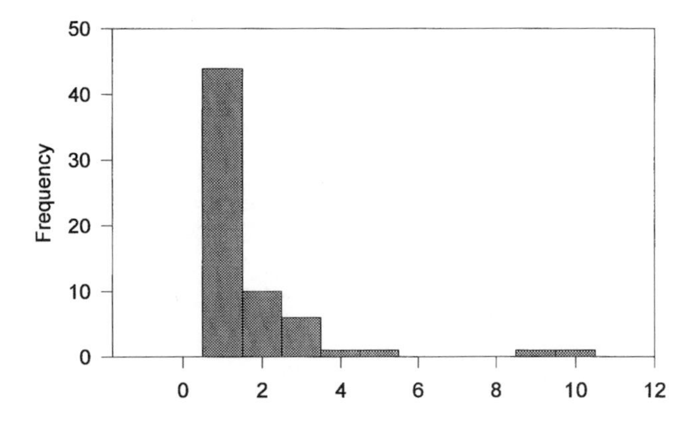

Figure 13.2. Histogram of Burkard spore trap counts.

Histograms. Histograms are bar graphs of frequency distribution tables containing quantitative data. Figure 13.2 is a histogram displaying frequency distributions of fungal spore counts from air samples collected with a Burkard spore trap in an office building over 6 months.

Bivariate frequency distribution can be presented as a two dimensional display of a three dimensional graph. Generally two horizontal axes are drawn in perspective perpendicular to one another. The vertical axis represents the frequency scale. Multivariate frequency distributions can be represented pictorially as a series of bivariate histograms.

Stem and leaf display. The stem and leaf display is similar to a histogram but the data is not grouped into intervals so that all the detail is retained. Data from Andersen culture plate samples in homes in Topeka, Kansas is shown as a stem and leaf display in Figure 13.3. The data are rounded to the hundreds for the display (i.e., 2429 shows as stem 2 leaf 4).

Scatter diagrams. Scatter diagrams give a good visualization of how quantitative bivariate data are distributed. These diagrams are often used with bivariate data because both the dispersion of each variable and the relationship between the two variables are displayed. Scatter diagrams should always be the first step in learning about the association between two variables. Bivariate scatter diagrams are constructed on a Cartesian graph where the horizontal axis represents one of the variables and the vertical axis represents the other. If there are repetitions of data points, the symbol used should reflect this fact. For example, the scatter diagram represented in Figure 13.4 is derived from a comparison of culturable fungi recovered by two different culture media from dust in domestic interiors, where X = the natural log of CFU/m^3 of fungi recovered on dichloran glycerol agar, and Y = the natural log of CFU/m^3 of fungi from the same dust recovered on malt extract agar.

```
          STEM AND LEAF PLOT OF VARIABLE:      CFU    , N =    590

MINIMUM IS:          0.000
LOWER HINGE IS:        1286.000
MEDIAN IS:        2429.000
UPPER HINGE IS:        4357.000
MAXIMUM IS:      42500.000

                     0    00111112222223334444
                     0    55555555555556666666666677777777777777777777777777777788888*
                     1 H  0000000000000000000000111111111222222222222222222222222223333*
                     1    555555555666666666666777777777777777777777777778888888888888*
                     2 M  00000000000111111111111222222222222333333333333444444444
                     2    55555555566666677777777777777777777888888888899 9
                     3    00000000000011111112223333344444
                     3    555555555566666677777788888888
                     4 H  0000000001111122222222222222333
                     4    5555555555666677777778
                     5    0000000111122222444
                     5    55555555558888999
                     6    000000222222334
                     6    5556667777999
                     7    0022244
                     7    555667788
                     8    00024
                     8    555777777
          ***OUTSIDE VALUES***
                     9    035778
                    10    5577
                    11    029
                    12    022
                    13    24
                    14    4579
                    15    07
                    16    9
                    26    58
                    27    7
                    38    1
                    41    8
                    42    5
```

Figure 13.3. Stem and leaf plot of culture plate sampling data from homes in Topeka, Kansas.

Examination of the scatter diagram can indicate the presence, or absence of a relationship between the two variables, and may suggest the need for calculating a correlation coefficient (see below).

Box plots. Box plots are extremely valuable tools in preliminary data analysis because they give an overview of the data distribution, the measures of central tendency (see below), and when more than one variable is plotted, a visual tool to indicate differences (or relationships) between variables. The relationship between a continuous variable and either a discrete or qualitative variable can be shown by constructing a box plot of the continuous variable within each value of the discrete or qualitative variable. Figure 13.5 is an example where airborne bacterial levels for terminals are compared with airborne levels during cruises for commercial airlines.

C. Summary Measurements

Qualitative data can be summarized numerically by using the relative frequencies or proportions of the categories of a variable. The mode (see

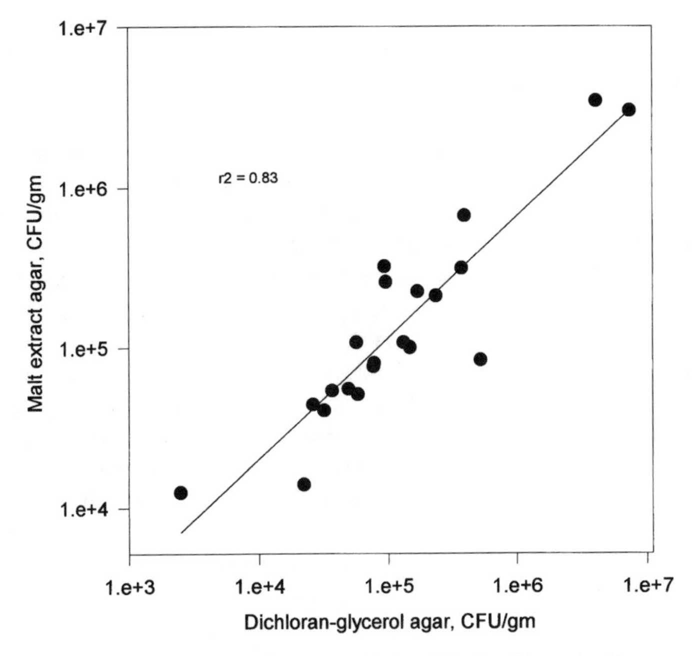

Figure 13.4. Scatter plot of data comparing two kinds of fungal culture media.

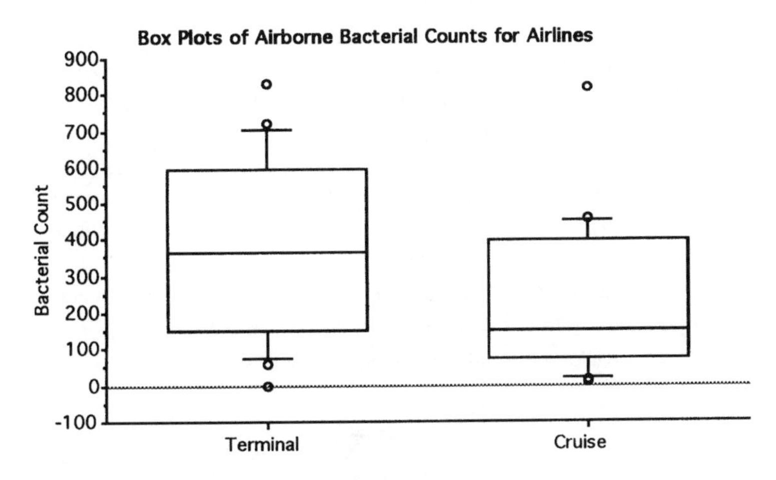

Figure 13.5. Box plots of airborne bacterial counts for terminals and cruises for commercial airlines.

below) can also be used for qualitative data. However, quantitative data lends itself more easily to summary measurements because it is inherently numerical. Summary measurements to describe central tendencies, variability, relative standing, and correlations are presented here, but the coverage is by no means exhaustive, and reference can be made to Daniel (1991) or a variety of other sources for further information.

Measures of central tendency. Measures of central tendency indicate where on the number line the data is centered. Commonly used measures include the mode, the median, the arithmetic mean or average, and the geometric mean.

The mode is used as a measure of central tendency for qualitative and discrete quantitative data, and is the most frequently observed category or count. When continuous data has been grouped into intervals, the mode equals the most frequently occurring interval.

The median is the value that divides the data into halves (see Figure 13.3). One must first order the data, which can be done by constructing a stem and leaf display. The median is generally used when an underlying population distribution shape is not assumed, and is the measure of central tendency used in nonparametric statistics (see below). It is also often used with data that has a highly skewed distribution, i.e., there exists a sizeable portion of the data with either extremely high values, or extremely low values relative to the rest of the data.

The arithmetic mean (commonly called the mean or the average) is by far the most commonly used measure of central tendency, and is usually designated by a horizontal bar over the variable (e.g., $\overline{x}$). It is the sum of all the data values divided by the sample size (n). The mean, in contrast to the mode and the median, is extremely sensitive to extreme values since every value in the data set is weighted equally. The term "mean" is used both to refer to the mean of a sample ($\overline{x}$) and the mean of a population (μ). It is the measurement of central tendency used in parametric statistics.

The geometric mean is defined as the antilog of the arithmetic mean of the logs of the data. In other words, each data point is transformed to log form, the arithmetic mean of these logs is calculated, and the geometric mean is the antilog of this value. Geometric means can only be used when all data points are positive. Actually, geometric means are often calculated by adding 1 to each data point, taking the log (either common or natural), finding the mean of the log values, deriving the antilog, and subtracting 1.

$$\overline{X}_{geometric} = anti\log\left(\frac{1}{n}\sum_{i=1}^{n}\log(x_i+1)\right)-1 \qquad (1)$$

The geometric mean is often used for bioaerosol data when the sample distribution is most accurately described as log-normal (i.e., when the logs of the data more closely approximate a normal distribution than the data itself).

Measurements of dispersion or variability. Everyone who hopes to interpret bioaerosol data must recognize that the value of a variable (or measurement) within a population varies, and that the reliability of any single data point or statistic as a predictor of a population parameter(s) is entirely dependent on this variability. Measuring the variability present in the sample gives indication of the variability present in the population that generates the data. The variability inherent in the data will indicate the accuracy of estimation and the power of testing (see section on power).

Various methods are used to measure the variability in a data set; the more frequently used measures are described below. We first describe the measures that are commonly (but not exclusively) used with nonparametric statistics (range and inner quartile range) and then go on to describe those used with parametric statistics (variance, standard deviation, and coefficient of variation). The basic difference between parametric and nonparametric statistics is that in parametric statistics we assume that we know the shape of either the population distribution, or of the sampling distribution. (These ideas are dealt with in more detail later in the chapter.)

One simple measure of the dispersion in the data is the range. The range is the number of units needed to span the data. As this is easily influenced by either extremely high or extremely low values, a more reliable measure is the inner quartile range (IQR) which is the number of units needed to span the middle 50% of the data. Note that, for either of these measures, we need to first order the data (as we did for the median). A percentile is a way to locate a data value relative to the rest of the data set or relative to another data set. For example, 95% of the data lies at or below the 95th percentile. The median is the 50th percentile, and the 25th, 50th, 75th percentiles are called the 1st, 2nd, and 3rd quartiles, respectively because they divide the data set into quarters. So now we can define the inner quartile range as: IQR = 3rd quartile – 1st quartile. The box plot (Figure 13.5) shows (from inside to outside) the median, the first and third quartile (within the box), the range of data falling within the inner fence (the whiskers), and data falling beyond the inner fence (open circles). The box part is drawn from the first quartile to the third quartile. The inner fences are a distance of 1.5 IQRs out from the quartiles.

Variance and standard deviation are based on the assumption that the arithmetic mean is the appropriate measure of central tendency. The variance is the average of the squared distances from the data values to the mean. The sample variance is:

$$S^2 = \frac{1}{n-1}\sum_{i=1}^{n}\left(x_i - \bar{x}\right)^2 \qquad (2)$$

The sample standard deviation is the square root of the sample variance. The standard deviation has the same units as the data, and its magnitude depends on the variability of the data. If two data sets are centered at a similar location, the data set with the larger standard deviation will have more variability in the data. Population variance is symbolized by σ^2, and the population standard deviation by σ. (The formula for population variance differs slightly in that we divide by N, the size of the population, rather than N-1.)

The coefficient of variation is the standard deviation divided by the mean. This value is usually multiplied times 100 for convenience:

$$V = \frac{S}{\overline{X}} \cdot 100 \tag{3}$$

The coefficient of variation has no units and is not affected by magnitude. The coefficient of variation is extremely useful for bioaerosol data where the standard deviation tends to increase as measurement values increase.

Measurements of relative standing. These measurements indicate where a datum lies with respect to a data set (either the data set from which it arose or another data set). As mentioned above one such measure is percentile. The percentile of a datum indicates what percent of the data set has values less than or equal to that datum. When using the mean and the standard deviation as measures of central tendency and variability, respectively, the Z-score is also used as a measure of relative standing. The Z-score measures the distance (and direction) from the specific datum to the mean of the data set in units of standard deviations:

$$Z = \frac{(x_i - \bar{x})}{SD} \tag{4}$$

Thus, Z-scores are close to zero near the center of the data and far away from zero at the extremes of the data. When the data is assumed to have come from a normal distribution (bell shaped) the percentile of the specific datum can be found by using the Z-score and tabled percentiles of the standard normal distribution (often this can be requested from a statistical computing package as well).

Outliers. Outliers are unusual data values. They lie far away from the center of the data set. Data values that lie outside the outer fences (i.e., three inner quartile ranges away from the center) on a box plot are outliers. Data values falling between the inner and outer fences are called mild outliers. When the data has a histogram which is bell shaped we often use the Z-score as an indicator of outliers. Z-scores that are larger than 3.00 in absolute value are considered outliers. If outliers are present in the data set one must first determine their validity. If possible, the original data should be verified to insure that a recording error is not responsible for the outlier(s). If a true outlier is found, one must ask whether or not the value should be discarded as a random case that has a very slight chance of re-occurring, or whether the value indicates a potential new trend in the system.

D. Relationship between Variables

Graphical methods. As discussed above, scatter diagrams can be examined for apparent relationships between two continuous variables, and should always be the first step in exploring such associations (see Figure 13.4). Box plots also provide a visual method for examining the relationship between two variables, one continuous, the other discrete or qualitative (see Figure 13.5).

Summary measures. Covariance measures the average degree of common variation between the two variables in a bivariate data set. The bivariate data point (x,y) deviates from the mean of the X's and the mean of the Y's by the deviations $(x - \bar{x})$ and $(y - \bar{y})$, respectively. The covariance is the average of the products of the variable's respective deviations from their means. The formula for the sample covariance is:

$$COV(X,Y) = \sum_{i=1}^{n} \frac{(x_i - \bar{x})(y_i - \bar{y})}{n} \tag{5}$$

Pearson's correlation coefficient. Correlation is commonly used to measure the strength of an association (dependency) between two variables. Pearson's correlation coefficient for the population is denoted ρ and it measures the strength of the linear relationship between two variables in the population. Its value falls between –1 and +1 inclusive. An absolute value of 1 indicates a perfect linear relationship (i.e., all the data in the population lie on a straight line). If ρ is positive, then X and Y increase (or decrease) together, if ρ is negative then when X increases Y decreases (or vice versa). The closer ρ is to either + or –1, the stronger the linear relationship is between the two variables. If ρ is 0 there is no linear relationship between X and Y and the variables are called (linearly) uncorrelated. Note that this does not imply independence. Independent variables are uncorrelated, but it is possible to construct a pair of uncorrelated variables that are not independent. Pearson's sample correlation coefficient, denoted r, measures the strength of the linear relationship between two variables in the sample:

$$r = \frac{COV(X,Y)}{SD(X) \cdot SD(Y)} \tag{6}$$

Statistical inferences based on r assume that the data are independent observations from a bivariate normal distribution (see Section V). Basically, even when we do not assume the "ideal" bivariate normal distribution we use r as an indication of the strength of the linear relationship between X and Y inherent in the population.

Spearman's correlation coefficient. Spearman's sample rank correlation coefficient (r_s) measures the strength of association between two variables in a bivariate sample of continuous (or at least ordinal) data and is the most frequently used nonparametric correlation coefficient. It can be applied to count data, and to continuous data which obviously deviate from the normal distribution. It is calculated on the relative ordering of X and Y, respectively (the ranks of data are discussed in Section V, Part E). The properties of Spearman's rank correlation coefficient include:

- r_s falls between –1 and +1 inclusive
- $r_s > 0$ indicates that large values of X tend to occur with large values of Y.
- $r_s < 0$ indicates that small values of X (or Y) tend to occur with large values of Y (or X).
- $r_s = 1$ if the ordering of the Xs and Ys is the exactly the same within each pair (x,y).
- $r_s = -1$ if the ordering of the Xs and Ys is the exact opposite within each pair (x,y).

Spearman's correlation coefficient is especially useful for bioaerosol data because such data is often not normally distributed and is likely to contain extreme values. It is also extremely useful for comparing rank ordering of taxa between populations (done by assigning a number to each taxon). The reader is referred to a standard text on nonparametric statistics, such as Daniel (1991) or Walsh (1962), for more details on the calculation of r_s and the inferences about Spearman's rank correlation coefficient for the population.

IV. RANDOM VARIABLES AND THEIR DISTRIBUTIONS

A. Characteristics of Variables

A random variable is a measurable characteristic of the elements of a population. A data set consists of observed values of a random variable (or random variables). Random variables can be discrete (assuming a countable number of values) or continuous (assuming values in an interval or intervals, i.e., an infinite number of values). Because discrete random variables only have a countable number of values, each specific value attains a probability between 0 (if it never occurs) and 1 (if it always occurs). Hence, the statement: "The probability is 0.06 that we will find 150 of the 350 houses having mold on interior surfaces." makes sense. On the other hand, continuous random variables take on an infinite number of values so that the probability of one specific value occurring is always infinitely small, essentially 0. Often a discrete variable may take on so many values that it is regarded as continuous. For example, counts of biological particles (e.g., pollens or spores) in an air sample are technically discrete but are often treated as continuous. Therefore, the question: "What is the probability of measuring a microbial concentration of exactly 478.23 CFU/m^3?" is inappropriate, because the answer will always be extremely small (for truly continuous data, the answer would always be 0). However, the question: "What is the probability of measuring a microbial concentration between 475 and 480 CFU/m^3?" is more appropriate and will result in an answer between 0 and 1 inclusive.

B. Distributions of Random Variables

Distributions are, essentially, descriptions of the frequency distributions discussed above. Data plotted in a histogram, or a stem and leaf plot, take on a shape that is known as the distribution. Some of these shapes are extremely common and important in statistical analysis. Technically, the distribution of a discrete random variable in some manner "lists" all the possible values of the variable and the corresponding probabilities (chances) that the variable will realize each of the values. The distribution of a continuous random variable shows all the possible values for the variable and the probabilities that the variable will take on a value within intervals.

The normal distribution. The normal distribution is a continuous distribution which often occurs when the variable being measured represents a cumulative response to many effects. It is a continuous, symmetric distribution with a smooth bell shape. The two parameters for the normal distribution are the arithmetic mean, indicating the central location of the distribution, and the variance (or the standard deviation which is the square root of the variance), indicating the spread of the distribution. Normal distributions are symmetric about the mean. The standard normal distribution has a mean of 0 and a standard deviation of 1. Any normally distributed random variable can be transformed to a standard normal distribution by computing the Z-score. Thus, any probability question about a normal random variable can be transformed into a question about a standard normal random variable. The probabilities of a standard normal random variable being between two values are tabled extensively and easily accessed. Because a normal random variable is a continuous random variable, only probability statements concerned with an interval

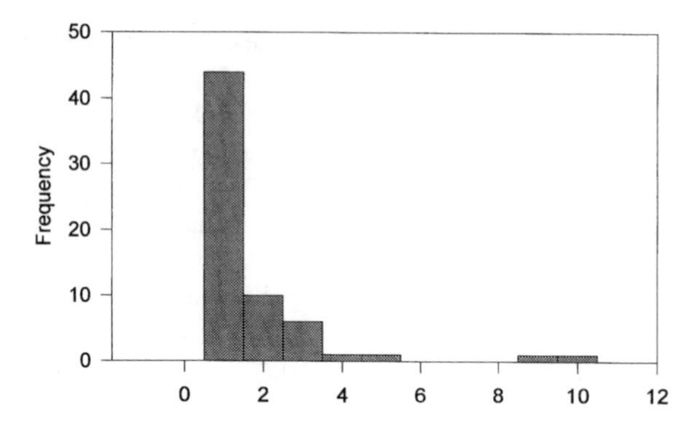

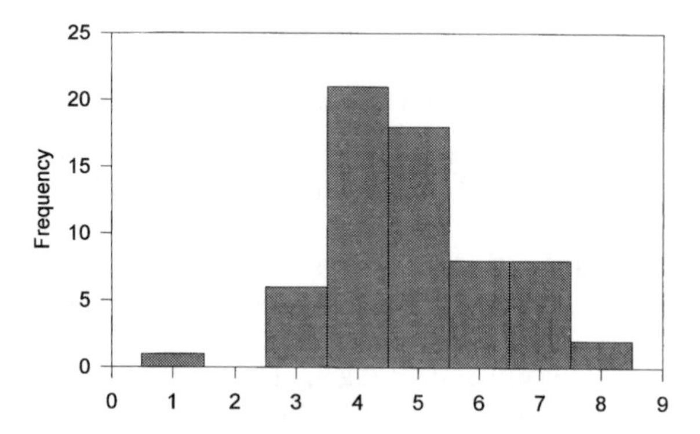

Figure 13.6. (A) Histogram of data from 64 culture plate air samples. (B) Histogram of data from part A after log transformation.

of values, rather than a single value, will make sense. The normal distribution plays a key role in parametric statistics, as the reader will witness.

Logarithmic distributions. Most bioaerosol data, when plotted in a histogram, does not form a symmetric bell-shaped distribution characteristic of the normal distribution. Further, graphical presentations of such data are often impossible on an arithmetic scale because of the prevalence of extreme values. Often, the use of a log scale will render the data more accurately. In many cases, if each data point is changed (transformed) into its log value, and these log values are plotted into a histogram, the resulting distribution is much more symmetric and bell shaped. Often, the transformed data has a distribution which is close enough to the normal distribution so that parametric statistical methods can be used. Figure 13.6A presents the histogram of the frequency of colony forming units of fungi per cubic meter of air measured using the

Burkard personal culture plate impactor in homes (Su et al., 1992) and Figure 13.6B shows a histogram of the same data after it has been transformed. If log transformation results in a normal distribution, then the original data are said to have a log-normal distribution.

C. Sampling Distributions

Introduction. A sampling distribution is the theoretical distribution of a sample statistic. Recall that a sample statistic is a numerical measure of a sample (e.g., the sample mean). For any given sample size, n, there are many possible different samples that can be selected from the population (the population remains fixed). Hence, theoretically, we can create a population of possible values of the sample statistic of interest. The sampling distribution is the distribution of this population. For example, suppose that we have a population of the numbers {1, 2, 3, 4, 5, 6, 7, 8, 9, 10} and we are interested in the population mean (we know that $\mu = 5.5$ but we use this as an illustration of the principles). If we are to take samples of size 2, with replacement of values, then there are 100 possible samples each with its own sample mean (e.g., a sample of (9,4) would have a sample mean of 6.5 and sample of (7,3) would have a sample mean of 5). If we were to list all of the possible values that the sample mean could take on and their respective probabilities then we would have the sampling distribution of the sample mean for this population when the sample size is $n = 2$. If we choose a different sample size then we would get a different sampling distribution. So, when we speak of sampling distributions we keep the sample size constant. Note also that no matter which sample we select, the population parameter is fixed (in our example, $\mu = 5.5$). Any random variable, including a summary statistic, has a corresponding probability distribution specifying each and every possible value the variable can take on and their respective probabilities. The distribution of a sample statistic is called a sampling distribution.

Properties of sampling distributions. Sampling distributions have the same properties as the distributions discussed above; this is because a sample statistic is a random variable. A statistic and its corresponding sampling distribution will usually have a mean and a variance. The mean of the statistic, as for other random variables, is the average of all possible values that the statistic can assume (recall that we hold the sample size constant). The variance of the sample statistic is a measurement of the variability in the sampling distribution. It is the average of all the squared deviations from the mean of the statistic. Again, the standard deviation is the square root of the variance. We refer to the standard deviation of a statistic as the standard error (SE) because, if the statistic is unbiased for a parameter (defined below), the standard deviation is measuring the "average error" of the statistic in estimating that parameter (see section on estimation). In other words, the standard error is a measurement of the accuracy of the sample statistic.

Usually a statistic is used to estimate the value of a population parameter. In our example above we would estimate the population mean to be about 6.5 with the first sample and estimate it to be about 5 with the second sample. When the mean of the sampling distribution of the statistic is equal to the population parameter then the statistic is called "unbiased" for that parameter (i.e., the statistic is "centered" at the value that we are estimating). In our example above, the average of all the possible sample means is 5.5, so the sample mean is unbiased for the population mean. When choosing between possible statistics (e.g., should the sample mean or the sample median be used to estimate the population mean) the criteria generally examined are the bias of the statistic and the variance of the statistic. It is desirable to use a statistic which is not only unbiased for the parameter but also has the smallest variation in distribution. In other words, we would like a statistic that is "centered" at the parameter and whose possible values do not stray very far from its center. In this manner we feel comfortable that our estimate will be close to the parameter.

Some specific sampling distributions. Before discussing the sampling distribution of the sample mean we present some distributions which are frequently encountered in parametric statistical inference. These distributions result from manipulations of normally distributed random variables (or statistics) and include the chi-square, Student's *t*, and F distributions.

The chi-square random variable is the summation of normal random variables after they have been squared. The chi-square random variable is always non-negative and the resulting chi-square distribution is asymmetric. This distribution depends on the sample size and is indexed by a quantity called degrees of freedom. An example of a chi-square statistic is the sample variance from a random sample selected from a normal population. For the sample variance the degrees of freedom are (n-1), the denominator of the sample variance formula (refer to Equation 2).

The Student's *t*-distribution was derived to show the distribution of a normal random variable divided by the square root of a chi-square random variable. Hence, if X is a normal random variable then the quantity:

$$T = \frac{x - \mu}{SD} \tag{7}$$

has a Student's *t*-distribution. The Student's *t*-distribution looks very similar to the standard normal distribution; it is symmetric about 0 and bell shaped, but it is a little more spread out than the standard normal distribution. The *t*-distribution also changes with degrees of freedom. As the sample size gets larger, and consequently the degrees of freedom get larger, the *t*-distribution looks more and more like a standard normal distribution. The *t*-distribution has n-1 degrees of freedom in the denominator.

The F distribution is another continuous distribution often used in parametric statistical inference, particularly in analysis of variance and in regression analysis. The F-statistic is the ratio of two chi-square random variables (usually the ratio of two variance estimates) after being divided by their respective degrees of freedom. It is always non-negative and is asymmetric. It too is indexed by its degrees of freedom but the F-statistic has degrees of freedom for both the numerator and the denominator (from the respective chi-square statistics).

Sampling distribution of the sample mean. As long as the sample is randomly selected from a fixed population (with a mean, μ, and a standard deviation, σ) the sample mean will be unbiased for the population mean (the average of all the possible $\overline{X}$'s = μ) and will have SE equal to the standard deviation of the population divided by the square root of the sample size (i.e., $SE = \sigma/\sqrt{n}$). If the original population (i.e., the population from which the data was selected) is normally distributed, then the sampling distribution of the sample mean will also be normally distributed, and the quantity:

$$Z = \frac{(\overline{X} - \mu)}{\sigma \sqrt{n}} \tag{8}$$

will have a standard normal distribution. If the population standard deviation is not known then we use the sample SD as an estimate of σ. The quantity:

$$T = \frac{(\overline{X} - \mu)}{(SD \sqrt{n})} \tag{9}$$

has a Student's *t*-distribution with (n-1) degrees of freedom. To use the *t*-distribution the sample data must have been randomly selected from a population with a normal distribution. If one cannot make this assumption, then the *t*-distribution is not appropriate.

D. The Central Limit Theorem

We have seen previously that if we are sampling from a normal distribution, then finding the sampling distribution of the sample mean is easy. However, often it is not reasonable to assume to state that the data came from a normally distributed population. The Central Limit Theorem (CLT), one of the most important theorems in statistics, addresses this problem. It basically states that as the sample size increases the distribution of the sample mean becomes closer and closer to a normal distribution, regardless of the distribution of the original data. The Theorem states:

> When a sample is the result of simple random sampling from a population with mean [μ] and standard deviation [σ], then, as the sample size n increases, the sampling distribution of the sample mean [$\bar{X}$] approaches a normal distribution with mean [μ], and standard deviation [$\sigma/\sqrt{n}$].

To understand this concept, think about taking many samples from a single population, each sample consisting of 100 individual data points. If you compare the distributions (histograms) of each of the samples, you would expect them to be similar and, in fact, you would expect them to reflect the distribution of the population. Likewise, you would expect the means and the standard deviations of each of the samples to be similar. Recall that we know that the sample means are centered around the population mean (and actually the sample variances are centered around the population variance). The CLT states that not only are the sample means centered around the population mean (with standard error $SE = \sigma/\sqrt{n}$) but also, the sample means have a distribution that is (close to) normally distributed. In other words, if we could collect all possible samples of size n = 100 from this population and draw a histogram of these sample means, this histogram would be bell shaped with mean μ and standard deviation $SE = \sigma/\sqrt{n}$.

Often statistical inference is concerned with estimating or testing the value of the population mean. Application of the CLT means that we can make parametric inferences about the population mean when we have taken a large sample from a population with a non-normal distribution. So, although we may not know the distribution of the population from which the data arose, we can assume that the distribution of the sample mean is close to normal when the sample size is large. Many texts choose n = 30 as a criterion to apply the CLT. However, as the skewness of the population increases, that sample size must increase as well to insure that the sample means approach a normal distribution. Practically speaking, we suggest a sample size of at least n = 100 for untransformed bioaerosol data before applying the CLT. With log transformations (see above) the distribution of the data becomes less skewed so that a smaller sample size may be appropriate.

When using the CLT the sample size will be large so that the sample SD is a fairly good estimate of the population standard deviation σ. Hence, the quantity:

$$Z = \frac{(\bar{X} - \mu)}{(SD\ \sqrt{n})} \tag{10}$$

is assumed to have a distribution which is fairly close to the standard normal distribution.

V. STATISTICAL INFERENCE

A. Introduction

Statistical inference is the process of using sample data to draw conclusions about a population parameter (or population parameters). Statistical inference takes two basic forms, estimation and hypothesis testing. There are also two basic approaches to inference, parametric and nonparametric. Parametric inference is based on knowing the form of the population distribution, or on the use of a CLT. Nonparametric inference does not rely on knowing the underlying distribution of the population. Estimation is described below as used in parametric statistics; hypothesis testing is described using each of the two approaches.

B. Estimation

Estimation is the process of deriving an educated guess for the value of an unknown population parameter from the sample data. One first takes a simple random sample from the population, calculates the analogous sample statistic, then uses this sample statistic as a point estimate of the population parameter, and finally determines the accuracy of this estimation. A point estimate constitutes a single guess, or estimate, of the population parameter; an interval estimate is the determination of a range of values that may cover the true value of the parameter. We call the latter a confidence interval for the parameter, because we can establish a statement of degree of confidence to go along with the interval estimation.

Estimating the population mean from a large sample. Generally, the sample mean is used for estimating the population mean. When using simple random sampling with a large sample size, n, then (based on the CLT) the sample mean ($\overline{X}$) has a normal distribution with mean = μ and standard error = $\sigma/\sqrt{n}$. The probability of a selected sample yielding a sample mean $\bar{x}$ exactly equal to the population mean μ is negligible, so a confidence interval is calculated. Here, the confidence interval is the point estimate, $\bar{x}$, plus or minus a multiple of the SE. Since the CLT is being applied, this multiple is based on the standard normal distribution. The commonly used 95% confidence interval is based on a distance of 1.96 times the SE. (i.e., 95% confidence interval for μ is: ($\bar{x}$ – 1.96 SE, $\bar{x}$ + 1.96 SE). Observe that the confidence interval for the population is symmetric about the sample mean.

Estimating the population mean from a small sample. We cannot apply the CLT when the sample size is small. Thus, to construct a confidence interval using a standard normal distribution, it must be established that the data were sampled from a normally distributed population (recall that the sample mean is always centered about μ, no matter what distribution the population has, so $\bar{x}$ is still a good point estimate). Providing that the population has a normal distribution, $\bar{X}$ will also have a normal distribution. If the population SD, σ, is known, then the confidence interval is constructed by finding the multiple from the standard normal distribution. However, σ is usually not known with bioaerosol data. In this case we use the sample SD in place of σ ($SE \approx SD/\sqrt{n}$) and find the multiple from the *t*-distribution with n–1 degrees of freedom. For the same confidence level (95% for example) the multiple found from the *t*-distribution will be somewhat larger than the analogous multiple from the standard normal distribution (i.e., a number slightly larger than 1.96 will be used in place of 1.96). Hence, the confidence interval for the population mean is slightly wider to make up for the inaccuracy of sample SD.

C. Hypothesis Testing (General)

Hypothesis testing is a kind of statistical inference wherein a hypothesis about a population parameter is stated and then tested. The hypothesis that states a specific claimed value of the population is called the null hypothesis (symbolized H_0) and the alternative hypothesis (H_A) is the complement. Hypothesis testing is a procedure which essentially puts the null hypothesis on trial. The null hypothesis, which claims a specific value for the parameter, can never be proven true without a full census of the population. However, the alternative hypothesis, which claims the parameter is within an open interval of values, can be proven to a set degree of doubt.

Sample data is used to test the null hypothesis. If the sample statistic, an analog for the parameter being tested, is close to the value claimed in the null hypothesis then the null hypothesis cannot be disputed; otherwise, one can conclude that the alternative hypothesis is more likely to be true. After a discussion of the types of decisions that can be made, some specific examples of hypothesis tests (both parametric and nonparametric) are given.

Decision making and error types. There are two possible errors that can be committed when making a decision between "not disputing the null hypothesis (H_0)" and "concluding that the alternative hypothesis (H_A) is more likely to be true than the null". Type I error occurs when one rejects a null hypothesis that is actually true. Type II error occurs when a false null hypothesis is not rejected (i.e., one states that the null hypothesis cannot be disputed when in reality the alternative hypothesis is true). Table 13.4 summarizes the possible situations in hypothesis testing. Recall that, in order to really know (without any doubt) whether the null or alternative hypothesis is true a census is needed.

Table 13.4. Decisions in hypothesis testing.

		Reality	
		H_0 is true	H_A is true
Decision	Reject H_0	Type I Error Probability = α (Significance level)	Correct decision Probability = $1 - \beta$ (Power)
	Do Not Reject H_0	Correct decision Probability = $1 - \alpha$	Type II error Probability = β

Note that, in reality, either the null hypothesis is true or the alternative hypothesis is true. Hypothesis testing is the procedure we use to decide between these two complementary hypotheses. Since the null hypothesis states a specific value for the parameter being tested then the distribution of the test statistic is known under the null hypothesis. For example, if the null hypothesis claimed that the mean of a population, μ, is 50, then we know that the sample mean has a distribution centered at 50 under the null hypothesis. Generally, the alternative hypothesis states that the parameter is within a specified open interval, not stating a specific value for the parameter, so under the alternative hypothesis we do not have a specific distribution for the sample statistic. To continue our example, if the alternative hypothesis claimed that μ is less than 50, then the sample mean has a distribution centered at a value less than 50 but how much less? The consequence of this set-up is that the probability of Type I error can be specified (i.e., we can set the level of α) and is called the significance level. It is the probability that the decision: "Reject the null hypothesis, we have proven the alternative hypothesis" is incorrect. This is the "set degree of doubt" referred to in the previous sub-section. Common values chosen for α are 0.10, 0.05, 0.01, and 0.005. One important fact to know is that when α decreases then β (the probability of Type II error) increases and we can only guess at how big this increase is. Other parts of the table above will be discussed in the section on Power and Sample Size.

Hypothesis testing. As mentioned above there are two basic approaches to hypothesis testing: parametric and nonparametric. Parametric testing requires a knowledge of the underlying distribution of the data or use of a statistical theorem such as the CLT (for sample means). Nonparametric testing assumes that the data are independently sampled from a continuous distribution but does not require knowledge of the shape of the distribution. It should be noted that nonparametric procedures are not a substitute for inadequate experimental design with resulting non-representative (or biased) samples. When testing the location of a distribution, it is best, whenever possible, to collect a large enough random sample so that the CLT applies. Distributions for most raw bioaerosol data are highly skewed and very large sample sizes (n 100) are necessary to insure the results of the CLT. Smaller data sets (n 30) may be appropriate when a transformation (such as logarithm) yields a transformed data set which is less skewed.

Sections D and E give examples of testing the location, or center, of a population, and also examples of comparing the centers of two populations. Within each of the examples the basic steps of hypothesis testing are followed:

Step 0. Determine the research question. The research question needs to be stated in terms of a population parameter in order to use hypothesis testing. Often this step is the most difficult.

Step 1. State H_0 and H_A. Recall that H_0 claims a specific value for the parameter being tested. Generally, H_A is the claim that the researcher would like to prove and H_0 is the currently accepted claim or the conservative claim.

Step 2. Determine the significance level. This is sometimes a field specific level (e.g., in Medical Statistics $\alpha = 0.05$ is commonly used.)

Step 3. Choose an appropriate test statistic for the parameter being tested. This choice will depend on whether a parametric or nonparametric test is used, on what assumptions can be made about the population distribution, and on the sample size.

Step 4. Determine, by the choice of the test statistic and the choice of α, how far the sample statistic can stray from the claimed value of the parameter (under the null hypothesis) before the decision is made to reject the null hypothesis and believe that the sample statistic is more consistent with the claim of the alternative hypothesis. This is done by determining the distribution of the test statistic under the assumption that the null hypothesis is true; this distribution will be referred to as the null distribution.

Step 5. Collect the random sample and calculate the value of the test statistic using the sample data. Make a decision according to the guideline set up in Step 4. This is the statistical decision. The decision is either "Do not reject the null hypothesis" (the sample data was consistent with the claim of H_0) or "Reject the null hypothesis" (the sample data gives significant evidence that the claim of H_A is more likely). Sometimes this decision is based on the "observed significance level" (or "p-value") which is a measure of how likely it would be to observe the observed sample test statistic (or one more extreme) from the null distribution. The null hypothesis is rejected if the observed significance level is less than α.

Step 6. Rephrase the statistical decision in terms of the research question. What conclusion (if any) can be drawn about the research question based on the hypothesis test?

D. Parametric Hypothesis Testing of the Population Mean

Parametric hypothesis testing is used when it is safe to make an assumption about the underlying distribution of the data, or when the sample size is large enough so that the CLT can be applied. The examples in this section either use the CLT (for the sample mean) or it is assumed that the data were sampled from a normal distribution.

Research problem. Suppose that past experience has indicated that the mean number of bacterial colonies recoverable from dust in a group of similar houses is 21.0 million/gram. (This value could have been determined from a large historical data set.) A local administrative board has proposed the use of a particular type of dehumidifier to reduce this figure. The administrative board would like to know if the dehumidifier is effective in reducing the density of culturable bacteria in the dust. The research question is: "Does the introduction and operation of the dehumidifiers reduce the levels of bacteria in dust?". This question will be addressed below with three differing experimental designs.

Example 1. Single location problem, large sample. One hundred (n = 100) houses are chosen at random from this neighborhood and the dehumidifier is installed. Sampling is conducted to determine the average number of bacterial colonies/gram of dust after operation of the dehumidifier.

Step 0. In this example the average (or mean, μ) number of bacteria/gram of dust with the use of the dehumidifier is the parameter which will answer the research question of whether the dehumidifiers are effective in reduction of bacterial density for this neighborhood.

Step 1. H_0: μ 21 million/gram of dust (the dehumidifier is either not effective, $\mu = 21$ million, or counterproductive, $\mu > 21$ million/gram of dust). H_A: <21 million/gram of dust (the dehumidifier is effective in reducing the density of bacteria). This is called a one-tailed test because we are only interested in a decrease in bacteria, not a change in either direction.

Step 2. Set $\alpha = 0.05$ (we are willing to decide that the dehumidifiers are effective in reduction of bacteria and be wrong about this decision 5% of the time).

Step 3. Since we have a large sample, n = 100, (by the CLT) the sample mean ($\bar{X}$) has a distribution which is approximately normal and it is appropriate to use the sample mean (transformed to the standard normal distribution) for the test statistic:

$$Z = \frac{\left(\bar{X} - \mu_0\right)}{\left(SD\ \sqrt{n}\right)} = \frac{\left(\bar{X} - 21.0\right)}{\left(SD\ \sqrt{100}\right)} \tag{11}$$

Note that this Z-score is based on the assumption that H_0 is true: that $\mu_0 = 21$ million is the true value of the average number of colonies of bacteria per gram of dust after the air filters are installed.

Step 4. The sample Z-score will be compared with the standard normal distribution (the Z-score above is centered at 0 and has standard deviation of 1 as long as the null hypothesis is true). If the observed sample Z-score, calculated from the observed $\bar{x}$, is –1.65 or less (Z –1.65) we will decide "Reject H_0". If $Z > -1.65$ then we will decide "Do not reject H_0". The value

–1.65 was chosen because it is the 5th percentile of the standard normal distribution ($\alpha = 0.05$), and only 5% of the samples will have a Z-score (if the null hypothesis is true) of –1.65 or less.

Step 5. After collecting the data we calculated the sample mean to be 19.5 million/gram of dust and the sample SD = 5 million/gram of dust. The observed Z-score is:

$$Z = \frac{\left(\overline{X} - \mu_0\right)}{\left(SD\ \sqrt{n}\right)} = \frac{(19.5 - 21.0)}{\left(5\ \sqrt{100}\right)} = -3.00 \tag{12}$$

Since the observed Z-score is less than –1.65 [and the observed significance level = P (Z   –3.00 = 0.0014)], the decision is "Reject H_0". (A significance level of 0.0014 means that, if the null hypothesis were true, we would observe a sample mean of 19.5 million or less in only 14 out of 10,000 samples.)

Step 6. We have conclusively shown that the average number of bacteria recovered per gram of dust is less than 21 million. If all other factors were equal in these houses except for the installation of the dehumidifiers, this is conclusive evidence that the dehumidifiers are effective in reducing the density of bacteria.

Note that, in this example, the random variable (count of bacteria colonies), can take on so many possible values that this measurement was considered close to continuous. Also, the sample size, n = 100, is considered large enough so the CLT was used in Step 3. No assumption about the distribution of count of bacterial colonies/gram of dust was made. Also, the large size of the sample allowed the use of the sample SD as a reliable estimate of σ, the population standard deviation. Consequently, a Z-score was used.

Example 2. Single location problem, small sample. Often it is not possible to take a large sample. Suppose that we have a similar situation to Example 1 in which we are testing the population mean, μ. To use parametric inference with a small sample the investigator needs to know the population distribution from which the sample arose. If the population has a normal distribution and the population standard deviation, σ, is known, then the Z-score (using $\sigma/\sqrt{n}$ as the SE) and the standard normal distribution can be used as the test statistic and the null distribution, respectively. If the population has a normal distribution and the population standard deviation, σ, is not known then the T statistic (using $SD/\sqrt{n}$ as an estimate of the SE) and the Student's *t*-distribution with n-1 degrees of freedom can be used for the test statistic and the null distribution, respectively.* (In this manner allowances are given for the inaccuracy of the SD as an estimate of σ with a small sample.)

* Recall that the *t*-distribution is a little more spread out than the standard normal distribution. Therefore, we are more conservative about rejecting the null hypothesis. In essence, the evidence that "proves" the alternative hypothesis has to be more convincing.

For example, suppose that it was only possible to sample 25 homes in Example 1 and that it had been determined that bacteria colony counts have a normal distribution. Hence, the T statistic is the appropriate test statistic and referring to the *t*-distribution with n-1 = 24 degrees of freedom, the 5th percentile (from $\alpha = 0.05$) is –1.711. The decision rule is to reject the null hypothesis if the sample test statistic is less than or equal to –1.711 (i.e., if T –1.711). In our sample of n = 25 we obtained a sample mean of 19.5 million bacteria/gram of dust and a sample SD of 7 million bacteria/gram of dust. Then the test statistic is:

$$T = \frac{\left(\bar{X} - \mu_0\right)}{\left(SD \;\; \sqrt{n}\right)} = \frac{(19.5 - 21.0)}{(7/5)} = -1.07 \qquad (13)$$

Since –1.07 > –1.711, we cannot reject the null hypothesis. Our conclusion is that we are unable to show that the dehumidifiers are effective in reducing the average density of bacteria. Note that the use of a small sample has several consequences:

1. The estimate of the population standard deviation, σ, is not as accurate as it would be with a large sample.
2. The estimated SE is larger because the divisor ($\sqrt{\text{n}}$) is smaller.
3. The cut-off value for deciding to reject the null hypothesis (–1.711 here) is further away from zero than it would be with the use of the standard normal distribution (–1.65 in Example 1).

Example 2 stresses the desirability of collecting a large sample whenever possible. This issue is also discussed on the section on Power and Sample Size.

Example 3. Comparing the locations of two independent populations, large samples. It may be that trusting the historical data for the current value of the population mean is not reasonable. To answer the research question we need to compare the levels of dust-borne bacteria in homes without the dehumidifier with the levels of bacteria in homes with the dehumidifier. We randomly sample 100 homes for measurement without the dehumidifier and then randomly select another 100 homes for measurement with the dehumidifier. The null hypothesis is still that the dehumidifiers are not effective in reduction of the average level of bacteria in the dust, so that the mean levels of bacteria for each population of homes is the same.

Step 1. $H_{0:}$ μ_w $\mu_{w/o}$ (either the dehumidifiers are not effective or they are counter-effective). $H_{A:}$ μ_w $\mu_{w/o}$ (the dehumidifiers are effective in reduction of the average level of bacteria in dust) μ_w symbolizes the population mean level of bacteria per gram of dust in all of the homes with an operating dehumidifier and $\mu_{w/o}$ symbolizes the population mean for all of the homes without the dehumidifier.

Step 2. $\alpha = 0.05$.

Step 3. The CLT can be used and we can transform the difference of the sample means to a Z-score (formula shown below).

Step 4. Same as in Example 1: decide to "Reject H_0" if Z -1.65.

Step 5. If we obtained the following sample results:

$\bar{x}_{w}$ = 19.5 million and SD_w = 4 million; $\bar{x}_{w/o}$ = 21 million and $SD_{w/o}$ = 5 million

then

$$\frac{\left(\bar{X}_w - \bar{X}_{w/o}\right)}{\left(SD_w/\sqrt{n_w}\right)+\left(SD_{w/o}/\sqrt{n_{w/o}}\right)} = \frac{19.5-21.0}{\left(4\ \sqrt{100}\right)+\left(5/\sqrt{100}\right)} = -1.67 \quad (14)$$

and the observed significance level: p-value = P (Z -1.67) = 0.0485. Hence, we would decide to reject the null hypothesis and conclude that, all other things being equal, the dehumidifiers are significantly effective in reducing the average density of bacteria in dust.

Example 4. Comparing the locations of two independent populations, small samples. If it is not possible to collect large samples from each of the two populations, a *t*-test can be used to compare the population means. The assumptions of the *t*-test are that independent samples are taken from two normally distributed populations and that the two populations have the same variance. The form of the test statistic (Step 3) is the same as it is for the test above (except that it is called T instead of Z) and we pool the estimates of the standard deviation. The T statistic is compared with a *t*-distribution with degrees of freedom = $(n_1 + n_2 - 2)$. If both of the populations are normally distributed but the variances are not equal then theoretical problems arise. There are approximate methods for this situation which use a T statistic (with a different denominator) but refer to a Student's *t*-distribution with an adjusted degrees of freedom.

Note that by using two samples (comparison of average bacterial counts between homes with air filters and homes without air filters) the danger that the bacterial count may have changed since the historical data set was taken has been removed. If the temporal problem is not a worry (i.e., it is assumed that the bacterial count per gram of dust is fairly consistent over time), but there is wide variation among homes, then the following approach to answer the research question may be more appropriate.

Example 5. Comparing two matched population locations. The approach here is to compare the bacteria count per gram of dust before, and after the introduction of the dehumidifier within each household. For each home chosen, a measurement without the dehumidifier would be taken ($X_{w/o}$); the dehumidifier would be operated and after a specified amount of time a second measurement would be taken (X_w). This design accounts for the wide variability in bacterial count from household to household by looking at the differences (D):

For home i: $D_i = X_{w,i} - X_{w/o,i}$, for i, …, n

$$\overline{D} = \frac{\Sigma D_i}{n} = \text{mean difference}$$

D measures the change due to the dehumidifier as long as the bacterial count is not changing over time due to some other cause. We have reduced a two population problem to a one population location problem and the research question becomes: “Is the mean of the differences less than zero?”. The null hypothesis assumes that the dehumidifiers are not effective: $H_{0:}$ μ_D 0 and the alternative: $H_{A:}$ $\mu_D < 0$, indicates that the dehumidifiers are effective in reduction of the density of bacteria. If the sample size is large, we use the method in Example 1; if the sample size is small and it is reasonable to assume that the differences have a normal distribution then we use the method in Example 2. For each case the form of the test statistic is:

$$\frac{\overline{D} - \mu_{D,0}}{\left(SD_D \ \sqrt{n_D}\right)} = \frac{\overline{D} - 0}{\left(SD_D \ \sqrt{n_D}\right)} \qquad (15)$$

Here, $\overline{D}$ represents the sample average of the differences, SD_D represents the sample variance of the differences, and n_D denotes the number of differences in the sample.

E. Nonparametric Hypothesis Testing of Location

In a nonparametric setting the basic assumption made about the data is one of independent sampling from a population that has a continuous distribution. When the interest lies in testing the location of the population, the median is used as a measure of location. Recall that, unlike the mean, the median is not sensitive to extreme values in the data set. Some commonly used nonparametric tests are described here in the context of an example.

Research Problem. From historical data it is known that for a particular office building the median level of culturable *Cladosporium* is 269 CFU/m^3 of air. Recently, in this building, there has been an increase in the number of complaints and an investigation is launched to find out if the level of culturable

Cladosporium in air has risen. A sample of 100 air samples is taken to answer the question: "Has the level of culturable *Cladosporium* in air risen?".

Step 0. We are using the median as a measurement of location and the research question translates into: "For this building, is the current median level of culturable *Cladosporium* above 269 CFU/m^3 of air?".

Step 1. H_0: population median 269 CFU/m^3 of air (i.e., the median level of culturable *Cladosporium* has either stayed at 269 CFU/m^3 of air or has decreased from this level). H_A: population median >269 CFU/m^3 of air (i.e., the median level of culturable *Cladosporium* in air has increased above 269 CFU/m^3).

Sign test. The sign test is a simple way of testing the null hypothesis. The test statistic for the sign test is to count the number of observations that are above the hypothesized median, in this case above 269 CFU/m^3: S = number of data values above the median claimed in H_0 (here, S = number of x >269). The sign test is based on the principle that the median is the 50th percentile. If the population median is truly 269 then we would expect about half of our sample values to be above 269 (i.e., about n/2 data values should be above 269). It's called the sign test because we assign pluses to data values above the hypothesized median and minuses to values below the hypothesized median and then count the number of pluses. (A technical note: If any of the data values fall exactly on the hypothesized median, here 269, we generally discard this data and decrease the sample size accordingly.) If there are many more pluses than minuses, we have too many values above 269 CFU/m^3 to believe that the population median has not increased. (If we had wanted to prove that the median level of *Cladosporium* has decreased from 269 CFU/m^3 (H_0: median 269 vs. H_A: median <269) then we would reject H_0 if we observed too few pluses.) How we measure whether or not there are too many pluses depends on the sample size. If the sample size is large (and large here does not depend on the distribution of the data, so even n = 25 is considered large enough) the CLT theorem, the Z-score, and the standard normal distribution can be used.

In our sample suppose that we found 80 observations above 269 CFU/m^3 of air. Then:

$$Z = \frac{S - n/2}{\sqrt{n/4}} = \frac{80 - 50}{\sqrt{100/4}} = \frac{30}{5} = 6 \tag{16}$$

The observed significance level would be: P(Z >6) <0.0001. This last statement tells us that if the true median level in the building were still 269 CFU/m^3 of air, then the chances of our collecting 80 out of 100 observations with levels above 269 CFU/m^3 of air is less than 1 in 10,000. Hence, the sample of 100 observations has conclusively shown that the median level of culturable *Cladosporium* in this building has risen above 269 CFU/m^3 of air. If the sample size is small, then we compare the number of pluses to a binomial distribution with parameters n, $p = 0.5$. Refer to Freedman et al., 1991 or Daniel, 1991, for a discussion of the binomial distribution..

Suppose we had only been able to collect a sample of 25 observations and had observed that 20 of these had levels above 269 CFU/m^3 of air. Then examining the binomial distribution with $n = 25$ and $p = 0.5$ yields an observed significance level (the chance that we would observe 20 or more observations with levels above 269 CFU/m^3 if 269 is the true median) of P(S 20) = 0.00204. Since this observed significance level is very small, we have again conclusively shown that the median level of culturable *Cladosporium* in this building has risen above 269 CFU/m^3 of air.

Sign test for paired data. The sign test can also be used for paired data. Using the home dust example from subsection D, we compare the level of bacteria in the dust before, and after the operation of the dehumidifiers. Here the test statistic would be the number of homes for which the level of bacteria was greater before the operation of the dehumidifiers. (The homes in which the "before operation" and the "after operation" values are equal are discarded from the sample and the sample size, n, is adjusted accordingly.)

Rank statistics. Quite often, nonparametric tests use the relative position of each data value in the data set. Ranks can be assigned to the data by first placing the data in numerical order and then noting which place each datum has; the "place" is the rank of that datum. If two (or more) data values are equal in value then the rank for each is the average of the ranks that they would have been assigned had they not been equal. For example, the rank statistics of the data set {–1, 2, 0, 7, 2, 10} is found by first ordering the data set {–1, 0, 2, 2, 7, 10} and then assigning ranks: {1, 2, 3.5, 3.5, 5, 6}. If we want to put the ranks in the original order of the data: {1, 3.5, 2, 5, 3.5, 6} are the ranks of {–1, 2, 0, 7, 2, 10}, respectively.

Wilcoxon signed rank test. The sign test has the disadvantage that each of the values above the hypothesized median has the same weight. So in our building example a reading of 365 CFU/m^3 would have the same "value" in the test as a reading of 785 CFU/m^3 (each would get a plus). Yet, obviously, the second reading gives much more justification to the alternative hypothesis than the first. With the additional assumption that the underlying distribution of the data is symmetric (i.e., the distribution of the data above the median is the mirror image of the distribution of the data below the median) the Wilcoxon signed rank test can be used to test the median. (One simple way to examine whether the underlying population distribution of the data is symmetric is to compare the sample mean with the sample median. If the population distribution is symmetrical then the sample mean should be close to the sample median.)

The Wilcoxon signed rank test looks at the absolute differences of each datum from the hypothesized median (in the example 269 CFU/m^3 of air) and ranks these differences. In this manner data values that are further away from the hypothesized median get larger ranks. The procedure then gives the rank of each datum either a plus or a minus depending on whether the original data value was above or below the hypothesized median, respectively (the method used in the sign test). So, now each datum not only has a plus or a minus, but it also has a rank indicating distance from the hypothesized median. Specifically the reading of 785 CFU/m^3 would get a higher positive rank than the reading of 365 CFU/m^3. The test statistic has several equivalent forms, one of which is the sum of the positive ranks. If the sample size is small then this statistic is compared with tabled values for the Wilcoxon signed rank test (indexed by sample size); if the sample size is large then a CLT can be employed and the appropriate Z-score is compared with the standard normal distributions.

Wilcoxon-Mann-Whitney U-Test. This test is used to compare the locations of two independent populations with continuous (or at least ordinal) distributions. The assumptions are more rigid than those of the two tests presented above: each of the populations must have the same shape distribution and the same variance. Hence, the two population distributions are only allowed to differ by location. In addition, each of the random samples are drawn independently of one another. The null hypothesis is that the medians of the two populations are the same. This test is very sensitive to departures from the assumptions and the null hypothesis may be rejected, not because the medians of the populations differ, but because the shapes of the two population distributions differ. The procedure is to rank the data of the two independent samples together and then compute the sums of each sample's ranks. The steps are outlined below:

1. Rank the observations of the combined sample, keeping track of which observation is from which sample.
2. Let R_1 = the sum of the ranks of sample 1, and R_2 = the sum of the ranks of sample 2.
3. Let $U_1 = (n_1n_2) + 0.5\,[n_1(n_1 + 1)] - R_1$ and $U_2 = (n_1n_2) + 0.5[(n_2(n_2 + 1)] - R_2$. Here n_i = size of sample i.
4. The test statistic U = the minimum of U_1, U_2.

The decision to "Reject the null hypothesis" is made by comparing U with tabled critical values of U. These values are indexed by n_1, n_2, and α, the significance level. If U is less than or equal to the critical value then the null hypothesis is rejected and the alternative hypothesis, that the two medians are not equal, is accepted. The reader is referred to references (Daniel, 1990; Walsh, 1962; Sachs, 1984) for presentation of this test and tabled critical values. Again, this test has different forms and care should be taken when using tables to make sure that the form of the statistic that is computed agrees with the form of the statistic as it is tabled.

F. Testing Locations of Three or More Continuous Distributions: Analysis of Variance

The analysis of variance is a parametric test which is used to compare the means of three or more independent populations. The populations must be normally distributed and each have the same variance, σ^2. The use of the CLT is not appropriate because the test is based on comparing the variation within each of the three samples with the variation between the three samples. This means that transformation will be necessary for most bioaerosol data.

The approach of the test is to compare two estimates of σ^2: the first estimate is pooled from the individual *within* sample variances and the second estimate is based on the distance *between* the individual sample means and the overall mean. If the ratio of these two estimates is close to one, then the data is consistent with the hypothesis that the population means are all equal. If the *between* sample estimate is much larger than the *within* sample estimate then this implies that the populations differ by location since their respective centers (estimated by the sample means) are far away from each other relative to the variation within each of the populations (estimated by the pooled within variance). The null hypothesis states that all the population means are equal; the alternative hypothesis is that at least one of the population means differs from the others. The reader is referred to Daniel (1990) or a text on parametric statistical procedures for a more detailed description and for details on sampling designs.

Many statistical packages perform analysis of variance procedures and report the observed significance level (sometimes called the p-value); if the observed significance level is less than α, then reject the null hypothesis.

A note here to remind the reader of the assumptions of the analysis of variance test:

1. Independent random samples are taken from k independent populations.
2. Each of the k populations has a normal distribution.
3. Each of the k populations has the same variance, σ^2. (As stated earlier, this does not mean that the sample variances will be equal but they should be close to one another.)

For bioaerosol data, it is almost essential to perform some kind of transformation (perhaps the log transform described earlier) to meet the second assumption.

We have described the basic one-way analysis of variance for the completely randomized design (i.e., each of the samples were randomly and independently drawn from independent normal populations with common variance). The paired-data *t*-test can be expanded to a one-way analysis of variance for randomized block designs. With this design and analysis the investigator is comparing k locations of matched data. For example, this analysis can be used to compare before, during, and after scores taken on the same element. The analysis of variance technique has been extended to other designs as well, but within each, the assumption of randomly drawing from normal populations with the same variance remains.

Analogous nonparametric tests exist for various designs and analysis of variance tests. The test which is analogous to one-way analysis of variance for completely randomized design is the Kruskal-Wallis test. For a design with randomized blocks (i.e., the data are "matched" in k-tuples and the comparison is between the k populations) the Freidman test is used. In general, these nonparametric tests determine whether the k populations have the same distribution against the alternative hypothesis that at least one of the k distributions differs from the others. Hence, these tests are not only for detecting a difference in location, but a difference in shape of distribution as well.

VI. POWER AND SAMPLE SIZE

When planning a statistical experiment the investigator will initially have the question: "How large a sample should be taken?". Another question that may also arise is: "What are the chances of detecting a specific change?". These two questions are linked together; one cannot be answered without considering the answer to the other one. Recall that the mean of a sample of size n has standard error, SE $= \sigma/\sqrt{n}$. The SE is a measure of how far the sample mean varies from the population mean; it measures the accuracy of the sample mean in estimating the population mean. As the sample size increases the SE decreases, which implies that the sample mean gets closer to the population mean. Considered in the extreme this makes sense: if the largest possible sample is taken, i.e., a census, then the sample mean equals the population mean, SE = 0; if the smallest possible sample is taken, n = 1, then the sample "mean" is an individual observation and the SE = σ. Consider the answer to the two questions above in the context of hypothesis testing and in the context of an example.

Referring back to the research problem of the home dehumidifiers (Section V, Part D), suppose that historical data has shown us that the average density of culturable bacteria is 21.0 million/gram of dust with a standard deviation of 4.5 million/gram of dust. Suppose further, that the administrative board has decided it would be advisable to recommend introduction of the dehumidifiers if they can reduce the average density to 20.0 million/gram of dust. Then the administrative board has defined the "specific change" that we would like to detect. There are two ways to approach answering the power and sample size questions for this situation. For a given sample size and level of significance, α, we can answer the question: "What are the chances of detecting an average of 20.0 million/gram of dust?". Or, a decision can be made that we would like to detect an average of 20.0 million/gram of dust with 80% chance when testing with $\alpha = 0.05$ and then ask "How large a sample do we need to do this?". The latter approach is the one that is usually taken. We set the probability of detecting a specific change (80% for detecting an average of 20.0 million/gram of dust) at a given significance level ($\alpha = 0.05$) and then answer the question: "How large does the sample need to be?" Referring to Table 13.4, the power $(1 - \beta)$ is the probability of rejecting the null hypothesis when the alternative hypothesis is true. In our example, we want power of 80% to detect a population mean of 20 million/gram of dust. Power depends on sample size: we need a larger sample size to have 80% power to detect a population mean of 20 million/gram of dust than we would need to have 80% power to detect a population mean of 19 million/gram of dust. In other words, if our measurement tool, the SE, is smaller then it will be more sensitive to small changes in the population mean. (One cannot tell the difference between a person who is 110.75 lb and a person who is 110.25 lb if one is measuring on a scale that is only accurate to the pound; a more sensitive scale is needed.)

To complete our specific example, we want to know the sample size, n, needed to detect a population mean of 20.0 million (when H_0: $\mu = 21.0$, and $\sigma = 4.5$ million) with 80% power when testing at = 0.05. Our decision rule was to reject the null hypothesis for a sample Z –1.65. That is, we want 80% of the $\overline{X}$s that come from a population with mean 20.0 million/gram of dust to have Z-score less than –1.65 as calculated by:

$$Z = \frac{\left(\overline{X} - \mu_0\right)}{\sigma \sqrt{n}} = \frac{\left(\overline{X} - 21.0\right)}{4.5 \sqrt{n}} \leq -1.65 \qquad (17)$$

For the $\overline{X}$s that come from a population with mean 20.0 million/gram of dust the 80th percentile is:

$$\overline{X} = 1.28 \cdot \left(4.5 \sqrt{n}\right) + 20.0 \qquad (18)$$

Since z = 1.28 is the 80th percentile of the standard normal distribution, we can solves Equations 18 and 19 simultaneously to get:

$$n = \frac{(1.65 + 1.28)^2 \cdot 4.5^2}{(21.0 - 20.0)^2} \approx 173.84 \qquad (19)$$

and since parts of homes cannot be sampled, we sample 174 homes. Thus, to detect a change to $\mu_1 = 20.0$ with 80% chance (at $\alpha = 0.05$) we need to sample 174 homes. In general, if power $1-\beta$ is desired to detect an alternative mean of

μ_1 when testing H_0: $\mu = \mu_0$ at significance level α, in a population with standard deviation σ, then the sample size needed is:

$$n = \frac{\left(z_\alpha + z_\beta\right)^2 \cdot \sigma^2}{\left(\mu_0 - \mu_1\right)^2} \tag{20}$$

and if the result is not a whole number then it is always rounded up. Here z_α is the $(1 - \alpha)\cdot 100^{th}$ percentile and z_β is the $(1 - \beta)\cdot 100^{th}$ percentile of the standard normal distribution.

More complicated formulae exist for calculating the sample sizes needed to detect specific differences between locations of two populations for a given power and significance level and for the sample sizes needed in analysis of variance for detecting specific alternatives. All of these formulae are based on the same principle: A larger sample size will yield more accurate results (i.e., a smaller SE). An exaggeration of this principle can be misleading as well. If an extremely large sample is taken, then the SE will be very sensitive and will detect a change that may not be meaningful. To continue our home dehumidifier problem, suppose that the administration decided that 10,000 homes should be used in the sample. Then the SE is $4.5/\sqrt{10{,}000} \approx 0.045$ and we would have 80% chance of detecting an average of μ_1 20.87 million/gram of dust which may not be a meaningful change, and yet it could honestly be reported to the board that the average level has decreased. This is how one can "lie" with statistics.

VII. MORE ON RELATIONSHIPS BETWEEN VARIABLES

A. Linear Regression

Linear regression is a method for deriving an equation that describes the relationship between two variables: one variable is called the response or dependent variable (Y), the other is called the predictor or independent variable (X). The goal of linear regression is to find a linear model in X which will yield more accurate inference about Y than is possible with the Y's alone. The basic assumption is that a linear relationship between X and Y exists in the population and can be characterized by:

$$y_{ij} = \beta_0 + \beta_1 \cdot x_1 + \varepsilon_{ij} \tag{21}$$

This equation says that the jth observation of Y taken at the ith observation of $X = x_i$ can be found by multiplying x_i by β_1 (called the slope parameter), adding β_0 (the intercept parameter), and then adding a random error, ε_{ij}. We use the sample to initially decide if a linear model seems plausible and if so, to then estimate the values of β_0 and β_1, and consequently infer values of Y for given values of X. The basic assumptions behind linear regression analysis are

1. The model (Equation 22) is appropriate.
2. The values of X are fixed, or measured without any error.
3. The average of all the Y's at a given value of X = x is $y = \beta_0 + \beta_1 \cdot x$.
4. For each given value of X, the variance of the Y's is the same.
5. For each given value of X, the Y's have a normal distribution.

The first step in fitting a linear model to a given sample is to plot a scatter diagram. If a linear trend seems apparent on the scatter diagram, then fitting the above model is probably appropriate. If the scatter diagram appears to be randomly scattered and/or show a form other than linear, then fitting the above model is probably not appropriate. Data which appear to be related through a non-linear function can sometimes be transformed. For example, the data may appear to fit a shape "$y = x^2 + e$" so that if we let $W = X^2$ then Y is linear in W rather than X. Also, the variance of the Y's should be fairly consistent for each value of X, so that the spread of Y's should be about the same as X changes. If the spread of the Y's tends to increase as X increases (decreases) then assumption of equal variances may not be met. This, too, can often be handled with a transformation. A "football" shape to the scatter diagram is a desirable result. Generally, a sample which has a scatter plot which has a "football" shape is not inconsistent with the assumptions needed for linear regression.

Usually, this model is fit through a statistical computing package by the method of least squares. Essentially, the computing package will find the line that comes "closest" to all of the sample data points at once. The results of the process are usually presented in a table and will provide the investigator with estimates of β_0 and β_1 as well as other information to make inference about the parameters, β_0 and β_1, and the value of Y for any specific X.

Statistical models, such as linear regression, are used to model association. Association does not imply causation. So, even though knowledge of the value of X may help in inferring a value of Y this does not mean that the value of X "causes" Y to be a specific value. The reader is referred to *Statistics*, Freedman, et al., 1991, for a very readable discussion on the concepts of regression and correlation.

The Pearson correlation coefficient, r, is used as a relative measure of fit to the linear model. Recall that r varies between –1 and +1 inclusive, and that an absolute value of 1 indicates all of the data lie on a straight line. There is no absolute scale for r, but rather a higher absolute value of r indicates a better linear fit. The value of ρ, the population correlation coefficient, and the value of β_1 are linked. If $\rho = 0$ then $\beta_1 = 0$, and vice versa. Recall, that $\rho = 0$ (and hence, $\beta_1 = 0$) means that X and Y are linearly uncorrelated. This means that X does not help in inference about Y. A common test used in regression analysis is to test H_0: $\beta_1 = 0$ vs. H_A: β_1 0. Rejection of the null hypothesis implies that the linear model is useful in inference about Y. The test statistic cited in statistical packages is the F statistic (we reject the null hypothesis for large values of F, much larger than 1.00) and usually the statistical package will print out an observed significance level (a p-value).

The inferential methods of linear regression are fairly robust. That is, the inferences will still be valid under small deviations from the assumptions listed above. However, bioaerosol data is often highly skewed, and it is not uncommon for the variance of a response variable to increase with the value of a predictor variable. Transformations of the original data can often correct these problems.

B. LOGISTIC REGRESSION

In linear regression the dependent variable, Y, is a continuous random variable. Often a researcher would like to model the presence, or absence, of some factor. For example, an investigator may want to model the occurrence of hay fever in the home on the level of *Cladosporium* in the air. In this case the response variable, Y, is dichotomous and is defined as 1 if hay fever is present, 0 if hay fever is not present, and X is the measure of the level of *Cladosporium* in the air. Logistic regression envelopes the procedures used to model the probability that Y = 1 (hay fever is present) for a given value of the predictor variable, X. Let $\pi(x)$ = the probability that hay fever exists in a home with measurement X = x. Then logistic regression assumes that the following model exists in the population:

$$\ln\left(\frac{\pi}{1-\pi}\right) = \beta_0 + \beta_1 \cdot x \tag{22}$$

Standard statistical computing packages will fit this model from the sample data so that the investigator will have estimates of β_0 and β_1, and call them $\hat{\beta}_0$ and $\hat{\beta}_1$, respectively. Then the transformation below can be used to estimate the probability of hay fever at a given measurement of X = x_0:

$$\hat{\pi}(x_0) = \frac{\exp\{\hat{\beta}_0 + \hat{\beta}_1 \cdot x_0\}}{1 + \exp\{\hat{\beta}_0 + \hat{\beta}_1 \cdot x_0\}} \tag{23}$$

The reader is referred to *Applied Logistic Regression,* Hosmer and Lemeshow (1984), for a more detailed explanation of this model and its uses in inference.

VIII. CONCLUSIONS

Our aim has been to present a brief introduction to statistics, not so that the reader can use our paper alone to conduct statistical analyses, but to stimulate an interest in applying statistical principles to data interpretation (and, more particularly) experimental design in even small, problem-solving studies. In addition, we hope this treatment will assist beginners in interpreting the print-outs that come with some statistical packages. Needless to say, there is a great deal more to statistics than is presented in this brief paper. For example, there are tests that allow one to make inferences about correlation coefficients and slopes of regression curves that are extremely useful in analyzing bioaerosol data. As pointed out above, we have not included a detailed discussion of logistic regression, which is also a powerful and valuable tool in bioaerosol research. These and other topics are covered in many excellent statistics textbooks, including those listed in the references of this paper.

REFERENCES

Daniel, W. W., *Applied Nonparametric Statistics*, C. W. Ken, Boston, 1990.

Daniel, W. W., *Biostatistics: A Foundation for Analysis in the Health Sciences*, 5th ed., John Wiley & Sons, New York, 1991.

Freedman, D., Pisani, R., Purves, R., Adhikari, A., *Statistics*, 2nd ed., Norton, New York, 1991.

Hosmer, D. W., Lemeshow, S., *Applied Logistic Regression*, John Wiley & Sons, New York, 1984.

John, P., *Statistical Design and Analysis of Experiments*, MacMillan, New York, 1971.

Sachs, L., *Applied Statistics: A Handbook of Techniques*, Springer-Verlag, New York, 1984.

Walsh, J. E., *Handbook of Nonparametric Statistics*, Van Nostrand, Princeton, NJ, 1962.

Zar, J., *Biostatistical Analysis,* 2nd ed., Prentice-Hall, Englewood Cliffs, NJ, 1984.

Index

A

B

C

D

E

F

G

H

I

N

O

P

U

V

W

X

Y

Z